Water Quality and Its Control

Water Quality and Its Control

James C. Lamb, III
University of North Carolina at Chapel Hill

John Wiley & Sons
New York Chichester Brisbane Toronto Singapore

Library of Congress Cataloging in Publication Data:

Lamb, James C.
Water quality and its control.
Includes bibliographies and indexes.
1. Water quality.
I. Title.

TD370.L36 1985 363.7'394 84-20945
ISBN 0-471-80047-3

Printed in the United States of America

10 9 8 7 6 5 4 3 2 1

To Pat

Preface

The past two decades have produced sweeping changes in how we approach water quality problems. There have been major developments in technology along with improved implementation and enhanced funding of pollution control programs. Most important, the attitudes of professionals and the public have undergone radical shifts that reflect national demands and expectations for improved water quality.

Remarkable progress has been made, but it is now clear that some past program activities were ambitious but unsound, zealous but wasteful, well intentioned but ill conceived. During the 1960s and 1970s, we devoted great effort to developing legal structures and technological tools, but too little attention to methods for using them wisely. The flood of federal money supplied to expedite water quality programs, which often exceeded our capability to use it wisely, has caused the gains in technology to outstrip our ability to apply it soundly. The result is that water quality control activities are sometimes viewed with suspicion by review agencies, politicians, and the public, producing increased pressures to justify our positions and actions. This is especially true of expenditures for treatment facilities, which face tougher competition for funds by other important social needs.

The explosion of information (and misinformation) over the past two decades about water quality has produced strong views, many of which are fundamentally sound. Unfortunately, others are based on erroneous assumptions, faulty data, defective reasoning, and illogical conclusions. A critical need now exists for accurate facts and rational concepts to provide a valid approach for putting our house in order. It is a time to rethink our positions and improve our methods. We must give greater attention to the rationale behind our actions, based on better understanding of goals, tools available for attaining them, and techniques for deciding among alternatives. This requires careful attention to basic concepts to provide a sound foundation for improving regulatory structures and modifying the design and operation of facilities to enhance efficiency and economy in water quality evaluation and control.

The goal of this book is to present facts about water quality that are accurate in detail, balanced in perspective, and that can be accepted as reliable by readers, whatever their points of view.

This text is an outgrowth of an introductory course that I have developed and taught for 20 years; it is motivated by a conviction that professionals entering the water quality field cannot effectively apply water technology and management techniques without first acquiring a sound grasp of the fundamental rationale behind the activities in which they participate. This requires an accurate understanding of key facts and issues underlying the public, legal, political, and technological goals and decision-making processes.

The course content evolved gradually into a structure calculated to provide a solid foundation for professional development of scientists, engineers, and others who one day may occupy leadership positions in the field. Its goal is not to teach the technological details necessary to prepare one for professional practice—that logically follows in other courses and books—but to convey an understanding of the nature of that technology, the rationale behind its application, what it can accomplish, and its limitations. The book is based on the premise that, in this type of social undertaking, the "why" behind our actions is at least as important as the "how" of the facilities that we design, construct, and operate.

An early problem in the course was that no text was available to satisfy its needs. The few that seemed close were oriented too heavily toward technology and engineering details, with only minimal attention to background issues and fundamental concepts underlying program development. Accordingly, reading materials for the course were selected from current literature, reprints, reports, and specially prepared outline notes. The arrangement was unwieldy, and, as often happens in the academic world, the notes ultimately became the outline for this book.

Chapter 1 begins with a brief introduction to water quality, the historical development of quality problems, the significance of those problems, and the evolution of control programs. It also outlines recent trends and current needs in water quality control. Because of a need to express water quantities and quality characteristics in numerical terms whenever possible, this chapter also introduces the complex systems of measurements and units used in this field. The appendix explains how to make conversions among the various units.

Chapter 2 describes some characteristics of water that are helpful in understanding its roles in nature and its utility to humanity. Chapter 3 outlines some concepts of aquatic biology that are important in understanding certain parts of following chapters.

Water quantities and quality are intimately related in several ways. Accordingly, Chapter 4 presents certain elements of hydrology, that is, the water resources, or "assets," available to us. Chapter 5 describes how we use water and the amounts we use for those purposes in the United States.

Having laid that groundwork, Chapter 6 develops some fundamentals of water quality, beginning with the concept of quality and the systems used to measure it. The identities and concentrations of constituents often found in natural waters are outlined in Chapter 7.

Chapters 8, 9, and 10 are devoted to the dark side of water quality—the problems commonly encountered with it. First, several types of constituents that can cause quality problems are discussed in Chapter 8, including the nature of their impacts and sources from which they may be introduced into watercourses. Two of the most important specific types of problems—health effects and oxygen relationships in streams—are discussed in detail in Chapters 9 and 10.

This is followed by three chapters on practical water systems. Chapter 11 addresses the facilities often employed in municipal supplies to collect, transport, treat, and distribute water for drinking and other purposes, including the rationale for their selection and development. Chapter 12 discusses systems for collecting, treating, and disposing of municipal wastewaters. Industrial water uses and wastewater characteristics are outlined in Chapter 13, including the similarities and differences between industrial and municipal systems.

Two of the remaining three chapters deal with management techniques used to control water quality in the United States. Chapter 14 describes the types of standards (and the advantages and disadvantages of each) used to regulate water quality. Chapter 15 outlines the present legal and regulatory structures at federal, state, and local levels. Chapter 16 discusses several major current issues, problems in existing programs, alternatives proposed to correct them, and trends anticipated for the next decade.

The information in this book is important in establishing a necessary foundation for practicing engineers, scientists, and managers in the water quality field. The facts and concepts form a framework within which technological and managerial decisions can be made rationally.

Furthermore, it should be useful as a self-study course for practicing professionals who wish to review fundamentals, new professionals who need to develop background in the field, industrialists who require accurate knowledge of the subject in dealings with regulatory agencies, and informed members of the public who wish to understand water quality issues better.

I am indebted to Dr. Donald Francisco of the UNC faculty for some of the ideas in Chapter 3 and for his advice and assistance. Of course, any shortcomings in the final presentations are mine.

JAMES C. LAMB III

Contents

CHAPTER 3 Some Concepts of Aquatic Biology / 27

CHAPTER 4 Characteristics and Availability of Water Resources / 47

CHAPTER 5 Water Uses / 81

CHAPTER 6 Methods of Evaluating Water Quality / 105

CHAPTER 1

Introduction

Questions relating to water resources almost always are controversial in some quarters—or soon will become so. This is especially true of water quality matters because of complex technological considerations and wide perceptual differences among individuals and groups interested in the subject.

For example, some view streams and lakes as valuable principally as resources that can be used to produce profits by supplying industrial water and economically disposing of wastes. Accordingly, their approaches to water quality questions are based on economic considerations and evaluation of cost-benefit relationships. At the other extreme are those who feel that the most valuable uses of water resources lie in the soul-restoring benefits derived through the observation and enjoyment of aesthetic values, which are rapidly disappearing in modern technological societies. These are emotionally important benefits and must be approached from that point of view.

The differences between the two schools of thought often produce very emotional and heated confrontations between seemingly irreconcilable positions. Accordingly, the wide range of viewpoints represented by them and many others related to them make it unsurprising that many questions concerning water quality have all of the elements required for strong disagreements.

Evaluation of water quality goes far beyond merely determining which consitutuents are present and at what concentrations. This is because identification of specific physical, chemical, and biological characteristics is never the primary goal in practical consideration of quality. It is just the mechanism by which suitability of water for attaining the real goal can be judged and endpoints of quality control activities defined.

Meaningful evaluation and control of water quality must include technological aspects, but it is also necessary to consider fully the wide variety of desires and needs of various interested persons and groups. Thus, water quality matters always involve the balancing of public interests; ultimately, many questions can be settled only through legal and political actions. For that reason, political and bureaucratic organizations have become more involved in water quality matters during the past decade or two, often with substantial beneficial impact, but sometimes with an unfortunately negative impact.

This book deals with many technological and other aspects of water quality. I try to develop and present them realistically, within the framework and constraints of sociological, political, and economic situations often encountered in the United States today. It is not my objective to promote or defend materialistic-utilitarian approaches to water quality, nor is it my goal to extoll the beauty and other desirable aesthetic characteristics of water resources so well known to one who has spent many delightful hours boating, fishing, swimming, or just watching and enjoying the incomparable scenes of coastal waters. My goal is to present accurate information that readers can rely on in developing their own personal and professional rationales for understanding, evaluating, and resolving water quality issues.

This information makes it possible for newcomers to the field to develop viable, pragmatic, and defensible positions, whatever their positions might be. Individuals already experienced in water resources often can review some of the concepts to good advantage in updating and reevaluating their own positions.

DIFFERENCES ABOUT WATER QUALITY

Water quality means different things to different persons because of the various perspectives from which they approach it. Generally, it refers to characteristics or attributes of water, good or bad, that relate to its acceptability for certain purposes or uses.

From a technological point of view, those characteristics usually are defined in terms of appropriate physical, chemical, and biological parameters, preferably ones that can be measured quantitatively and reproducibly to avoid ambiguity in reporting or discussing them. In this context, water quality is relatively simple and straightforward, at least conceptually, because accurate quantitative data compiled in accordance with sound scientific principles form an excellent basis for discussion and agreement, even among those who might differ radically in other respects.

It is in the interpretation of water quality data that wide disparities exist among interested persons and organizations—that is, in determining whether the data indicate that quality is acceptable or what corrective actions might be necessary. These questions, of course, relate to potential utility of the water and often produce wide differences of opinion, even in the face of full agreement on specifics of technical water quality characteristics. For example, it is not difficult to understand that views of a manufacturer seeking an industrial supply and a stream to receive wastewaters from manufacturing operations could differ widely from those of an angler desiring to minimize interference with game fish. Both might disagree with others who desire to use the water for still different purposes.

Thus, water quality (what characteristics might be satisfactory or unsatisfactory) can mean very different things to different persons. This is the basis for most disagreements about current conditions of our streams and lakes and the validity of national and state pollution control programs. It also is the basis for much of the misinformation on water quality issues often encountered in news media, articles, and books. Many (most?) spokespersons seem compelled to take the position of strong advocate for some point of view, with little or no sense of responsibility either for recognizing differing but valid arguments or for adhering to facts and balanced perspective to avoid misleading the public and its representatives.

EVOLUTION OF WATER QUALITY PROBLEMS

Most water quality problems originate either with people or with industrial operations. Accordingly, many centuries ago, when populations were low and widely dispersed and major industrial discharges were almost nonexistent, there were few quality problems.

Nevertheless, they did occur when people congregated in cities and produced large flows of wastewaters without safe and sanitary means for disposing of them. Those conditions often resulted in deadly epidemics of cholera, typhoid fever, and other waterborne diseases, such as the ones that swept London, Hamburg, Philadelphia, Pittsburgh, and Lawrence. These problems became especially acute in the nineteenth century, in response to increasing urbanization.

Stream pollution problems other than those involving health impacts also were encountered, sometimes with dramatic impacts. For example, consider the conditions in London in 1858–59 resulting from pollution of the Thames River, as documented in the classic book *Typhoid Fever* (Budd, 1874):

> For the first time in the history of man, the sewage of nearly three millions of people had been brought to seethe and ferment under a burning sun, in one vast open cloaca* lying in their midst.
>
> The result we all know. Stench so foul, we may well believe, had never before ascended to pollute this lower air. Never before, at least, had a stink risen to the height of an historic event. Even ancient fable failed to furnish figures adequate to convey a conception of its thrice Augean foulness. . . .
>
> . . . At home and abroad, the state of the chief river was felt to be a national reproach. "India is in revolt, and the Thames stinks," were the two great facts coupled together by a distinguished foreign writer, to mark the climax of a national humiliation.

Consider also the graphic statement quoted by Clark on the main drainage works of Boston, in 1885 (Fair et al., 1966):

> Large (surrounding) territories were at once, and frequently, enveloped in an atmostphere of stench so strong as to arouse the sleeping, terrify the weak, and nauseate and exasperate everybody.

These and other problems also were experienced in many areas where there were major concentrations of industrial wastewater discharges.

THE ERA OF PUBLIC HEALTH CONCERN

During most of recorded history, the terrible epidemics were attributed to impure air, vapors arising from swamps, odors from decomposing animal or vegetable matter, acts of God, or admittedly unknown causes. Not until the late nineteenth century was the germ theory of disease developed and extended to provide an accurate understanding of the origins of waterborne disease outbreaks. This new information gave concrete evidence of the relationship between water and health as well as knowledge that could serve as the basis for controlling disease outbreaks by improving municipal water quality. Thereafter, under pressure from the public and their leaders, progress to that end became rapid.

The period around 1890–1910 marks the beginning of major concern about water quality, especially drinking water. This concern led to rapid advances in the development and application of water treatment technology as well as extensive evaluation of the resulting impacts of those improvements on human

*A sewer or cesspool.

health. During this golden period of the water supply field, extending to perhaps 1925, the key processes of filtration and chlorination were applied broadly to drinking water supplies, with dramatic impacts on the incidence of waterborne diseases, as outlined in Chapter 9. It also was during this period that wastewater treatment was first recognized as an important adjunct to the water supply industry and was initiated in some communities.

As noted, the rapid progress in water quality control was fueled by concern about public health. It therefore follows that after radical reductions in incidence of waterborne diseases during the first quarter of the century, public interest in water quality also diminished, since water safe for drinking and other purposes was taken for granted. For the next quarter century, progress continued at a steady but slow pace, with relatively minor improvements in technology, consolidation of knowledge, and extension of the benefits of modern water and wastewater systems to increasing segments of the population.

THE ENVIRONMENTAL MOVEMENT

After World War II, the rapid expansion in industrial activities and increased urbanization in the United States produced more adverse impacts on our water resources and caused rapid growth in water pollution problems. This situation combined with a new national affluence and an increased interest in water-based leisure activities to make the public more aware of undesirable watercourse quality and more vocal in their objection to it. Public pressures and recognition of water pollution as a national issue of major importance caused increased political attention and passage of more stringent laws and regulations.

The movement for improved regulation of water quality in streams quickly focused on industrial sources as the first line of attack for many reasons, including rapid industrial growth and general lack of progress in earlier years, when the primary target had been municipal wastewater sources. The trend for increased attention to water pollution as a part of a broad environmental movement continued with growing momentum through the decades from 1950 until the present, reaching a crescendo—sometimes even approaching a panic state—in the 1970s.

This period of increased concern about water quality was marked by attention principally to matters other than public health impacts of drinking waters. This is reflected in myriads of regulations developed specifically to protect fish, other aquatic life, and recreational values, and to prevent nuisances, as well as in pronouncements of public figures about goals of pollution control programs. For example, widely publicized statements by President Johnson and others ultimately resulted in a national policy, expressed in Public Law 92-500, of making all streams in the United States fishable and swimmable by the early 1980s—a goal clearly motivated by factors other than health (U.S. Congress, 1972). Although admirable as an expression of intent to expedite cleanup of the nation's waters, the differences in natural characteristics of streams throughout the nation and competing needs for use of those resources make simple and sweeping preemption of *all* streams for fishing and swimming of dubious wisdom and clearly unrealistic in setting national water quality goals.

The low priority given to human health matters in the 1970s, on the other hand, is illustrated, for example, by virtually negligible funding in the Environmental Protection Agency budgets for research on human health effects, as contrasted with much larger appropriations for other research within a total agency budget of billions of dollars.

A RENEWAL OF HEALTH CONCERNS

It is interesting to see that the situation has changed substantially in recent years because of new concerns about potential health impacts of toxic substances in the environment generally and in water especially. These concerns stem from disclosures about potential carcinogens and teratogens that have been identified and uncertainty about their effects on the health of humans who drink waters containing them over many years or over their entire lifetimes. The validity of many of the concerns is highly controversial, and in most instances the present state of knowledge is inadequate to provide reliable and clear answers. Nevertheless, they have generated a resurgence of interest in human health effects and a more cautious attitude about eliminating uncertain risks whenever feasible by preventing the presence of suspect chemicals in drinking waters.

The net effect is that in the 1980s an increasing part of the motivation for water pollution control is fueled by health issues, with somewhat less attention in research and technical literature to aquatic life and other aspects that are better understood and more amenable to control by using current technology. This suggests that the principal water quality control concerns of the next decade or two almost certainly will return to the public health arena, with special attention to toxic chemicals found in drinking waters and foods derived from aquatic life. These matters will be discussed in more detail in Chapter 9.

SYSTEMS OF MEASUREMENT

Anyone desiring to consider water quality seriously must understand the quantitative units used to describe it and also should be able to make simple calculations for constituents of water. Their concentrations and quantities present are vital factors in assessing the quality of water and its suitability for various purposes. They also form the basis for selecting specific treatment methods for correcting quality deficiencies and determining the costs of those steps.

The units in which water quantity and quality characteristics are measured and reported have always been complex and confusing to newcomers to the field because of the mix of English and metric units and because of unusual units often employed by various businesses and professions with vested interests in water. Now the situation is even worse because the units are in a state of transition from the commonly used English-metric mix to the SI (International System) units. This is part of the current trend toward metrication of all measurements in the United States.

The result is that reports, books, and journals now are divided along somewhat different lines with respect to the units employed—that is, what basic system of measurements is given editorial priority. Some remain mostly in the English system, especially those that deal with water quantities, stream flows, water uses, and the like. Others rely almost exclusively on the SI system, especially organizations and individuals dealing principally with water quality characteristics, analytical procedures, and research activities—activities in which the metric system has been employed for many years anyway. Several journals have elected to resort to dual units; one system is selected as the primary one for reporting data, and units in the other system are presented parenthetically for readers more attuned to them.

The situation presents a dilemma here because this book is aimed at readers with varied backgrounds and interests. A conviction that most persons in the United States still think principally in terms of the English system and are most familiar with those units of measure has been instrumental in deciding to continue usage of many English units in this book. Another important consideration is that much or most data on water quantities and uses still are reported in those units.

On the other hand, the metric system has been used almost exclusively for many years to report water quality data because most of them are based on laboratory measurements, and the metric system has been the language of science for many years. Accordingly, those types of data are reported here in metric units, which usually are the same as or closely related to units of the SI system. Where it seems to be desirable for effective communication with readers, both English and SI units are given.

Recognizing that many readers of this book are not familiar with units used in the water resources field, the appendix presents information on the SI system and methods for converting data between English and SI units. The appendix also contains information on how to make other types of calculations on water quantities and quality characteristics. Readers with solid academic backgrounds should be able to follow the information presented there without undue difficulty. Many will be able to meet their needs merely by scanning the appendix as a review or refresher of facts learned long ago. Others might experience difficulty in understanding some of the topics and should pursue those subjects further in encyclopedias, appropriate high school or college texts, or other references until the material is clearly understood.

The study questions at the end of this chapter include several questions and problems dealing with these matters and can be used as a self-test by readers. If difficulty is encountered in answering them, the appendix and other publications may be consulted as required.

A few years from now enough of the population of the country probably will have been "retreaded" to the SI system to make today's confusion on this subject a thing of the past. The painful, gradual movement in the United States to the SI system eventually will allow us to adopt entirely that more logical, correct, and convenient approach to measurement and data handling.

STUDY QUESTIONS

1. Identify several groups having different vested interests in water quality and uses of streams and lakes.
2. For each group in question 1, itemize the uses and quality characteristics of greatest concern to it and the basis for those concerns.
3. List other organizations with which each of these groups would be likely to disagree about water quality and outline specific policy areas in which those disagreements could be expected to occur.
4. Identify the seven basic measurement units of the SI system and write the proper symbol for each.
5. Identify the relationships among Kelvin, Celsius, and Fahrenheit temperatures. Be able to solve problems converting temperature measurements among those systems.
6. Identify the correct SI units, symbols, and base units for each of the following: concentration, area, density, velocity, volume, energy, work, force, power, and pressure.
7. What is the correct multiplier for each of the following SI prefixes: nano, giga, milli, tera, micro, kilo, pico, mega?
8. Identify each of the following conversion factors:
   pounds weight per cubic foot of fresh water ____________
   pounds weight per gallon of water ____________
   gallons per cubic foot ____________
   grams per pound ____________

millimeters per inch ________
liters per gallon ________

9. Given that the population of a town is 40,000 persons, the per capita wastewater flow is 100 gal/person/day (378 liters/person/day), and the concentration of phosphorus in it is 10 mg/liter: (*a*) calculate the pounds (kilograms) of phosphorus that could be recovered from this source daily, assuming 90 per cent efficiency of the recovery process; (*b*) calculate the tons of wastewater that must be processed daily to recover that phosphorus.
10. Stream flows commonly are reported in cubic feet per second. For a flow of 5 ft^3/sec, set up equations and convert that flow rate to the following units: million gallons per day (mgd), gallons per minute (gpm), liters per second (l/s); and cubic meters per day (m^3/d).
11. The volume of a drinking water reservoir is 500 million gal (1.89×10^6 m^3). Assuming that the permissible level of some potentially toxic chemical is 0.5 $\mu g/l$, and assuming uniform distribution of it throughout the reservoir, how many pounds (kilograms) of the chemical must be added to the lake to exceed that concentration?
12. Laboratory studies determine that a chlorine dosage of 4.0 mg/liter would be adequate for disinfection of some specific wastewater. With a plant flow of 5.5 mgd (million gal/day) (or 241 liters/sec), how much chlorine would be used monthly at the plant?

REFERENCES

Budd, W., *Typhoid Fever: Its Nature, Mode of Spreading, and Prevention,* London (1874). Reprinted by the American Public Health Association, Washington, D.C. (1931).

Fair, G. M., J. C. Geyer, and D. A. Okun, *Water and Wastewater Engineering: Volume 1, Water Supply and Wastewater Removal,* Chapter 1, Wiley, New York (1966).

Marshall, B. R., and J. C. Lamb, "Evolving Water Quality Regulations," in *Proceedings, Conference on Water and Energy: Technical and Policy Issues,* American Society of Civil Engineers, New York (1982).

U.S. Congress, "An Act to Amend the Federal Water Pollution Control Act," Public Law 92–500, 92nd Congress (1972).

CHAPTER 2

Some Important Characteristics of Water

Water is a vitally important substance, with properties that are unique, interesting, and wide-ranging in potential impact. An understanding of its nature and behavior requires knowledge of certain fundamental facts about its chemical and physical characteristics. Several of those characteristics play major roles in determining water quality and causing changes in it, either through natural phenomena or impacts brought about by humans and their activities.

BASIC PHYSICAL CHARACTERISTICS

Physical State

Perhaps the most unique characteristic of water is that it occurs on earth as solid, liquid, and gas within the range of climatic conditions commonly encountered. At low temperatures, it is present as ice, a solid in which the molecules form relatively rigid circular structures. In this state, the molecules have little motion and possess only low energy as compared with the other two states.

With the addition of heat, the molecules have higher rates of motion and more energy. They are still close to each other but no longer are fixed in a rigid structure. The resulting fluid moves readily to fit the shape of any container in which it is placed or flows from one location to another in response to the application of only slight force.

With the addition of still more heat, the molecules achieve even more rapid motion and higher energy levels. They are much more widely separated into a gaseous state and disperse readily among the molecules of other gases occupying the same volume, mixing thoroughly with them. This form is commonly referred to as water vapor.

The three physical states of water are not mutually exclusive—they can coexist and often do. Also, the portions in each state are not fixed or independent of each other, but molecules can exchange between them continuously. For example, when water is in contact with air, there is constant movement of molecules across the interface between the two phases. Some water molecules pass through the surface from liquid into air, becoming vapor while, at the same time, some vapor molecules move across the interface to become part of the liquid. This aspect of water will be discussed more fully shortly.

Density and Specific Gravity

The density of water depends on its temperature and chemical content. Density also is influenced by pressure, but close spacing of the liquid molecules requires very high pressure to cause even slight changes in density. Thus, in most practical water supply and wastewater applications, those effects are usually ignored and it is assumed that water is "incompressible."

The effect of temperature on water density also is small, but it is especially important in determining its behavior in many instances; for example, small density changes can produce effects of great importance in reservoirs and treatment systems. Table 2.1 shows the relationship between temperature and density for pure water. An especially important point about that relationship should be noted. The density of water increases with a reduction in temperature down to 39°F (4°C), but further cooling causes a slight decrease, producing its maximum density at that temperature (see middle column of Table 2.1). The density changes are small, but they are enough to have vitally important effects on freezing of water bodies, as will be discussed subsequently.

The effects of consitutuents in water on its density depend on the types of materials involved and their concentrations (quantity per unit volume of water). The density of "fresh" water, including most wastewaters, is usually cited in practical water resources literature as 62.4 lb/ft^3 or 8.34 lb/gal (1000 kg/m^3). Seawater usually has a density of about 64 lb/ft^3 (1025 kg/m^3).

The specific gravity of a substance is merely the ratio of its density to that of pure water at 39°F (4°C). Table 2.1 shows that the specific gravity of water is numerically equal to its density, when expressed in grams per cubic centimeter, because the density of water at 39°F (4°C) is 1.00000 g/cm^3.

TABLE 2.1
Density and Specific Gravity of Pure Water

Temperature (°C)	Density (kg/m^3)	Specific Gravity (or Density in g/cm^3)	Density (lb/ft^3)	Temperature (°F)
0	1000	0.99987	62.42	32.0
2	1000	0.99997	62.43	35.6
4	1000	1.00000	62.43	39.2
6	1000	0.99997	62.43	42.8
8	1000	0.99988	62.42	46.4
10	1000	0.99973	62.41	50.0
15	999	0.99912	62.38	59.0
20	998	0.99823	62.32	68.0
25	997	0.99707	62.25	77.0
30	996	0.99567	62.16	86.0
35	994	0.99406	62.06	95.0
40	992	0.99224	61.96	104
45	990	0.99025	61.82	113
50	988	0.98807	61.69	122
60	983	0.98324	61.38	140
70	978	0.97781	61.04	158
80	972	0.97183	60.67	176
90	965	0.96534	60.27	194
100	958	0.95838	59.83	212

EFFECTS OF HEATING AND COOLING

Specific Heat

The "specific heat" of a substance defines the relationship between the amount of heat transferred to or from it and the change in temperature produced by that exchange. For example, water requires the addition of 1.0 Btu to each pound to raise its temperature by one degree Fahrenheit, or 1.0 calorie to each gram to raise the temperature by one degree Celsius. Accordingly, its specific heat is 1.0 Btu/lb/°F or 1.0 cal/g/°C (4187 J/kg-K).

Note that specific heat is the same numerically (1.0 for water) in both the English and metric systems. That is because the British thermal unit is, by definition, the amount of heat required to raise the temperature of one pound of water one degree Fahrenheit, and the calorie is defined as the amount of heat required to raise the temperature of one gram of water one degree Celsius. The units in the SI system, of course, are substantially different (4187 J/kg-K).

Water has the highest specific heat in nature, except for very few substances, including ammonia. Examples of other values for specific heats include: 0.2 Btu/lb/°F (837 J/kg-K) for sand and 0.107 Btu/lb/°F (448 J/kg-K) for iron. This means that the transfer of a given amount of heat to or from water produces far less temperature change than it would in most other substances. That large heat capacity accounts for the stabilizing effects of large water bodies on atmospheric temperature and climate near them.

Vapor Pressure

As indicated earlier, the movement of molecules through an interface between water and air proceeds constantly in both directions. The rate of that exchange depends on molecular velocity which, in turn, is a function of the temperature of the liquid and gaseous phases and several other factors.

Consider the situation in which water is added to a vessel containing dry air and the vessel is closed to prevent further additions or losses of contents. The rate at which water molecules pass from liquid into the air is constant and depends on temperature. Initially, of course, no water molecules can move from the dry air into the liquid. As time passes, however, the concentration of water vapor in the air rises, causing an increase in the rate at which vapor molecules may reenter the liquid. In this example, the *net* rate of transfer from liquid to air must decrease as the two opposing processes continue.

Ultimately, the rate at which vapor molecules transfer from the air into liquid becomes equal to that at which they pass from liquid into air and *net* interchange between the two phases stops, although actual movement continues unabated in both directions. This produces a dynamic equilibrium in which concentration of water vapor in the gas phase is characteristic of the temperature and depends on several other factors, including impurities in the water.

One way to describe that equilibrium is by reporting the "partial pressure" of vapor—i.e., that portion of total pressure of gas phase within the vessel that is exerted by water vapor. That is called the "vapor pressure" and, for pure water, it varies with temperature, as shown in Table 2.2. The vapor pressure rises with temperature in the range of 32 to 212°F (0 to 100°C). At the freezing point, water still exerts a slight vapor pressure; in fact, even ice at substantially lower temperatures does so and can release water molecules as vapor.

At the upper end of the temperature scale in Table 2.2, the vapor pressure reaches 14.7 lb/in.^{2} (101 kPa), which is equal to atmospheric pressure at sea level. That produces spontaneous formation of vapor bubbles in the liquid (boiling) if pressure in the system is equal to 1 atm. At higher altitudes, water boils at lower temperatures because its vapor pressure reaches the reduced atmospheric pressure (the boiling point) at a lower temperature, as shown in Table 2.2.

TABLE 2.2
Vapor Pressure of Pure Water

Temperature (°C)	Vapor Pressure			Temperature (°F)
	mm Mercury	kPa	lb/in.^{2}	
0	4.58	0.61	0.088	32.0
2	5.29	0.70	0.102	35.6
4	6.10	0.81	0.118	39.2
6	7.01	0.93	0.136	42.8
8	8.04	1.07	0.156	46.4
10	9.21	1.23	0.178	50.0
15	12.8	1.71	0.248	59.0
20	17.5	2.33	0.338	68.0
25	23.8	3.17	0.460	78.8
30	31.8	4.24	0.615	86.0
35	42.2	5.62	0.816	95.0
40	55.3	7.37	1.07	104
45	71.9	9.58	1.39	113
50	92.5	12.3	1.79	122
60	149	19.9	2.88	140
70	234	31.2	4.53	158
80	355	47.3	6.87	176
90	526	70.1	10.17	194
100	760	101.3	14.70	212

The practical significance of vapor pressure in natural water-air systems includes the fact that it is vital in determining rates of evaporation from lakes and streams. Also, it is important in designing pumping systems because it limits the amount of negative pressure (vacuum) that can be applied by a pump without causing boiling. That places a practical limitation on how far the pump may be located above the surface of water to be moved.

Evaporation and Condensation

Applying information developed in the preceding section, whenever the vapor pressure of water exceeds its partial pressure in the atmosphere in contact with it, there is a *net* transfer from liquid to gas. That process is called evaporation. The opposite process—condensation—refers to a *net* transfer of molecules from gas to liquid, which occurs when partial pressure of water vapor in the gas exceeds vapor pressure of the water in contact with it. These processes provide the basic mechanisms by which water undergoes changes in physical state between liquid and gas in natural or other systems.

Evaporation and condensation must involve simultaneous changes in energy of molecules undergoing exchange between the two phases because those in the gaseous phase have much higher velocities (and therefore higher kinetic energies) than those in the liquid. Accordingly, the change of water from a liquid to a gas (evaporation) requires the addition of energy to the molecules in the amount of about 970 Btu per pound of water (2.26 MJ/kg), even without a change in its temperature. The magnitude of this "latent heat of vaporization" may be realized more fully by recognizing that only 180 Btu (190 KJ) of heat would be required to raise the temperature of that pound of water from freezing point to boiling point at sea level.

Because molecules of liquid that are vaporizing must absorb energy from (and cool) their surroundings, evaporation is referred to as a "cooling" process. Conversely, that same amount of heat is given off by molecules during condensation, causing it to be termed a "heating" process.

Natural evaporation and condensation processes are of special importance in the water resources field. Evaporation from land and water areas provides a continuous supply of vapor to the atmosphere, and subsequent condensation of that vapor as rain and snow supplies most of our freshwater needs. On the other hand, evaporation also can be an important debit; through that process, substantial amounts of our freshwater resources are lost to the atmosphere from lakes, streams, and vegetation. In treatment technology, the combination of controlled evaporation and subsequent condensation in distillation systems permits production of fresh water from ocean and other saline sources. Practical aspects of evaporation and condensation will be discussed in more detail in subsequent chapters.

Freezing and Melting

The interchange between liquid and solid phases also requires exchanges of energy between molecules and their surroundings. Changing from ice to liquid—the melting process—requires increases in molecular energy; therefore, heat must be added, similar in principle to the latent heat of vaporization described in the preceding section. In melting, however, there is less change in energy—only about 144 Btu per pound of water (336 kJ/kg). This is referred to as the "latent heat of fusion."

When water freezes, the molecules readjust in configuration to form a rigid structure in which they are more widely separated than they are in the liquid phase. This causes a decrease in density to about 57 lb/ft^3 (910 kg/m^3), which accounts for the fact that ice floats and has great practical significance. For example, as lakes and streams cool in the winter, the formation of ice at the surface provides a solid, fixed barrier between atmosphere and water and interferes with further cooling and freezing. That reduces the extent to which rivers and lakes freeze and the resulting impact on aquatic life in them.

Another practical effect of the density change is that the expansion of water when ice formation occurs in confined areas can produce great structural forces, which can cause bursting of metal pipes, cracking of swimming pool tanks, and accelerated disintegration of wet stones and concrete.

When the molecular arrangement of water changes during freezing, there is a strong tendency to exclude impurities from the crystalline structure of the ice. That provides the basis for processes that produce fresh water from seawater through freezing. The ice crystals formed in those processes are mostly fresh water and can be separated and melted to provide drinking water. The seawater salts excluded during freezing accumulate in the remaining brine, which is discarded as a waste stream.

Summary

Figure 2.1 summarizes the relationship between heat and temperature for one pound (0.454 kg) of pure water, as discussed in the preceding sections. Beginning at absolute zero temperature, −460°F (−273°C or 0°K), the temperature of ice rises as heat is added. The relationship is a curved line because the specific heat of ice varies somewhat with temperature.

When the melting (freezing) point is reached, at 32°F (0°C), the further addition of heat produces no net change in temperature until enough energy has been added to satisfy the latent heat of fusion and convert the ice to liquid water. Thereafter, the addition of heat results in a temperature rise until the boiling point is reached, at 212°F (100°) for one atmosphere pressure.

No net temperature change occurs during vaporization of the water because the energy satisfies the latent heat requirements until all of the water has been converted from liquid to vapor. Subsequently, the temperature rises again in response to heat additions, but the curve is steeper than that for liquid water because the specific heat of steam is lower.

CHEMICAL CHARACTERISTICS

Water's unique physical characteristics, discussed earlier, can be accounted for by molecular structure, which differs from most other compounds in important ways. In this section, we examine some chemical characteristics that can have great influence on water quality and its changes. The subjects will not be developed in detail here because of space limitations, but they are discussed further

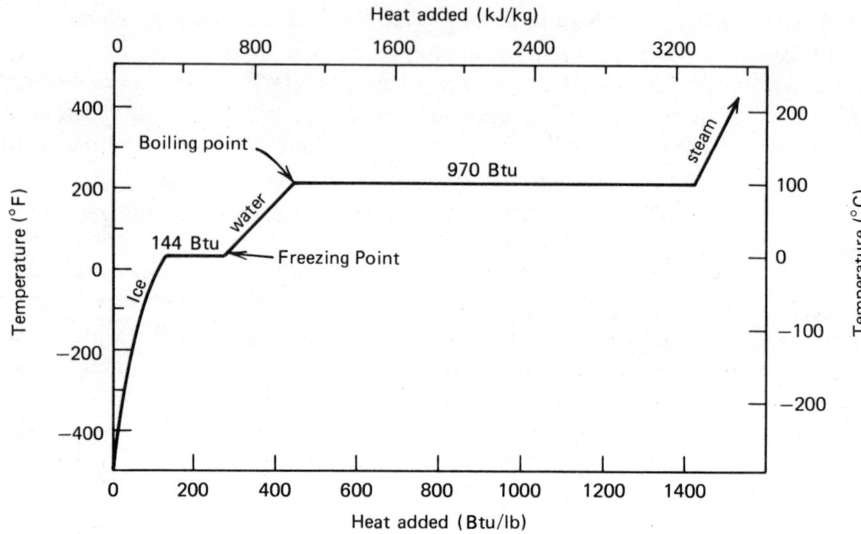

FIGURE 2.1
Relationship between heat added to water and its temperature.

in other references (Davis and Day, 1961; Leopold and Davis, 1966; Wetzel, 1975).

Molecular Structure

It is generally known that a water molecule is composed of three atoms: one oxygen and two hydrogen. Functionally, this description is convenient and adequate for most purposes, although it is oversimplified.

The hydrogen and oxygen atoms are joined in a water molecule through a linkage that is called a "covalent bond." Each hydrogen atom needs one additional electron in its structure to make it chemically stable, and each oxygen atom needs two additional electrons. In water, the two hydrogen atoms and one oxygen atom share electrons, thereby satisfying the needs of each and forming an exceedingly strong bond between them. The molecule is very tight in structure and highly stable, accounting for some of water's unique characteristics.

The water molecule is not symmetrical because its hydrogen atoms are attached to the oxygen atom at points separated by an angle of 105°. This lopsided physical arrangement causes an unbalanced distribution of electrical charges because the oxygen atom, on one side, is negatively charged and each hydrogen atom, on the other side, has a positive charge. These produce a "polar" molecule—similar in effect to a magnet—that is affected by electrical fields, including those exerted by nearby similar molecules. The electrostatic interaction causes hydrogen atoms in one molecule to be bonded to the oxygen atom in an adjoining molecule, connecting them to each other very firmly and producing structural continuity throughout the liquid. This type of interaction is referred to as "hydrogen bonding."

The effects of having covalent bonds within water molecules and hydrogen bonding between them account for the unusual characteristics of this compound. The high strength of its bonding system requires expenditure of more than the usual amount of energy, for a molecule of its weight, before it can escape from the liquid into the gas phase. This causes water to have an unusually high latent heat of vaporization. The bonds also account for water's high latent heat of fusion and high heat capacity.

Its bonding strength also is responsible for the unusual cohesive nature of water, producing very high surface tension—high enough, for example, to support insects on its surface. Hydrogen bonds are formed with glass and other compounds that water "wets," producing forces that are responsible for its capillary action. This makes it possible for water to move through soils, roots, plant structures, and the human body.

The strong intermolecular forces (hydrogen bonding) cause ice crystals to form expanded structures, decreasing the density of water in freezing and making ice float. It also causes the slight expansion of water between 39°F (4°C) and the freezing point because, in that temperature range, bonding forces overcome the kinetic energy of some molecules and begin formation of a semirigid structure. That produces the maximum density of water at a temperature above freezing.

The molecular structure of water really is more complex that depicted here. There are three isotopes of hydrogen, which have the same chemical characteristics but different atomic weights, and three isotopes of oxygen. Furthermore, some water molecules dissociate slightly into several different charged ions, compounding the types of chemical structures that are present. Thus, there are many different combinations of atoms, molecules, and ions, even in the purest water. Details of these factors, however, are beyond the scope of this discussion but are covered in other publications dealing with water characteristics and chemistry.

Ionization and pH

Some water molecules dissociate into electrostatically charged ions, as illustrated by the following reaction:

$$H_2O \leftrightarrow H^+ + OH^-$$

The amount of dissociation is very small—only about 10^{-7} moles* each of H^+ and OH^- ions are present in a liter of pure water. Nevertheless, this characteristic is vitally important because the effectiveness of water in dissolving certain constituents is influenced greatly by the concentrations of hydrogen and hydroxyl ions.

An increase in acidity is reflected by a higher concentration of H^+ and a corresponding lower OH^- concentration. The addition of a base has the opposite effect. The character of dissociation is such that the product of the concentrations of the two types of ions is constant at a given temperature; in this example, that product would be equal to about 10^{-14} at about 25°C (77°F).

Because of very small numbers involved in the ionic concentrations and their wide ranges in practice, a logarithmic scale has been adopted to describe the extent of water acidity. This "pH" scale is defined as the logarithm of the reciprocal of the hydrogen ion concentration, expressed as before†. In pure water, for example, the hydrogen ion concentration is 10^{-7} moles/liter. The reciprocal of 10^{-7} is 10^7, and the logarithm of that number is 7. Therefore, a pH of 7 indicates that the hydrogen ion concentration is the same as that for pure water.

The presence of a higher concentration of hydrogen ions (more acid condition)—say 10^{-6} moles/liter—would produce a lower pH (6 in this example). Thus, a lower pH indicates that water is more acid, while a higher pH means reduced acidity. Note that because the pH scale is logarithmic, a change of one pH unit actually reflects a *tenfold* change in acidity. In the preceding example, a pH of 6 means that the water is ten times as acid as pure water (pH = 7) would be. A pH of 5 would represent another tenfold increase in acidity.

WATER AS A SOLVENT

Water is as close as any compound to being the "universal solvent"; virtually all chemicals are soluble in it to some extent. This is one of the most important results of its molecular structure.

This characteristic often creates problems because it is practically impossible to produce and store pure water. Although high-quality water can be produced with modern equipment, maintaining that high quality is impossible because the purity of the "universal solvent" immediately begins to degrade by

---

*A "mole" is one gram molecular weight of the substance—in this example, 1.008 grams of hydrogen (H^+) or 17.008 grams of hydroxyl ions (OH^-).

†Actually, hydrogen ion "activity" is used instead of concentration, but the difference is of no concern here.

slowly dissolving whatever container is used to hold it. That can be a serious problem in delicate laboratory analyses.

Atmospheric precipitation (distilled water) dissolves elements and chemicals in the air during rainfall, adding significant amounts of many constituents to the water before it even reaches the ground (see Chapter 7 for examples). Subsequently, during passage over and through the earth, it dissolves other organic and inorganic substances. Seawater contains hundreds of constituents, many at relatively high concentrations, and most are derived from the solution of substances contacted by the water en route over or through the soil in its journey to the sea.

The atoms in many substances are bonded primarily through simple electrostatic attraction, based on their having electrical charges of opposite sign. These types of bonds are much weaker than the covalent bonds in water and are weakened even further by the polar nature of the water molecule, which cancels some of the electrostatic attraction. This permits atoms of the substance to separate, become surrounded by water molecules, and exist as "ions," which are electrically charged atoms or groups. Accordingly, atoms joined by these electrostatic forces—ionic bonds—readily dissolve in water. On the other hand, the water molecules retain their basic structure because of covalent bond strength and hydrogen bonding. For that reason, most substances dissolved in water do not change the basic chemical character of the water itself.

Volume changes involved in preparing solutions of various substances are not additive; the final volume of the solution cannot be predicted from the volumes of water and solute. In some instances, the volume of the water increases after the addition of solutes, in others it decreases, and sometimes it does not change at all because the solute molecules are interspersed between water molecules.

Many chemicals (such as sulfuric acid) liberate large amounts of heat on solution in water, but others (such as potassium nitrate) absorb heat, producing a cooler solution. The solubilities of most materials in water increase with a rise in temperature, but there are important exceptions. For example, calcium carbonate decreases in solubility as the temperature rises. This is important in practice because heating water that contains calcium carbonate can produce scale in pipes and water heaters, leading to a loss of capacity or premature failure. On the other hand, this relationship sometimes is used beneficially by heating water to be used in boilers to enhance treatment efficiency in removing calcium carbonate.

WATER AS A BIOLOGICAL FLUID

The role of water in the evolution of life now is widely recognized. Primitive forms began in the oceans, and living plants and animals that exist today evolved from those origins, adapting as necessary to meet special needs generated by the environment in which they are located. Frequent note is made of similarities in saline content of human body fluids and the oceans from which we evolved, leading to the observation that we carry our ocean environment within us.

The importance of water in maintaining plant and animal life cannot be overstressed; it is the one truly vital fluid, for many reasons. Its solvent action makes it possible for water to serve as the medium by which nutrients are dissolved from soil and carried to various locations throughout plants. Movement of water through plant membranes provides osmotic pressure that forces the fluid and its dissolved nutrients from the roots to high levels in tall plants. This is enhanced by hydrogen bonding, which permits water to "wet" the walls of tiny channels within the plant and is the basis for the capillary action that moves the liquid through them. The hydrogen bonding also provides the internal strength of water needed to withstand negative pressures generated during those movements.

The human body contains 50 to 70 percent water, averaging about 65 percent. Water permeates all parts of the body, from 22 percent of bone weight to 83 percent of the blood and kidneys; a loss of only about 15 percent of that liquid could cause death. The chemical reactions involved in human and other animal metabolisms are carried out in an aquatic environment. Water is a major vehicle for moving nutrients within the body and removing salts and other wastes from it. Its large latent heat of vaporization forms the basis for cooling by evaporation from sweat glands. Water even provides the mechanism for cleansing eye surfaces, lubricating them to ease movement of the eyelid (Leopold and Davis, 1966).

Water plays similarly important roles in microbiological life processes. It provides the basic environment within which organisms live and contains the elements and chemical compounds that make their survival and growth possible. It is the medium by which metabolites are transported from remote locations to supply the needs of organisms and the means by which nutrients are transported through cell walls and wastes products are removed later.

Water is the original elixir of life.

WATER AS A GEOLOGICAL TOOL

Since its first appearance on earth, water has constantly rearranged the physical nature and location of its landscape through several mechanisms that will be outlined briefly here.

First, the solvent action of water—a result of its molecular structure—has been of major importance. As indicated earlier, virtually everything is soluble in water to some extent. Its solution processes begin during precipitation, even before reaching the ground, and transfer constituents from the atmosphere to the earth below. Subsequently, during passage over and through the soil, water dissolves many organic and inorganic chemicals contacted by it, including even the stones over which it flows. The burden of dissolved substances is revealed in chemical analyses of natural waters, which are discussed in more detail in Chapter 7.

The impact of raindrops on soils and stones has been likened to small bullets because of their physical strength and high velocities of travel. The result is that centuries of hammering raindrops gradually wear away even the toughest stones through erosion, slowly changing their sizes and shapes. The materials eroded sometimes may dissolve in the water, but in other instances they may be removed as particulate matter. The rates of solution and erosion may increase markedly and be strikingly rapid when rain contains corrosive materials. For example, the impact of acid rain on marble statues and concrete structures is especially damaging in many areas today.

Water also has a major geological impact through transporting dissolved and suspended matter to other locations. This may be accomplished by flow across land and in streams, in which suspended particles are further reduced in size by abrasion against each other and the streambeds over which movement occurs. Large quantities of suspended and dissolved matter are transported constantly from land areas to lakes and oceans (see Chapter 7). This movement accounts for the loss of soils from land areas, sometimes in alarming amounts, and the gradual accumulation of those materials in the lakes and oceans of the world.

Finally, sedimentation of much of the removed matter occurs after reaching quiescent stretches of streams, lakes, and oceans. Suspended solids of substantial size settle readily after reaching a water body in which there is little turbulence. This is a major problem in many artificial impoundments because the accumulation of solid deposits causes a reduction in water storage capacity. Smaller materials, even of colloidal size, often are coagulated or precipitated by salts and settle when they reach the oceans. Through these mechanisms deltas are formed, for example, in the sea near the mouths of rivers.

STUDY QUESTIONS

1. Describe the relationship between temperature and density of water, identifying important ways in which it differs basically from most other substances.
2. A drum has a volume of 55 gallons (208 liters). What is its weight when filled with fresh water at 50°C (122°F)?

3. What is specific gravity? If water in an estuary has a specific gravity of 1.013, what is its density in (*a*) pounds per cubic foot, and (*b*) kilograms per cubic meter?
4. What is specific heat, and what is its practical significance?
5. How much energy, in joules and British thermal units, must be added to five pounds (2.27 kg) of water to raise its temperature from 15°C to 75°C?
6. What is vapor pressure?
7. What is meant by the term "boiling?" Identify the conditions that could cause water to boil at a temperature of 80°C (176°F) and explain why.
8. Water at 75°C (167°F) is added quickly to half fill a vented tank containing dry air at 75°C (167°F) and 14.7 lb/in.^{2} (101.3 kPa), which then is tightly capped. What will be the pressure in the tank several hours later, at a temperature of 75°C (167°F)?
9. Explain why evaporation is referred to as a cooling process.
10. What are "latent heat of vaporization" and "latent heat of fusion?" Give reasonable values of each for water. How do these values differ from those for most other substances? Why?
11. Explain why ice floats and describe three important practical results of that phenomenon.
12. How many joules must be added to 50 pounds of ice at 0°C to convert it to water vapor?
13. Describe the molecular structure of water and explain the role and significance of its covalent bonds and hydrogen bonding.
14. Explain the relationship between the molecular structure of water and its unusual heat-temperature characteristics.
15. What is the effect on the hydrogen ion concentration of a change in pH from (*a*) 8.0 to 9.0, and (*b*) 6.0 to 4.0?
16. Why is water so effective in dissolving substances that it often is referred to as the "universal solvent?"
17. Explain the importance of water in the life processes of microorganisms and humans.
18. Describe the mechanisms by which water rearranges global landscapes.

REFERENCES

Davis, K. S., and J. A. Day, *Water, the Mirror of Science,* Anchor Books, Garden City, N.Y. (1961).

Feth, J. H., *Water Facts and Figures for Planners and Managers,* Circular 601-I, U.S. Geological Survey, Washington, D.C. (1973).

Leopold, L. B., and K. S. Davis, *Water,* Time, Inc., New York (1966).

Leopold, L. B. and W. B. Langbein, *A Primer on Water,* U.S. Geological Survey, Government Printing Office, Washington, D.C. (1960).

Swenson H. A., and H. L. Baldwin, *A Primer on Water Quality,* U.S. Geological Survey, Government Printing Office, Washington, D.C. (1965).

UOP Inc., Johnson Division, "Everything You've Always Wanted to Know About Water . . . ," *The Johnson Drillers Journal,* (May–June 1972).

U.S. Geological Survey, *What Is Water?* (pamphlet), Washington, D.C. (1982).

U.S. Geological Survey, *Why Is the Ocean Salty?* (pamphlet), Washington, D.C. (1984).

Weast, R. C., and M. J. Astle, Editors, *CRC Handbook of Chemistry and Physics,* 63rd Edition, Chemical Rubber Publishing Company, Boca Raton, Fla. (1982).

Wetzel, R. G., *Limnology,* Saunders, Philadelphia (1975).

CHAPTER 3

Some Concepts of Aquatic Biology

Water quality and aquatic biology are closely interrelated. The types of biological life that develop in a watercourse depend heavily on its quality characteristics, making the populations of organisms in aquatic communities measures of water quality (Greeson, 1982). The other side of that coin is that the quality of water may be greatly influenced, for better for worse, by the species and numbers of organisms in it. Clearly, then, an understanding of water quality and the factors affecting it requires at least a reasonable acquaintance with certain aspects of aquatic biology.

The origins of life in watercourses may be divided into two broad categories. First, there are organisms that are not native to that environment but are added to it in land runoff, wastewater discharges, dustfall from the atmosphere, and other mechanisms. These organisms include many important forms of life, such as pathogens capable of causing human disease (discussed in more detail in Chapter 9). Although they may survive in water for extended periods, most of them do not thrive in that environment and do not multiply or even remain viable indefinitely.

Other organisms are native in that they often are present in watercourses and find that environment favorable, or even preferable, to others for continued existence and growth. These include, for example, many types of algae commonly found in lakes, which play important roles in determining the quality of the water. The native organisms are almost all nonpathogenic to humans, although a few may have direct or indirect health significance. Native organisms will be the principal subject of the aquatic biology discussions that follow in this chapter.

SOME FACTORS THAT AFFECT SURVIVAL AND GROWTH

Whether or not a specific type of organism thrives in a watercourse depends on temperature, chemicals available, pH, light, other organisms present, and many other environmental factors. These are explained in bacteriology, biochemistry, and ecology texts, but the details are beyond the scope of this discussion.

However, for an example, consider that a simplified description of a microorganism's composition may be depicted by a chemical formula summarizing its atomic content, such as $C_{106}H_{180}O_{45}N_{15}P$. When analyzed in detail, organisms also are found to contain lesser amounts of other trace elements, which are vitally important to them. Slack (1982) reports that 40 to 50 essential plant nutrients are known; Table 3.1 lists 19 elements that are essential for the growth of freshwater plants (Kimmel, 1982).

For growth (multiplication) to occur, necessary elements must be available to the organism in its food supply or in the water; otherwise, they cannot build new cell materials. Of course, there is much more to the matter than just the presence of appropriate atoms. They also must be available in the types of chemical structures that can be assimilated and incorporated into cells by that type of organism. The specific elements and chemical structures that are required vary among different species. Also, various species have different needs, or preferences, with respect to temperature, light, pH, and other environmental factors. Thus, the quality characteristics of a watercourse may be highly favorable for some organisms but less desirable, or even unfavorable, for others. That is one reason why most pathogens, for example, do not fare well in watercourses; they are favored by environmental conditions present in the bodies of humans, not those in natural waters.

An important environmental characteristic of natural water environments is that they are extremely dilute. It is not uncommon for the food supply and required nutrients of an organism to be present in water at concentrations of only a few milligrams per liter or less, and often they occur at concentrations of only a few micrograms or nanograms per liter. Thus, it is common for food or some trace element to be the factor that limits rates and amounts of growth for many species in natural waters. A corollary is that striking changes in rates of growth and in the biological balance among species often can be initiated by adding only very small amounts of specific chemicals to the system.

TABLE 3.1
Elements Essential for Growth of Aquatic Plants[a]

Element	Symbol	Plant Content (%)
Oxygen	O	80.5
Hydrogen	H	9.7
Carbon	C	6.5
Silicon	Si	1.3
Nitrogen	N	0.7
Calcium	Ca	0.4
Potassium	K	0.3
Phosphorus	P	0.08
Magnesium	Mg	0.07
Sulfur	S	0.06
Chlorine	Cl	0.06
Sodium	Na	0.04
Iron	Fe	0.02
Boron	B	0.001
Manganese	Mn	0.0007
Zinc	Zn	0.0003
Copper	Cu	0.0001
Molybdenum	Mo	0.00005
Cobalt	Co	0.000002

Source: Kimmel (1982).

[a]Living tissues of freshwater plants.

THE ECOLOGICAL NICHE AND FOOD WEB

An organism must operate within limits established by its genetic characteristics. Especially important are the enzymes that it can produce, which in turn identify the specific chemicals that it can metabolize, the degree to which they can be utilized, and the ultimate degradation by-products of the reactions. Accordingly, organisms in ecosystems may be viewed as occupying "niches." That is, each species plays a specific role based on what it may accomplish metabolically and the environmental conditions under which it functions.

Microorganisms are highly specialized; they are limited in the types of chemicals that they can use as substrates and in the reactions that they may carry out on those chemicals. Thus, one organism may be able to use only a portion of a complex organic compound, leaving the remainder as a by-product, or waste. Unless removed from the organism's environment, accumulations of that waste ultimately could inhibit its further activity. A different organism in the community may be unable to use the original chemical as a substrate but can metabolize the by-product of the first species, taking the degradation another step and releasing yet another waste material to the environment. In turn, that may be assimilated by a third species, carrying the chemical breakdown still further.

This type of biochemical chain reaction carried through many steps may degrade a complex chemical completely, supplying the metabolic needs of many different types of organisms in the process. In this type of system, each organism has an important symbiotic relationship with its neighbors; it degrades the wastes of those above it in the chain, making possible their continued functioning, and supplies substrate that can be used by others below it in the chain. In short, it occupies a specific and specialized role, or niche, within that ecological system.

Note that the niche of an organism does not depend solely on its food supply but is also a reflection of other environmental conditions affecting its survival and growth. Furthermore, ecological niches also include predators—organisms that use others as sources of food—and those predators themselves may be victims of different predators (Cassell et al., 1966).

Actually, niches of organisms are not as clearly defined as the preceding discussion might suggest. There may be hundreds or thousands of chain reactions in a biological system that extends from bacteria to fish. The varied food chains often are cross-linked with each other because the biological niches of organisms participating in them can overlap in many ways and to different degrees with respect to usable food supply and environmental conditions favorable to growth. These overlaps often produce direct competition among organisms for food supplies, further complicating the dynamics of the community organization and activities. For that reason, the term "food web" frequently is used to summarize the reactions underway and suggest their complexity and interactions (Campbell, 1977; Odum, 1971; Warren, 1971, 1976).

It is clear that an ecological system such as that normally present in a stream or lake, with its many substrates and organisms, is an extraordinarily complex assembly of biochemical activities. Detailed information on these subjects is available in many references on aquatic biology and ecology (Campbell, 1977; Hynes, 1979; NAS, 1969; Odum, 1971; Warren, 1971, 1976; Wetzel, 1971; Whitton, 1975) but is beyond the scope of this discussion.

POPULATION DYNAMICS

The fact that food and nutrients usually are present only in very low concentrations and frequently are growth-limiting factors produces a situation in which growth in natural watercourses often occurs mostly in response to additions of those substances. The types of substrates present and the environmental conditions determine which species of organisms are favored, and the amounts of necessary chemicals available to them dictate the extent of growth for each type.

That is responsible for establishing the types of biological balances often found in relatively clean watercourses. Most streams and lakes contain many types of chemicals, largely derived from natural sources, but only low concentrations of each. Accordingly, many species of organisms develop in response to the varied chemical supply, but only a few organisms in each species can be produced because of limitations in substrate availability. This type of biological balance—many species, but few organisms in each—is termed a "diverse" population and is viewed as a desirable watercourse characteristic because it results in increased community stability (Odum, 1971; Warren, 1971, 1976). Typical polluted streams, on the other hand, have less diversity because they contain far fewer species, with relatively large numbers of those favored by the organic and nutrient additions to the water. The concept of diversity can be summarized by various expressions, a simple one of which follows, and has been used as an index of water quality, as will be discussed later (Averett, 1982; Warren, 1971, 1976; Odum, 1971).

$$\text{Diversity} = \frac{\text{Number of Species}}{\text{Number of Individuals}}$$

An important characteristic of microorganism communities in watercourses is their variability. The types of species and numbers of each shift in response to changes in food supplies and chemical availability, as indicated by the considerations discussed earlier. Beyond that, changes in environmental conditions also can produce wide variations in communities. As temperature, sunlight, pH, stream flow, and other factors change, seasonally and otherwise, watercourse communities are impacted, with some species gaining a competitive edge and others suffering in comparison. These can produce wide swings in the biological balance of the ecosystem from time to time.

Furthermore, although overlooked sometimes even by professionals in the field, all complex ecosystems are dynamic. That is, the balance among various groups of organisms is always changing, even though they may receive constant supplies of substrates and nutrients under constant environmental conditions. This can be attributed to interaction among the organisms through competition for limited system resources and to the presence of species that are predators for some groups (Cassell et al., 1966; Warren, 1976). Accordingly, communities in streams, lakes, and biological treatment systems are never constant in composition, making the "steady-state" balance of organisms that usually is assumed in deriving and applying mathematical models for them nonexistent in nature!

TROPHIC STRUCTURE

The trophic structure of an ecosystem pertains to its nutritional aspects. Of course, as suggested earlier, any type of biochemical reaction involves the growth of microorganisms and that, in turn, requires the presence of adequate chemicals and trace elements to supply their needs in building cell materials. Because carbon is a key element in nutrition of all organisms and growth of cells, it deserves special attention. Processes involving transformations of carbon compounds may be divided into two broad groups: photosynthesis and respiration.

Photosynthesis results in biochemical production of organic compounds and oxygen, using inorganic carbon and other chemicals, with sunlight as the source of energy.

$$CO_2 + H_2O + \text{light} \rightarrow (CH_2O)_n + O_2$$

Respiration, on the other hand, is the group of biochemical processes in which organic matter is utilized as the substrate, consuming oxygen and releasing carbon dioxide, water, and energy.

$$(CH_2O)_n + O_2 \rightarrow CO_2 + H_2O + \text{energy}$$

Energy released in respiration reactions is that in the chemical bonds between atoms in organic compounds—often the same energy that was stored there during photosynthesis.

Organisms capable of photosynthesis commonly are referred to as "primary producers" because they can use relatively simple inorganic chemicals and energy derived from sunlight to produce organic matter in the form of additional living organisms. Those, in turn, can serve as a food supply for other forms of life that are incapable of photosynthesis. On land, for example, trees, bushes, and grass are primary producers, generating vegetation that is used for food by a wide variety of life forms.

As illustrated in Figure 3.1, in watercourses this role is filled by various types of green plants, including phytoplankton (algae, which are microscopic floating plants), various attached forms of algae, and higher rooted or floating aquatic plants (see Box 1). Also a few bacteria can use inorganic chemicals, without photosynthesis, to produce cell material (chemosynthesis). All of the producers serve as food supplies for herbivores, or grazers, which are referred to as "primary consumers" (Box 2). They include protozoa, which feed on bacteria and small algae, and rotifers and crustaceans, which feed on protozoa and algae. A few species also feed on rooted or floating plants, such as the Israeli carp, which may be useful for controlling hydrilla growths. Small fish serve as secondary consumers (Box 3) by feeding on protozoa, crustaceans, and rotifers,

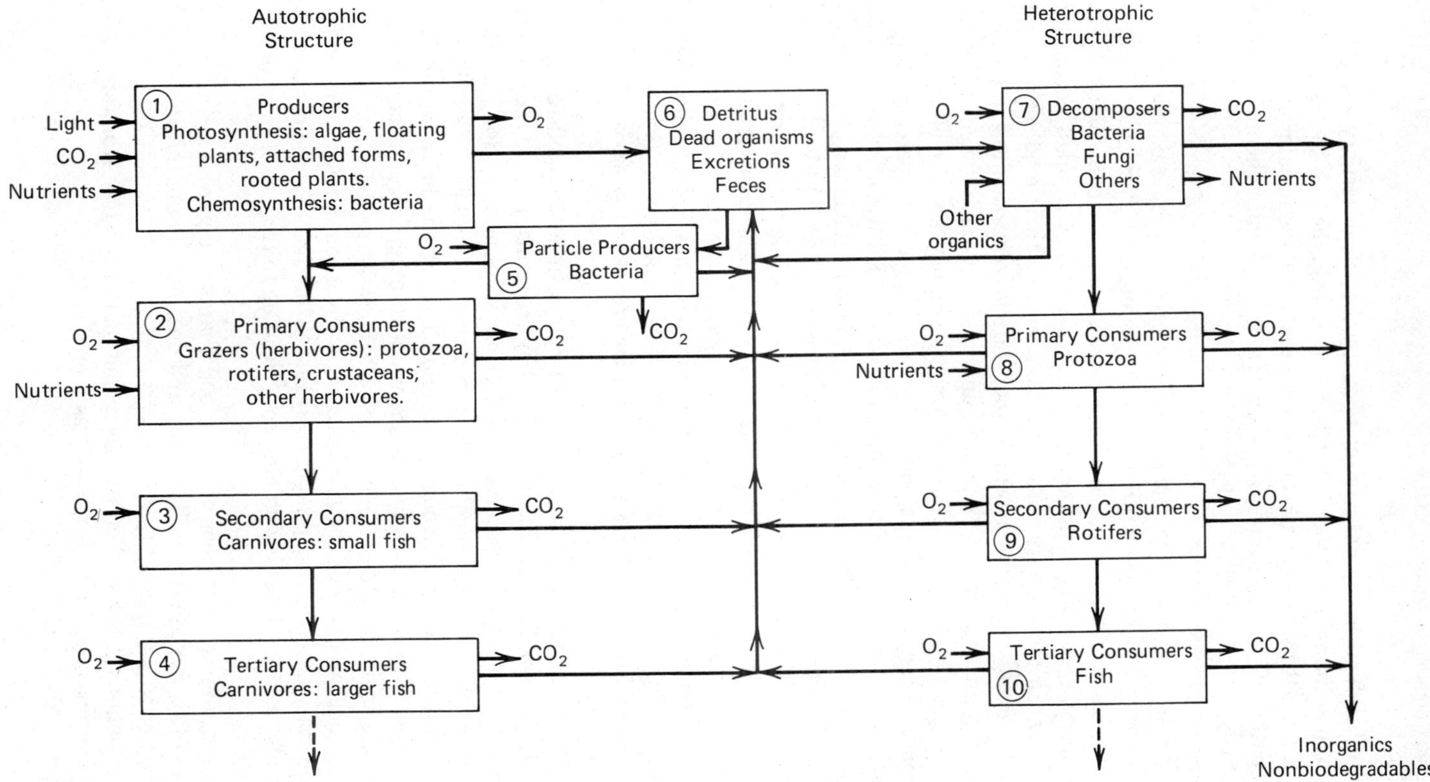

FIGURE 3.1
Illustrations of trophic structures.

while large fish feed on smaller ones and occupy the role of tertiary consumers (Box 4). When humans or other animals consume fish, they fill the role of yet another level in this trophic structure. Because most food materials in this "grazer pathway" are derived originally from inorganic chemicals, it is referred to as an "autotrophic" structure.

The photosynthetic primary producers use some portion of the incident sunlight energy in converting the inorganic chemicals to organic compounds in cells. Less energy is used in converting those organisms to primary consumers because efficiencies of the biochemical processes are only perhaps 10 to 30 percent. Similarly, secondary consumers effectively use only a small portion of energy present in the primary consumers that serve as their food supply, and the available energy is reduced even further during conversion to the tertiary forms. The significance of this in determining overall system efficiency may be seen by using a simple hypothetical example, assuming that the efficiency of energy conversion in each step along the trophic ladder is 10 percent. Beginning with energy imparted by the sunlight, only 0.1 of it would appear in primary producers, 0.01 in primary consumers, 0.001 in secondary consumers, and 0.0001 in tertiary consumers. The remainder is released as heat.

As shown in Figure 3.1, all of the forms release organic matter to the water through the excretion of waste products or on death. These materials collectively are known as detritus (Box 6). Some bacteria can assimilate dissolved organics that cannot be used by the primary consumers, building cell material that subsequently can be used by them. These serve as special types of producers and sometimes are referred to as "particle producers" (Box 5) (Kimmel, 1982).

Detritus is decomposed by bacteria (primary decomposers) and forms the basis for another similar food chain (Box 7). Higher forms that use the decomposers for food include protozoa, which are primary consumers (Box 8). Rotifers can be secondary consumers, utilizing bacteria and protozoa (Box 9), followed by small fish as tertiary consumers (Box 10) and, perhaps, larger fish. All of the consumers release organics to the water through excretion and death, which recycle as detritus to provide for more bacterial growth.

Ultimate end products of the processes include carbon dioxide, other inorganic chemicals, and nonbiodegradable organic by-products. These conversion processes also are relatively low in efficiency, leading to the loss of most of the energy as heat.

The basic difference between the two food chains in Figure 3.1 is that the one on the left is fueled principally by sun and does not require the addition of organic matter to the system, that is, it is "autotrophic." The one on the right is "heterotrophic"; it is based on using organic matter, derived from an autotrophic system or other sources (Box 7) as a food supply.

STREAM CHARACTERISTICS AND EFFECTS ON AQUATIC BIOLOGY

Water bodies have physical and chemical characteristics that greatly influence the aquatic environment and affect types and numbers of organisms that develop. In this section, we will discuss some characteristics of streams and how they can influence biology.

There can be major differences among various streams, or even along the length of a single one. For example, consider a stream that is the mountainous origin of a major river. Because its upland watershed is small and relatively isolated, stream flows are low and little sediment enters from the predominantly forested area. The steep slopes of the streambed cause high velocities, and bottom scour removes silt and clay, leaving rocks, gravel, and sand, depending on the velocities actually attained. Accordingly, organisms that populate this zone must be able to cling to stones in the presence of high velocities, and the absence of soft bottom deposits make this a poor environment for burrowing organisms.

The water is cool and clear and contains little organic matter because the stream has received neither discharges of municipal or industrial wastewaters nor runoff from farmlands or urban areas. Concentrations of nitrogen and phosphorus are low and limit the production of phytoplankton, making the water low in biomass content. Also, low concentrations of organics limit heterotrophic growths. Overall, the communities are diverse but sparse in numbers.

As the stream flows into piedmont zones, it gains increased drainage area, producing higher flow. Its slopes are flatter, reducing velocities and bottom scour, and its depth and width are greater. Runoff from farmlands and, possibly, the addition of municipal and industrial wastewaters add suspended matter, nutrients, and organics to the stream. Some sediment settles to the bottom in pools with reduced velocity, producing deposits of fine sand, silt, and organics. Temperature and turbidity are higher, decreasing suitability for some types of fish, but burrowing organisms now can develop in the soft bottom, allowing the development of different types of fish that feed on them.

Increased quantities of organics, nitrogen, and phosphorus added in runoff and wastewaters result in more growth of phytoplankton and greater accumulations of biomass in the water as well as changes in species of organisms present. If the additions of organics are great enough, dissolved oxygen may be reduced and cause an additional impact on the aquatic community balance.

As the stream approaches the coast, its drainage area is still larger and its slopes are even flatter, producing a slower and deeper flow. In the lower reaches of the estuary, where stream meets ocean, there is a pronounced tidal action with a periodic reversal of flow. The reduced velocities, flow reversals, and differences in specific gravity allow seawater to penetrate upstream in a wedge along the river bottom, often for many miles above its mouth. Higher salinity eliminates suitable conditions for some freshwater organisms but allows the development of different communities favored by that change in environment.

The additions of farm and urban runoff and wastewaters may greatly increase the quantities of sediment, organics, nitrogen, phosphorus, and other chemicals in the water. These, combined with reduced light penetration through the deeper and more turbid water and increased salinity, create a radically different environment for aquatic life. The result is an ecological balance in the estuary that is substantially different from those in unidirectional, freshwater flows upstream.

Differences notwithstanding, all streams have certain common characteristics that are basic and important in determining the type of ecosystem that develops. Perhaps most significant is the fact that it is a lotic (running water) system. The impacts that this can have on aquatic life depend on whether the organisms are fixed in location (attached) or move with the water (plankton).

Looking first at attached growths, consider a simple example. Beginning at a point where a food supply (in runoff or a wastewater, perhaps) is introduced, attached organisms downstream of that point are supplied with it by transport in the water flowing past. Waste products that they release after food utilization are transported downstream, where they become food supplies for different species of organisms located there; in turn, these organisms release their byproducts into the moving water. Repeating this pattern, the types of attached growths in the watercourse change in a downstream direction to reflect the different environmental conditions existing in those locations (Bartsch and Ingram, 1959; Warren, 1971; Hynes, 1970).

This situation provides a good environment for the development and maintenance of fixed growths. They are supplied with needed food and chemicals and merely extract them from the flow moving by. Also, their waste products are removed promptly, making possible their continued development.

On the other hand, the arrangement has substantially different effects on plankton, which are suspended in and move with the water. Usually, relatively few organisms that can use the added food are present in the flowing water at the point of addition, making assimilation slow. As the food and organisms move along together, its use results in the multiplication of plankton and reduces the food concentration until the environment finally becomes unfavorable for the further growth of those species. Other organisms that can use waste products of the first group develop in response to the increase in their food supplies, also reaching peak concentrations and running out of food later. The pattern is repeated for other organisms in the food web, none of which are maintained in a constant environment for long. This produces sequential development of different species as the flow passes downstream and their environment changes, as illustrated so clearly by Bartsch and Ingram (1959). Thus, flowing streams are somewhat less favorable for phytoplankton than for fixed growths, each species of which is exposed to more constant environmental conditions and food supply.

Growths in streams tend to be largely heterotrophic because photosynthesis plays a less important role than in lakes. That can be attributed to relatively short times within which growth must take place in the moving water, as opposed to longer detention times in lakes. Also, streams usually carry higher sediment loads and have turbulent conditions less favorable for settling, producing more interference with light transmission through the water to supply needs of photosynthetic organisms.

LAKE CHARACTERISTICS AND EFFECTS ON AQUATIC BIOLOGY

Lakes differ from streams in that they are lentic (still water) bodies, and some of their physical and chemical characteristics produce significantly different environmental conditions and ecological systems. One should not take too literally their classification as still waters, however, because there can be substantial movements within lakes. These are induced by stream flow that passes into and out of most lakes, Coriolis force, wind action, density currents, and other vertical and horizontal movements brought about by various causes. Thus, their classification as lentic bodies is relative; velocities are generally much lower and less unidirectional than those in most streams.

With low velocities and long detention times—varying from days to years—an obvious effect is that the quiescent conditon of most lakes is favorable to the settling of sediment. This results in the movement of most of the incoming suspended matter and much of the organic content, as well as constituents attached to those particles, downward to accumulate in the bottom waters and deposits. Turbidity of the upper water is reduced; this allows sunlight to penetrate further and transmit energy to phytoplankton, synthesizing organics through photosynthesis. Thus, in most lakes the predominant reactions in upper levels involve the production of cell material from inorganic chemicals, using sunlight as the energy source. Assuming that adequate supplies of nitrogen, phosphorus, and trace elements enter the lake with runoff and stream flow into it, this results in higher populations of primary producers than would occur in streams under similar chemical conditions. That, in turn, produces greater growth of grazers and other forms higher in the trophic structure. In short, lakes are more productive via the autotrophic pathway than streams of similar chemical quality.

The downward movement of suspended matter and many of the phytoplankton produced in upper waters enrich the lower zones with organics that serve as the substrate for heterotrophic biochemical reactions. Accordingly, in strata below the level of light penetration adequate to maintain photosynthetic activity, the principal reactions involve respiration by bacteria and degradation of organics. These exert oxygen demands and often deplete the dissolved oxygen resources of bottom waters, with the potential for creating serious water quality problems.

Many of the organisms produced in both upper and lower strata ultimately die and settle to become part of the bottom deposits. There, they slowly decompose, and some organics, nutrients, and other products of their decomposition leach back into the overlying waters where they again can enter that ecological system. The exact extent and rate of recycling of organics, nutrients, and trace elements from bottom deposits to overlying waters is difficult to evaluate accurately and remains somewhat uncertain and controversial today.

The net effect of all of these reactions is the gradual accumulation of organisms and their by-products in the lake. This process is referred to as eutrophication and is the end product of reactions triggered by green plants using chemicals dissolved in the water and sunlight for energy. Over extended periods of time eutrophication can result in accumulations sufficient to change the character of a lake permanently, greatly increasing its organic content and community of organisms and ultimately converting it to marsh and land areas. Although it is a process that can proceed naturally without the intervention of humans, eutrophication may be accelerated drastically through the addition of key nutrients and other chemicals in wastewaters and land runoff, as outlined in Chapter 8. The increased eutrophication rate sometimes is referred to as "cultural eutrophication," in which the watercourse is enriched through human activities.

Many attempts have been made to develop classifications of lakes based on trophic state, as well as hydrologic and geologic classifications (Wentz, 1981). The details of proposed systems are beyond the scope of this discussion, but it is worth pointing out that the term "oligotrophic" refers to lakes that are on the nutrient-poor, unproductive end of the trophic scale. "Eutrophic" lakes are ones that are nutrient rich and highly productive.

A characteristic of a deep lake that is important in determining its ecological nature and water quality is thermal stratification. As shown in Figure 3.2, beginning during the winter there may be ice on the surface of the lake overlying denser water below at 39°F (4°C). When the weather warms during the spring, surface waters absorb heat from incident sunlight, melting the ice and warming the water. As the water at the surface continues to warm and decrease in density during the spring and summer, the difference in tempera-

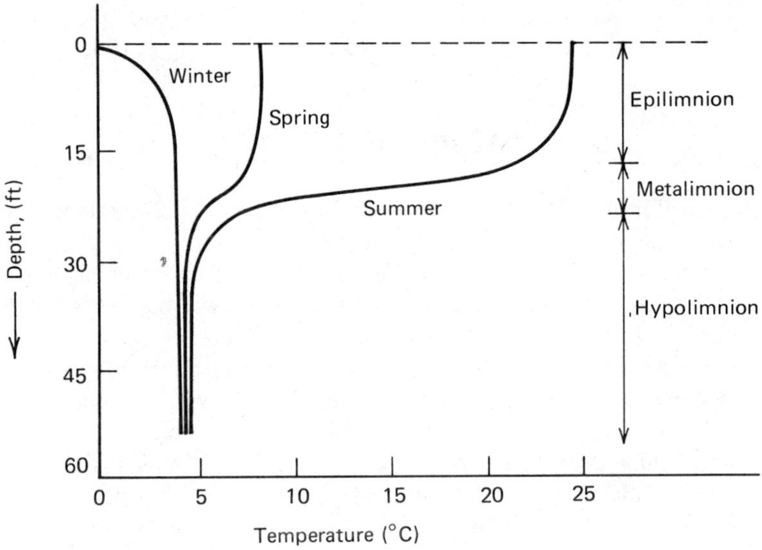

FIGURE 3.2
Examples of lake temperature profiles.

ture between surface and bottom waters increases, producing greater differences in specific gravity along a vertical profile.

This difference in density often is enough to inhibit mixing between upper and lower zones in deep lakes, leading to formation of three distinct layers. The upper one (epilimnion) may be 15 to 25 ft deep and contains the warmer water, as well as most of the phytoplankton, developed in response to available sunlight. The bottom layer (hypolimnion) is colder, receives little sunlight and, therefore, has little or no photosynthetic activity. Because it is enriched with organics, the respiration and consumption of dissolved oxygen can produce conditions unsuitable for sustaining fish life or many other aquatic organisms. Accordingly, that environment can be radically different from the epilimnion, causing an entirely different ecological balance. The transition zone between the two is known as the metalimnion, or thermocline.

As the lake cools in the fall, its surface waters, in the epilimnion, increase in density. Ultimately, the densities of the epilimnion and hypolimnion approach each other closely enough to permit top-to-bottom mixing of the lake by wind action. This usually occurs within a short period of time—a few hours—and is referred to as the "fall overturn" of the lake. At that time, of course, there is a radical change in the water quality in both the upper and lower strata and a corresponding upheaval in their ecological systems. This often causes temporary taste and odor and other problems in supplies that withdraw water from the lake, requiring changes in methods of treatment.

BIOLOGICAL IMPACTS OF POLLUTION

The biological impacts that result from adding potential pollutants to a stream or lake vary, depending on the life present in the watercourse, types of materials, amounts added, dilution available, temperature, other chemicals in the water, and many other factors. Some specific effects will be discussed later in other chapters, but a few of the principles are outlined here.

As pointed out earlier, the ecological system in a clean watercourse usually includes many species of organisms, but relatively few of any one type. This diversity is viewed as a stable and desirable system; in fact, many ecologists contend that any departure from that condition represents pollution of the stream and should be avoided. Controversial aspects of that position are discussed by Warren (1971, 1976) and in subsequent chapters of this book.

Pollution impacts on aquatic biology may be divided into two broad categories: (1) the generation of excessive growths in watercourses, and (2) the elimination, or major reduction, of some life forms. The first category sometimes results after the addition of large quantities of certain chemicals that can be used by some species to produce objectional growths. A specific example is the situation that occasionally has been observed downstream from some pulp and paper mills that discharge large amounts of wood sugars. Those chemicals can be used by filamentous organisms, called *Sphaerotilus,* to produce voluminous growths that are unsightly, clog fishing nets, and cause other water quality problems.

The second category may be illustrated by considering chemicals that are toxic to some aquatic organisms. Different species vary in their sensitivities to potentially toxic chemicals. Therefore, the addition of a specific chemical to a stream may sharply reduce, or even eliminate, some life forms that are sensitive to it, with little or no impact on other species. Progressively greater additions may adversely affect more organisms and exert greater impact on the biological balance by eliminating some members of the ecosystem entirely, decreasing its diversity. Depending on the position of affected organisms in the food web, there also may be secondary effects on other species that are not directly impacted by the chemical in question. For example, the elimination of a fish-food organism by a toxicant could adversely impact the fishery, even though the fish themselves may not be harmed by the chemical.

In many instances, wastewater discharges contain neither toxic chemicals nor ones that produce nuisance growths, but they still can exert major effects on diversity by modifying environmental conditions and upsetting the ecological balance. For example, a municipal wastewater discharge may contain large quantities of many organics, none of which are classified as toxic. The oxygen demand exerted by heterotrophic organisms using those chemicals as substrates may reduce the dissolved oxygen downstream to levels low enough to impact fish and other aquatic organisms. This can decrease populations of some species that are unfavorably impacted and increase the numbers of other species that can use the organics. This reduces system diversity both by decreasing the number of species and increasing the number of individual organisms.

BIOLOGICAL MONITORING OF WATER QUALITY

An understanding of the impacts of pollution on aquatic life has made it possible to use biological measurements to assess water quality in streams and lakes and to evaluate the effects of specific discharges on watercourses. The concept is simple, although its application can be highly complex. The basic idea is to examine biological life at various locations and compare the species and diversity found there with some "norm" for that type of watercourse, considering the season of the year, region of the country, characteristics of that and similar watercourses, and many other factors (Averett, 1982; Hynes, 1970; Odum, 1971; Slack, 1982; Warren, 1971, 1976; Wilhm, 1975).

One approach is to evaluate water quality based on populations of "indicator organisms" (Warren, 1971, 1976; Averett, 1982). This is based on the concept that various types of organisms have different sensitivities to water quality conditions and that the quality can be assessed by observing which and how many organisms are present and which are absent. For example, aquatic biologists have noted over many years that some forms of algae are very tolerant of pollution and occur frequently in polluted streams, while others are sensitive and seldom, if ever, found under those conditions. By observing the types of algae present in a watercourse and comparing them with compilations of information about their sensitivities to pollution, an estimate can be made of the overall water quality (Palmer, 1969.)

An examination of plankton in streams provides information covering a short time frame, because those organisms move with the current and indicate, principally, conditions in the specific mass of water from which the sample was taken. On the other hand, an examination of the attached or bottom growths provides an evaluation over a longer period because those organisms are mostly fixed in position in the moving flow and represent water quality conditions as integrated over an extended time. Sometimes, examining large bottom dwellers can provide even more information by evaluating the ages of living samples. Insects, larvae, birds, and other species can provide further information on secondary environmental impacts of pollution.

The indicator organism approach has been used by many past investigators but is viewed as more limited in value than other techniques now available because major variations in biological life can occur from time to time and place to place, even without the intrusion of pollutional effects on the system. Also, most species commonly found survive or thrive under a wide range of conditions, and this can be a serious limitation in trying to use them as indicator organisms (Warren, 1971, 1976; Averett, 1982). A preferable approach, according to Warren (1971), is to evaluate the overall composition of the community of organisms in the water, including species present and relative abundance of each. This is viewed as more reliable than merely determining which species are present or looking only at one or a few specific species as indicators.

The approach favorerd by many ecologists today is based on an evaluation of diversity, possibly combined with some elements of the preceding techniques. Various methods for defining species diversity have been proposed (Averett, 1982; Warren, 1971, 1976; Odum, 1971).

Several major industries have used these techniques to assess impacts of new manufacturing facilities on the local environment. One approach is to conduct a comprehensive and detailed analysis of aquatic and other life near the plant site before its construction, followed by similar periodic surveys after it has been placed in operation. Comparisons of indicator species populations and diversity before and after start-up can provide valuable information about environmental impacts.

Another approach to evaluating the impacts of wastewaters on stream life is to identify the species present and the diversity above and below the points of discharge. Virtually identical populations above and below suggest that the wastewater has little or no impact on the watercourse. However, changes in diversity, or loss of sensitive species, between the two points would imply an adverse effect.

STUDY QUESTIONS

1. Identify the most important chemical elements required by microorganisms and explain the basis for their needs, based on production of new growth.
2. Explain why the addition of relatively small amounts of nitrogen or phosphorus, for example, to a stream or lake may result in rapid and extensive

growths of organisms that otherwise could continue at very low concentrations for extended periods.

3. What is "diversity?" Is the diversity in a clean stream high or low? Why?
4. Explain what happens to diversity in a stream after the addition of large amounts of pollutional materials.
5. Identify three factors that could cause variability in microorganism populations in streams and lakes and explain why each could cause changes.
6. What is the "food web?" Explain why complex food webs may be involved in the breakdown of organic chemicals.
7. What is meant by the "niche" within which an organism operates? Why is the concept of niches important in developing an understanding of metabolic action of microorganisms in streams and lakes?
8. Distinguish clearly between photosynthesis and respiration. Outline the role of each in watercourses.
9. What are "primary producers?" What is the relationship between primary producers and consumers in watercourses?
10. Outline the nature of trophic structure in a natural water.
11. What is the difference between autotrophic and heterotrophic biochemical reactions? What does each type accomplish in watercourses?
12. Identify and explain the watercourse characteristics that cause major differences between the ecologies of streams and lakes.
13. Explain the effect of stream slope on its physical characteristics and the types of biomass that could be expected to develop in it.
14. Explain the differences in phytoplankton activities between streams and lakes.
15. Why would the ecological balance in an estuary differ substantially from that in a free-flowing zone upstream?
16. What is thermal stratification of lakes? Explain how it happens and its significance.
17. What is the metalimnion? The epilimnion? The hypolimnion? How do their environments differ?
18. Why is oxygen often depleted from lower regions of lakes?
19. What is the fall overturn? Why does it occur? What is its significance?
20. Explain the effects of adding pollutional materials on the diversity of a watercourse.
21. Describe the rationale behind biological monitoring, using indicator organisms.
22. Identify and explain other approaches to biological monitoring of water quality.
23. Describe how the impacts of a specific wastewater discharge on a stream may be evaluated through biological monitoring.

REFERENCES

Averett, R. C., "Species Diversity and Its Measurement," in Circular 848-B, pp. B3–B6, U.S. Geological Survey, Washington, D.C. (1982).

Bartsch, A. F., and W. M. Ingram, "Stream Life and the Pollution Environment," *Public Works,* Vol. 90, p. 104 (July 1959).

Campbell, R., *Microbial Ecology,* Wiley, Halstead Press, New York (1977).

Cassell, E. A., F. T. Sulzer, and J. C. Lamb, "Population Dynamics and Selection in Continuous Mixed Cultures," *Journal Water Pollution Control Federation,* Vol. 38, pp. 1398–1409 (1966).

Greeson, P. E., "Why Biology in Water-Quality Studies?" in Circular 848-A, pp. A3–A5, U.S. Geological Survey, Washington, D.C. (1982).

Hynes, H. B. N., *The Ecology of Running Waters,* University of Toronto Press, Toronto (1970).

Kimmel, B. L., "The Ecological Role(s) of Aquatic Micro-organisms in Lakes and Reservoirs," in Circular 848-E, pp. E3–E12, U.S. Geological Survey, Washington, D.C. (1982).

McCoy, G. A., "Periphyton," in Circular 848-A, pp. A23–A26, U.S. Geological Survey, Washington, D.C. (1982).

McCoy, G. A., "Phytoplankton," in Circular 848-A, pp. A17–A22, U.S. Geological Survey, Washington, D.C. (1982).

National Academy of Sciences, *Eutrophication: Causes, Consequences, and Correctives,* Washington, D.C. (1969).

Odum, E. P., *Fundamentals of Ecology,* 3rd Edition, Saunders, Philadelphia (1971).

Palmer, C. M., "A Composite Rating of Algae Tolerating Organic Pollution," *Journal of Phycology,* Vol. 5, pp. 78–82 (1969).

Slack, K. V., "Drift Organisms in Streams," in Circular 848-A, pp. A27–A30, U.S. Geological Survey, Washington, D.C. (1982).

Slack, K. V., "Stream Biology," in Circular 848-A, pp. A7–A16, U.S. Geological Survey, Washington, D.C. (1982).

Tilley, L. J., "Family Chironomidae (Diptera)," in Circular 848-A, pp. A31–A36, U.S. Geological Survey, Washington, D.C. (1982).

Warren, C. E., *Biology and Water Pollution Control,* Saunders, Philadelphia (1971).

Warren, C. E., "Toward a Conceptual System for Water Pollution Biology," in *Handbook of Water Resources and Pollution Control,* edited by H. W. Gehm and J. I. Bregman, Van Nostrand Reinhold, New York (1976).

Wentz, D. A., "Lake Classification—Is There Method to This Madness?", in Circular 848-B, pp. B15–B24, U.S. Geological Survey, Washington, D.C. (1981).

Wetzel, R. G., *Limnology,* Saunders, Philadelphia (1975).

Whitton, B. A., *River Ecology,* University of California Press, Berkeley (1975).

Wilhm, J. F., "Biological Indicators of Pollution," *River Ecology,* edited by B. A. Whitton, University of California Press, Berkeley (1975).

CHAPTER 4

Characteristics and Availability of Water Resources

It is not intuitively obvious, but the evaluation of water quality must include careful consideration of available quantities as well. Many quality problems and their solutions can be influenced radically by the amounts and flows that are available, their variability, rates of renewals and losses, and amounts of water needed for the intended uses.

This chapter briefly describes some elements of hydrology that often are important in water quality evaluation and control. It is not a definitive presentation of the technology of that field, nor is it a summary of its concepts and methods. That is better left to the many texts and other publications on various aspects of hydrology, several of which are referred to specifically in the pages that follow. Concepts discussed here are accurate and well known, but some are developed in ways that differ from viewpoints to which the public often is exposed. Special efforts are made to explain the rationale behind them and provide an understanding of how they relate to water quality.

DISTRIBUTION OF WATER RESOURCES INVENTORY

For practical purposes, the amount of water on earth is constant. Table 4.1 summarizes its approximate distribution at any given time (Nace, 1967, 1976; Feth, 1973). About 97.2 percent of it is in the oceans, another 2.15 percent is

TABLE 4.1
Summary of Global Water Distribution

Location	Cubic Miles	Cubic Kilometers	Percent of Total	Average Depth[a]
Oceans	317,000,000	1,323,000,000	97.2	1.6 mi
Saline lakes, seas	25,000	104,000	0.008	8.0 in.
Icecaps, glaciers	7,300,000	30,500,000	2.15	196 ft
Groundwater	2,000,000	8,350,000	0.61	52 ft
Moisture in upper soil	16,000	67,000	0.005	5.1 in.
Freshwater lakes	30,000	125,000	0.009	9.6 in.
Rivers (average volume)	400	1,670	0.0001	0.13 in.
Atmosphere	3,100	12,900	0.001	1.0 in.
Others	90,000	375,000	0.028	28.9 in.
Total	326,000,000	1,362,000,000	100	

Source: Adapted from data presented by Nace (1967, 1976) and Feth (1973).

Note: Figures do not total precisely because of rounding.

[a]Computed as if spread uniformly over the entire earth.

immobilized in icecaps and glaciers, and 0.61 percent exists as groundwater. The remaining small fraction of 1 percent is divided among all other resources, including lakes and rivers.

Without minimizing the importance of oceans, it may be observed that our principal concerns currently relate to the freshwater resources that are suitable for drinking and other uses closely associated with human activities. Those include lakes, rivers, and a small fraction of the groundwater, all of which total less than 0.1 percent of the global water inventory. Most groundwater is not readily available as a practical resource or is unsuitable in quality for uses requiring fresh water. Glaciers and icecaps are used only infrequently for freshwater supplies and represent a negligible fraction of our practical resource.

About 13 percent of the total U.S. fresh water is in lakes (including the Great Lakes). Eighty-six percent is groundwater, but probably less than half of it is shallow enough to be available as a practical resource, and much of that is poor in quality (Nace, 1976; Koenig, 1963). The water in all U.S. river channels, from which we obtain 77 percent of our total fresh water for use in municipal systems, industries, and irrigation, represents only about 0.03 percent of our inventory at any given time.

If it were necessary to rely only on the fixed inventory of fresh water to meet our needs, the supply would be exhausted very quickly. For example, water in the 3 million miles of U.S. rivers averages only about 12 cubic miles, or 13 trillion gallons. That would be enough to supply our current total water uses (about 450 billion gal/day) only for about one month. If we include, in addition, all fresh water in U.S. reservoirs and lakes, that would increase life of the resource only to 25 to 30 yr.

We are able to maintain our present level of freshwater use only because the limited inventory is constantly renewed by the hydrologic cycle. This is crucially important if one is to understand the nature of freshwater resources and how best to utilize them. They are not simply fixed resources from which we make withdrawals, such as iron, coal, and petroleum, but must be recognized as dynamic systems. Our demands on them must be evaluated within the context of rates at which they are being renewed.

THE HYDROLOGIC CYCLE

Figure 4.1 summarizes the basic elements of the hydrologic cycle. Under favorable meteorological conditions, atmospheric water vapor can precipitate as rain or snow. Some of that which reaches the ground flows over its surface into streams and lakes and is termed surface runoff. Another portion infiltrates into the soil and moves downward to become part of the groundwater resource.

Water infiltrating downward ultimately accumulates in a region in which soil pores are completely filled with water—the saturated zone. Its upper limit is the surface of an underground stream or lake and is called the groundwater table. Water in this zone flows toward lower elevations and, eventually, much of it reaches the oceans, either directly by underground travel or by emerging from the ground to flow in surface streams. Groundwater characteristics will be discussed in more detail later in this chapter.

The amount of water vapor in the atmosphere is small (Table 4.1), but it is replenished continuously by evaporation at many locations. Some of the falling precipitation evaporates quickly and returns to the atmosphere as vapor, and part of that reaching the ground subsequently evaporates from soil and the surfaces of lakes, streams, and swamps. Vegetation plays a dual role in the process. Some precipitation is intercepted on the leaves of trees and plants and evaporates before reaching the ground. Also, plants extract water from the soil in their metabolic activities and release it through leaves into the atmosphere, a process called transpiration. A large proportion of all vapor in the atmosphere originates in evaporation from the oceans.

Atmospheric water vapor is moved by air currents and ultimately may precipitate in locations remote from where the evaporation occurred. This trans-

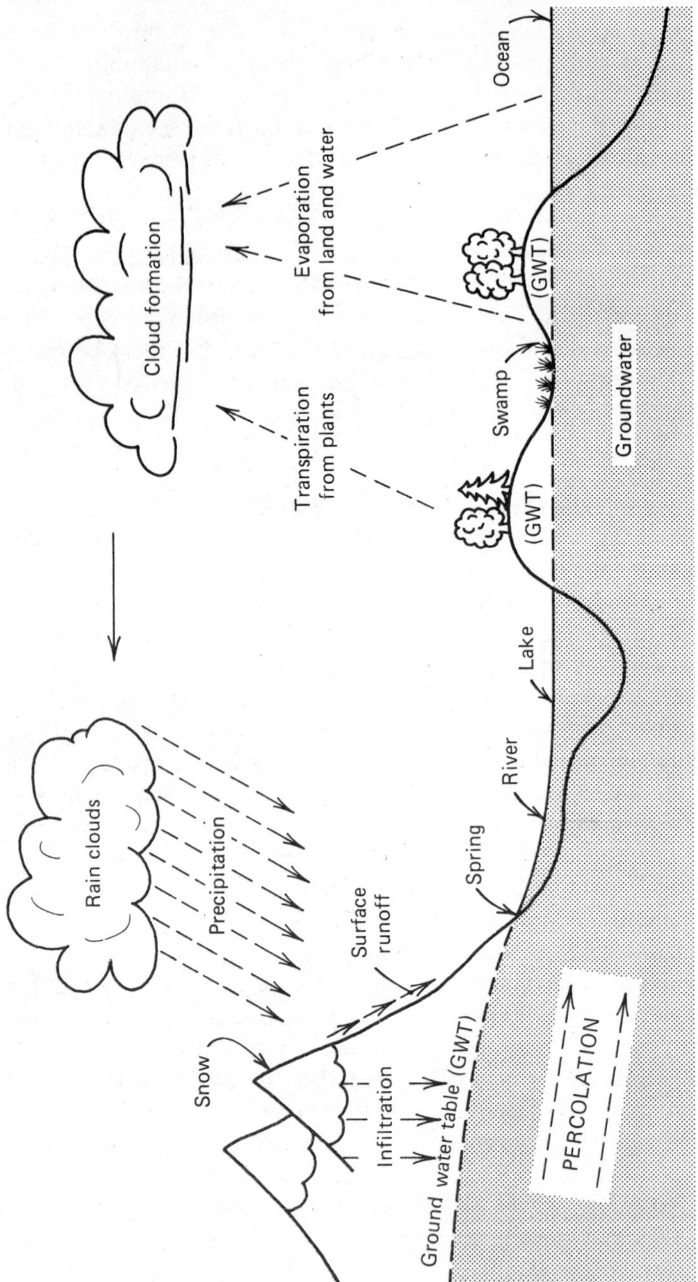

FIGURE 4.1
The hydrologic cycle.

ports water from oceans to inland areas and transfers it between land areas, completing the hydrologic cycle and providing our continuing supplies of fresh water.

Table 4.2 summarizes the worldwide movements of water by the hydrologic cycle. The total exchange is about 125,000 mi^3/yr (521,000 km^3/yr), or 0.038 percent of the global water resource listed in Table 4.1. Because the total water in the atmosphere approximates 3100 mi^3 (12,900 km^3) (Table 4.1), this produces a "turnover" in atmospheric water about 40 times per year, or an average of once each 9 days. Table 4.2 also reveals that 25,000 mi^3 (105,000 km^3) precipitates on land annually, but only 17,000 mi^3 (72,000 km^3) evaporates from land. The difference of 8000 mi^3 (33,000 km^3) annually is the total global replenishment of fresh water by net transfer from oceans to land areas.

Instead of cubic miles, or other conventional volume units, rainfall data often are expressed as depths to which water would accumulate if distributed uniformly over the area in question. Table 4.2 summarizes the average global water fluxes in inches per year and meters per year of evaporation or precipitation. The global evaporation averages 39.8 in./yr (1.01 m/yr), which must equal the global precipitation. The average evaporation from land areas is 18.9 in./yr (0.48 m/yr), while the precipitation on them averages 28.5 in./yr (0.72 m/yr). On the other hand, evaporation from oceans (48.4 in./yr or 1.23 m/yr) exceeds precipitation by 3.6 in./yr (0.09 m/yr).

The equivalent of about 40 to 50 trillion gal of water moves daily across the United States as atmospheric water vapor, only about 10 percent of which precipitates—4200 billion gal/day, or 30 in./yr (0.76 m/yr) (USWRC, 1978; Nace, 1976). The difference between water vapor incoming from the oceans

TABLE 4.2
Global Water Movements in the Hydrologic Cycle

Location	mi^3/yr	km^3/yr	Percent of Total	in./yr	m/yr
Evaporation					
Oceans	107,800	449,000	86	48.4	1.23
Land	17,200	72,000	14	18.9	0.48
World	125,000	521,000	100	39.8	1.01
Precipitation					
Oceans	99,800	416,000	80	44.8	1.14
Land	25,200	105,000	20	28.5	0.72
World	125,000	521,000	100	39.8	1.01

Source: Nace (1976).

and that leaving has been estimated at about 1680 billion gal/day or 12 in./yr (0.30 m/yr) (Nace, 1976). Accordingly, it is clear that much of the precipitation must originate in recycled water through evaporation from land areas and subsequent precipitation.

In addition to transporting vapor between ocean and land areas, the hydrologic cycle accomplishes another vital role in renewing our freshwater resources: treatment. When ocean water is evaporated, salts and many other impurities are left behind because they do not volatilize, making precipitation highly pure fresh water when it condenses. Accordingly, evaporation and precipitation in the hydrologic cycle comprise a massive distillation system, fueled by solar energy and highly effective as a water purification process.

EVAPOTRANSPIRATION

It usually is very difficult to identify separately water transferred to the atmosphere by evaporation and that transpired through plant metabolic activities. Accordingly, they often are combined and reported together as "evapotranspiration" in describing regional water balances.

However, the basic mechanisms and effects of variables on the two processes and the measurement techniques for evaluating them differ substantially. Also, methods used to control fluxes of vapor to the atmosphere require separate consideration of evaporation and transpiration. For those reasons, they are discussed separately in the following sections.

Evaporation

Vaporization occurs by the escape of rapidly moving water molecules through the surface into the air; condensation is the reverse process. The two processes always occur simultaneously, as described in Chapter 2. The combination of their opposing rates determines the net transfer between water and air and is referred to as evaporation.

Many variables affect evaporation rates, as outlined in Chapter 2 and discussed further in many books and articles (Chow, 1964; Linsley et al., 1982; Viessman et al., 1977). Some of the more important ones are outlined briefly here.

Temperature. Increased temperatures cause higher molecular velocities and higher equilibrium (saturation) vapor pressure. Those, in turn, increase rates at which water molecules exchange between the liquid and air, increasing net evaporation rates.

Humidity. An increase in the water vapor content of air increases the rate at which vapor molecules enter the liquid, decreasing the net rate of evaporation.

Area of exposed water surface. The rate of evaporation is proportional to the interfacial area between water and air.

Solutes. An increase in content of dissolved substances can reduce evaporation rates, but solute concentrations in fresh waters usually are too low to cause significant effects. Evaporation from seawater, for example, is only about 2 percent less than from fresh water under the same conditions.

Surface films. Substances that accumulate at the air-water interface can reduce evaporation rates significantly. Such agents—hexadecanol, for example—sometimes are added to reservoirs to reduce evaporation losses.

Wind velocity. The evaporation rate is proportional to the wind velocity. Higher wind velocities increase turbulence near the water surface and enhance movement of vapor from that region into the atmosphere above. That effectively reduces the relative humidity of air in contact with the water and disrupts films on the liquid surface, both of which enhance evaporation rates.

Barometric pressure. Reduced barometric pressures increase evaporation rates, but usually this is a small effect in practice.

Many formulas have been developed to describe evaporation rates as functions of variables that can be measured or estimated in the field. These largely empirical expressions include terms for major variables known to affect evaporation rates and constants determined through using field data to calibrate the expression. Many of these are similar in arrangement to the following type of relationship (Linsley et al., 1982).

$$E = (e_o - e_a)(a + bv)$$

where: E = evaporation rate

e_o = vapor pressure of water, or saturation vapor pressure of air in contact with it

e_a = actual vapor pressure of air at some specified height above the water surface

v = wind velocity at some specified height

a, b = constants determined through field study

Many specific formulas are presented in articles and hydrology handbooks and textbooks (Chow, 1964; Linsley et al., 1982; Viessman et al., 1977) and often differ widely in conceptual approaches and variables included. Most have been calibrated to meet conditions prevailing in specific water bodies or geographic areas and are of little value elsewhere.

The range of process variables encountered in practice result in wide variations in evaporation from place to place and time to time. This is illustrated by the data in Table 4.3, which summarizes average evaporation rates at several locations in the United States. The annual data vary from 16 in./yr (0.41 m/yr) at Eastport, Maine, to 100 in./yr (2.54 m/yr) at Yuma, Arizona, and illustrate geographic differences in average evaporation rates. The other data present evaporation rates for selected months and show wide seasonal variations at each location and differences in seasonal patterns among them.

Figure 4.2 further illustrates differences in patterns of variation by showing monthly average data at five locations. Although total annual evaporation rates are similar for Salt Lake City, Columbia, and Miami, their seasonal distributions differ greatly. Also, evaporation rates shown for Massachusetts, Eastport, and Miami are radically different, both in annual averages and seasonal patterns, although all are close to the Atlantic Ocean.

The data in Table 4.3 and Figure 4.2 give average evaporation rates from free water surfaces only, as determined through standard measurement techniques. They do not define the actual transfer of moisture into the atmosphere from either water or land areas or from regions with mixed water and land areas. Also, the average data are helpful for some purposes but have relatively little utility in solving practical local problems. Those usually require more detailed attention to actual rates likely to be encountered in that area, including maximum and minimum values.

TABLE 4.3
Average Evaporation Rates from Water Surfaces (in./mo)

Location	January	April	July	October	Annual
Yuma, Arizona	3.9	8.0	13.4	8.0	100
San Antonio, Texas	2.2	5.6	9.4	5.8	69
Oklahoma City, Oklahoma	1.5	4.7	10.2	6.3	66
Columbia, South Carolina	1.6	4.5	6.6	4.4	51
Salt Lake City, Utah	0.8	3.5	10.6	3.9	55
Eastport, Maine	0.8	1.1	2.0	1.6	16
Galveston, Texas	0.9	2.6	6.2	4.6	43
Miami, Florida	3.0	4.9	5.3	4.1	51
Gulf, off Texas	4.0	3.5	5.0	6.5	58
Ocean, off North Carolina	9.0	7.0	3.5	9.0	84
Ocean, off Massachusetts	3.0	1.5	1.5	3.0	28

Based on data from *The Water Encyclopedia,* D. K. Todd, copyright © 1970 by Water Information Center, Inc., Port Washington, N.Y., page 90. Reprinted by permission.

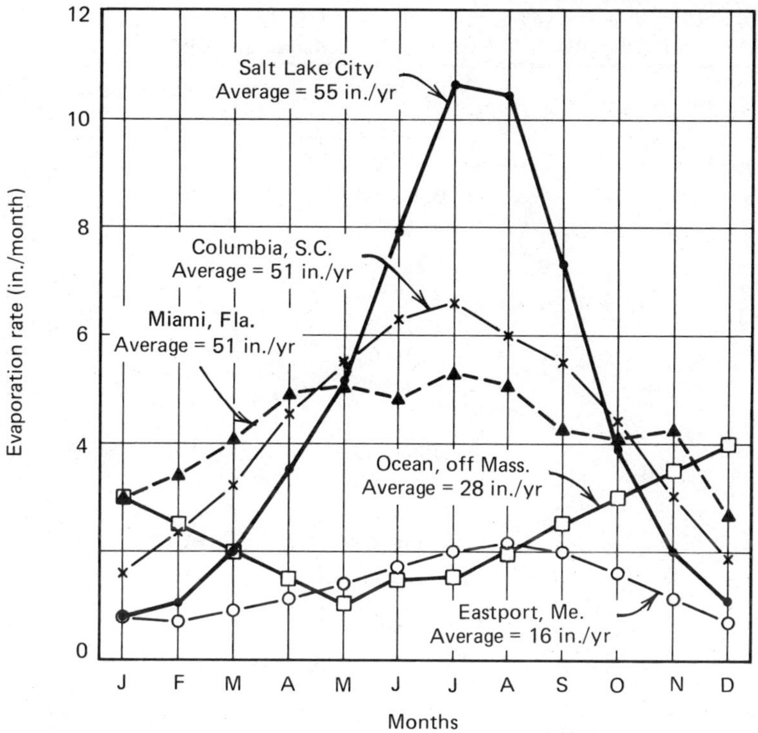

FIGURE 4.2
Seasonal distribution of evaporation from water surfaces in different locations.

Transpiration

Table 4.4 summarizes data on quantities of water transpired during the growth of different types of vegetation. Clearly, the rate of transpiration depends heavily on the type of plant involved. It is also influenced by many other variables, including temperature, humidity, type of soil, and the availability of soil moisture to the crop being grown (Wisler, 1959; Foster, 1948).

Evapotranspiration

Both evaporation and transpiration are affected by many variables, often simultaneously by the same ones. However, evaporation is a physical-chemical phenomenon, while transpiration is both physical-chemical and biochemical because living organisms are involved. Thus, the types and magnitudes of effects on the two processes by a given variable often differ greatly.

Evaporation rates for a given region depend heavily on the area of exposed

TABLE 4.4
Typical Quantities of Water Transpired in Production of Vegetation

Type of Plant	Pounds of Water per Pound of Dry Matter Grown	
	Range	Typical Average
Alfalfa	700–1000	850
Wheat	300–550	450
Oats	400–650	500
Corn	250–350	300
Rye	400–600	500
Barley	300–600	450
Rice	600–800	700
Cotton		600
Red clover	300–1000	800
Buffalo grass	250–350	300
Weeds	200–1000	600
Potatoes	300–600	500
Sugar beets	300–500	400
Larch		1150
Beech		1050
Elm		750
Norway spruce		250
Fir		100

Source: From *Hydrology,* Second Edition, C. O. Wisler and E. F. Brater, Copyright © 1959 by John Wiley & Sons, Inc., New York, page 218. Reprinted by permission.

water surface from which vapor may exchange into the atmosphere. That area usually is a mixture of surfaces of water bodies and soils covered in varying degrees with moisture. In practice, it is virtually impossible to define exactly the amount of water surface involved in the process.

Transpiration, on the other hand, is influenced predominantly by the types and amounts of vegetation covering the area of interest. There usually is a mix in types and density of vegetation present, even over relatively small areas. The problems involved in measuring accurately amounts of moisture released by that vegetation are immense and are compounded by the fact that evaporation and transpiration occur simultaneously over the same region.

This is why practical studies frequently combine the two into a single value called evapotranspiration. That seldom, if ever, can be determined accurately through direct measurements but often may be estimated through mass balances. Those can be accomplished by measuring precipitation on the area, runoff from it, and estimating groundwater inflow and outflow. These types of calculations for the United States, for example, produce average precipitation of 4200 billion gal/day (15.9 km^3/day) and runoff of 1330 billion gal/day (5.0 km^3/day) in stream flows and groundwater interchanges. That results in an estimate of 2870 billion gal/day (10.9 km^3/day) for average evapotranspiration (USWRC et al., 1978). Of course, the evapotranspiration varies widely geographically, seasonally, and annually in response to the effects of many variables affecting its two processes and variations in precipitation.

PRECIPITATION

Several factors determine the type, amount, and rate of precipitation, the most important of which is temperature. The process involves condensation of water vapor, which can occur when the air temperature drops below the dew point. That is caused principally by expansion and cooling of the gas in rising air columns, for which there are several mechanisms (Linsley et al., 1982; Viessman et al., 1977; Wisler and Brater, 1959). Of course, temperature also plays an important role in determining the type of precipitation, that is, whether it occurs as rain, snow, hail, or sleet.

One mechanism by which atmospheric masses rise and are cooled is the formation of convection currents by uneven heating of air near the earth's surface. That phenomenon frequently is observed through the development of towering cloud masses, especially during the summer.

Another mechanism is the chimney effect in large cyclonic (rotating) air masses. Air approaching the low-pressure center of the mass is deflected upward, expands, cools, and produces condensation. Much of the precipitation in the central United States can be attributed to this mechanism (Wisler and Brater, 1959; Viessman et al., 1977).

A third mechanism is "orographic" precipitation, which occurs when air is deflected to higher altitudes while passing over mountains, which causes it to expand and cool. For example, conditions favorable for orographic precipitation are found on the western mountain slopes of Oregon and Washington, where prevailing winds from the ocean rise as they pass over the ranges. This causes rainfalls averaging over 100 in./yr (2.54 m/yr) in many locations of that region. On the southern slopes of the Himalayas, in India, orographic precipitation causes annual rainfall in excess of 500 in./yr (12.7 m/yr).

As might be expected, the opposite situation prevails on the other side of the mountain range. Air reduced in moisture content by cooling and precipitation during the rise over the mountains then returns to lower elevations and is warmed, inhibiting further precipitation. For example, the dry or desert areas inland from the mountains along much of the west coast of the United States can be attributed to this phenomenon.

Wisler and Brater (1959) summarized some major factors that are especially important in determining the average rainfall in an area.

1. Distance from the ocean.
2. Direction of the prevailing wind.
3. Average annual temperature.
4. Altitude.
5. Topography of the land.

As shown in Table 4.2, the average global precipitation is about 39.8 in./yr (1.01 m/yr). The U.S. annual precipitation averages somewhat less than that—30 in./yr (0.76 m/yr). Data for average precipitation are helpful in providing general perspective, but more information is needed to solve most practical problems because only rarely are the averages actually experienced.

Variations in Annual Precipitation

Those who plan and design water resources facilities usually must pay close attention to variability, especially maximum and minimum precipitation and flow rates. For example, where one must provide reliable water supplies for municipalities or industries, minimum expected rainfalls and stream flows are of special interest because those conditions dictate the amount of storage required to insure adequate "safe yield" of the system. On the other hand, when the principal concern is flood control, rainfall and stream flow data near the maximum end of the spectrum are of greatest concern.

Evaluating the probabilities for experiencing minimum and maximum occurrences of precipitation or flow requires sophisticated statistical techniques, beyond the scope of this discussion. They are described in many hydrology texts and other publications, some of which are among the references for this chapter (Bernard, 1949; Chow, 1964; Riggs, 1968a, 1968b; Linsley et al., 1982; Viessman et al., 1977).

Table 4.5 illustrates some geographic variations in U.S. precipitation. The annual rainfall varies among the selected locations from 4.1 in./yr (0.104 m/yr) in Las Vegas to 140 in./yr (3.56 m/yr) in some areas of the Olympic Mountains of Washington. Most of the data in Table 4.5 are for widely separated geographic areas, but sometimes precipitation may vary drastically over distances of only a few miles. For example, rainfall on the windward side of one mountain in Hawaii averages 460 in./yr (11.68 m/yr), but just 15 mi away, on the downwind side, it is only 18 in./yr (0.46 m/yr). That can be attributed to the orographic effect and prevailing winds, as described earlier.

TABLE 4.5
Average Annual Precipitation at Selected U.S. Locations (1939–78)

	Average (in./yr)	Minimum (in./yr)	Maximum (in./yr)
Albuquerque, N.Mex.	8.3	4.1	15.9
Asheville, N.C.	48.1	30.3	64.9
Amarillo, Tex.	20.4	9.6	37.2
Fargo, N.Dak.	21.0	8.8	30.4
Wichita, Kans.	30.0	12.2	50.5
Indianapolis, Ind.	39.9	28.4	55.7
Daytona Beach, Fla.	49.3	31.4	79.3
Las Vegas, Nev.	4.1	0.6	10.7
Miami, Fla.	58.9	37.0	83.4
Portland, Maine	42.2	25.3	57.6
Seattle, Wash.	39.5	23.8	55.1
Spokane, Wash.	16.2	10.6	26.1
Fairbanks, Alaska	11.2	5.6	17.3
Fresno, Calif.	9.7	3.6	19.1
Yakutat, Alaska	132.0		
Phoenix, Ariz.	7.4	2.8	16.3
Olympic Mountains, Wash.	140.0		

Source: U.S. Weather Service.

At any given location, variations in precipitation occur from time to time. In many areas of the United States, for example, it is not uncommon for the minimum annual precipitation to be 25 to 50 percent of the average for that region, while precipitation for the maximum year may be 150 to 200 percent of the average. The data in Table 4.5 show the range in annual precipitation observed at each location over a 40-year period.

Seasonal Variations

Important seasonal variations within each year are superimposed on the annual ones just discussed. Data for four locations in Figure 4.3 illustrate seasonal patterns and how they can vary from place to place. The bar graphs were developed for each city by averaging precipitation in each calendar month over a period of several years. Weekly or daily data would show even greater variability.

The data for Seattle indicate that minimum rainfall usually occurs in the summer, with above-average precipitation from October to March. Miami, on

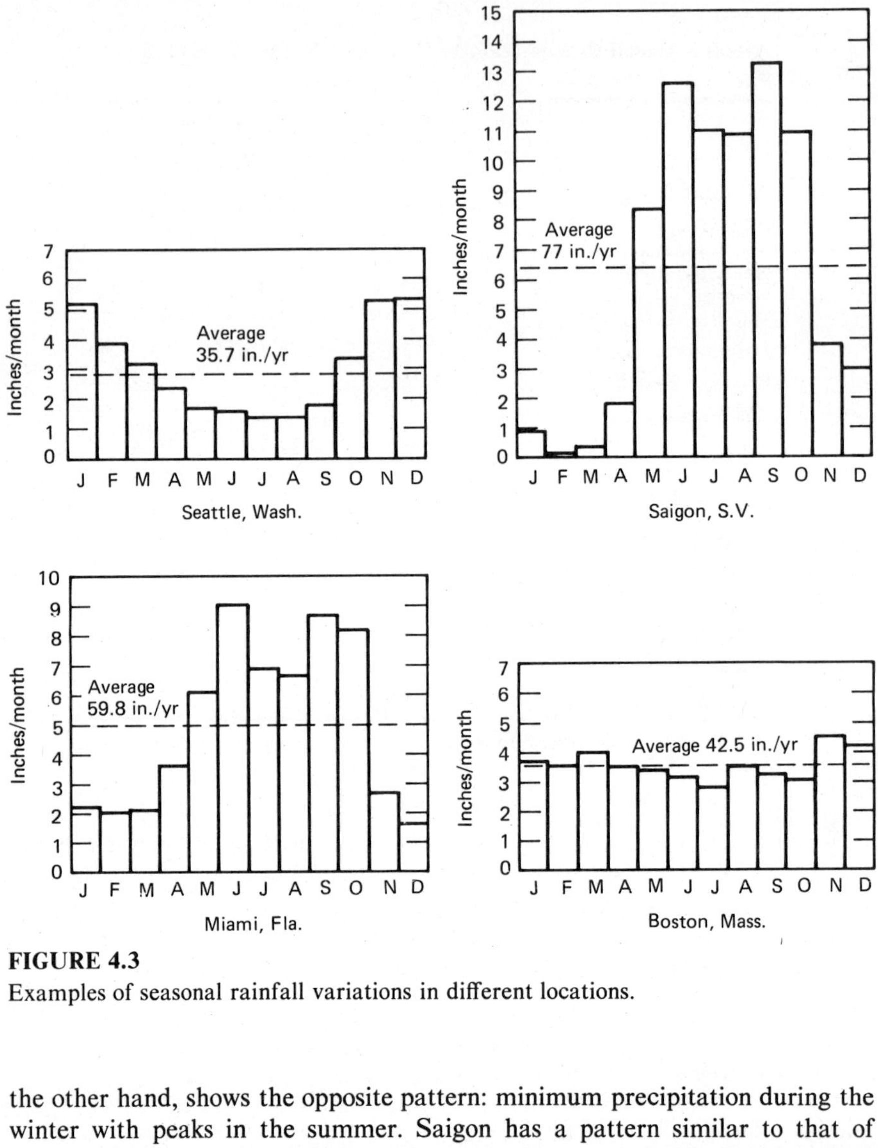

FIGURE 4.3
Examples of seasonal rainfall variations in different locations.

the other hand, shows the opposite pattern: minimum precipitation during the winter with peaks in the summer. Saigon has a pattern similar to that of Miami, but with more extreme values: virtually no rainfall in February and extremely high precipitation in the summer. Finally, the data for Boston show little seasonal variation in precipitation, based on monthly averages (NOAA, 1983).

Clearly, seasonal variations may be substantial and the patterns of variation may differ widely at various locations. Accurate planning and design of water resources facilities that depend on knowledge of precipitation require information on both average rates and variability. Accordingly, it usually is necessary to evaluate expected annual, seasonal, and even daily variations at that specific location and often to determine the probabilities of maximum and minimum occurrences.

Graphical Presentation of Precipitation Data

Often statistical analyses of rainfall data are summarized graphically as "rainfall frequency curves," similar to that shown in Figure 4.4 (Hammer and MacKichan, 1981). This figure shows the frequency of annual rainfall at Lincoln, Nebraska, based on data collected from 1884 to 1976. This presentation technique provides a convenient way for visualizing: (1) the probability of experiencing an annual rainfall at or above some given amount, or (2) the annual rainfall that might be expected to occur at some specified frequency. For example, Figure 4.4 shows that annual rainfall equal to or exceeding 490 mm might be expected to occur in about 90 percent of the years (nine years

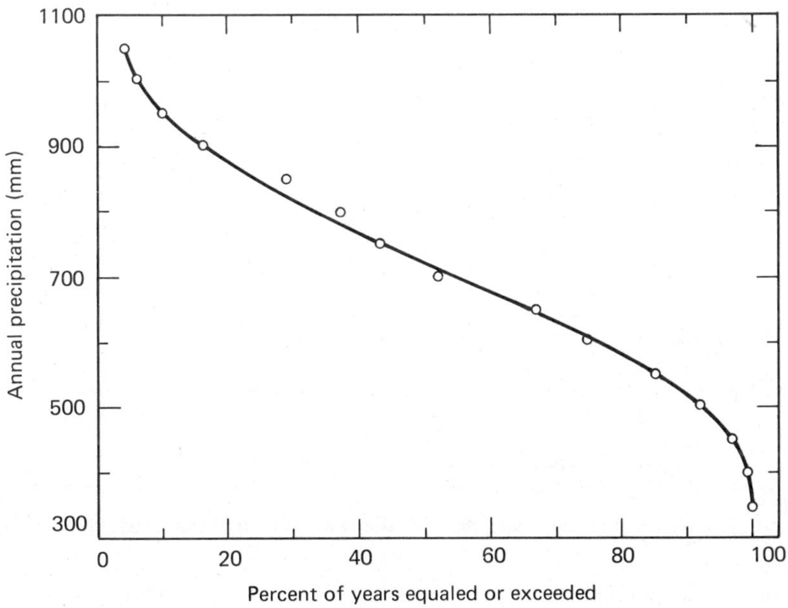

FIGURE 4.4
Probability of annual precipitation at Lincoln, Nebraska (1884–1976). From *Hydrology and Quality of Water Resources,* M. J. Hammer and K. A. MacKichan, Copyright © 1981 by John Wiley & Sons, Inc. New York, page 156. Reprinted by permission.

out of ten). Also, at a frequency of one year in twenty (5 percent probability), the rainfall would be expected to equal or exceed about 970 mm/yr.

Figure 4.5 shows the same data plotted on a special type of graph paper often used for this purpose. In Figure 4.5, the "probability" scale, on the abscissa, is divided into intervals derived by solving a mathematical expression commonly used to describe a certain type of probability relationship—in this instance, a "normal" statistical distribution. Graph paper also is available, or can be prepared, to produce straight lines for data that approximate other types of statistical distributions.

Figure 4.6 shows a different type of statistical analysis of rainfall data. Instead of summarizing average annual rainfall data, it describes statistical expectations for rainfall intensities and durations in relatively short storms in Boston, Massachusetts. This graph provides the type of data needed to design

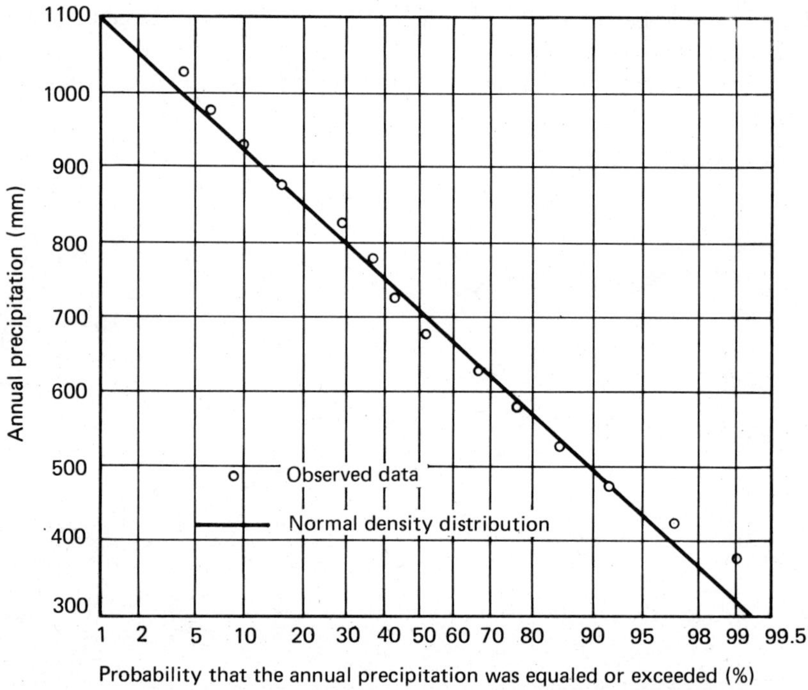

FIGURE 4.5

Annual precipitation probabilities at Lincoln, Nebraska (1884–1976), plotted on probability graph paper. From *Hydrology and Quality of Water Resources,* M. J. Hammer and K. A. MacKichan, Copyright © 1981 by John Wiley & Sons, Inc., New York, page 156. Reprinted by permission.

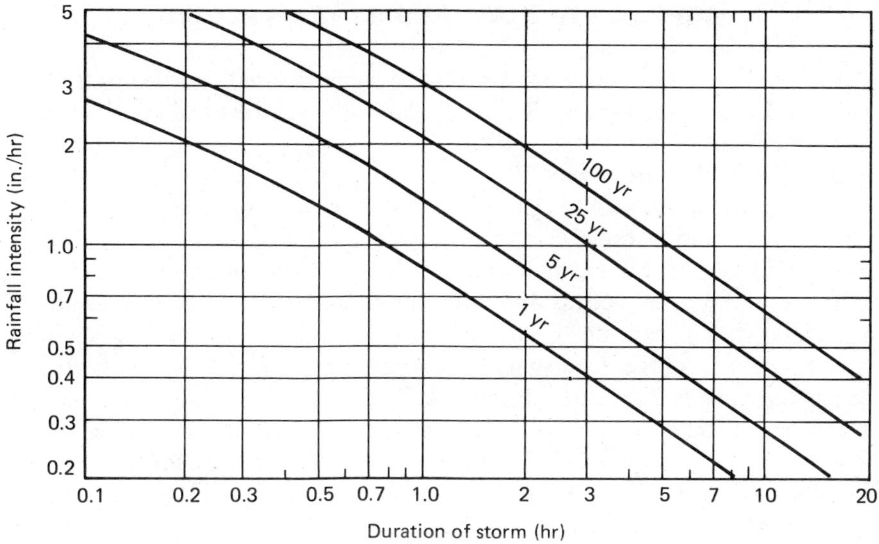

FIGURE 4.6
Intensity of precipitation for various durations and frequencies at Boston, Massachusetts.

runoff channels, storm sewers, and storm water detention facilities. It summarizes the relationships between intensity of rainfall, its duration at that intensity, and the probability of occurrence for that combination of intensity and duration. Figure 4.6 shows, for example, that in Boston a rainfall of 1.0 in./hr or more, lasting for 0.8 hr or more, could be expected to occur at an average frequency of about once per year. On the other hand, a rainfall of 0.8 hr duration with an intensity of 2.5 in./hr or more could be expected to occur on the average only once in about 25 years.

These types of curves have been compiled for many areas by the U.S. Weather Bureau and often are used by engineers to design sewers or storage facilities for storm runoff. A policy decision might be made, for example, that a storm sewer must not be overloaded more often, on the average, than once in five years. The sewer, then, would be designed to handle runoff from storms having any combination of rainfall intensity and duration represented by the five-year curve on Figure 4.6.

Similar information sometimes has been summarized as an empirical equation based on rainfall data for the area in question. Examples of these equations may be found in some hydrology texts and other publications (Chow, 1964).

SURFACE RUNOFF AND STREAM FLOWS

Precipitation reaching the earth's surface may be divided into two broad categories: (1) that which runs off over the land into streams and lakes, and (2) that which infiltrates into the ground.

Sometimes the term "runoff" is used in hydrology as a synonym for discharge in surface streams to distinguish that flow from water leaving the same area underground or through evapotranspiration. In this context, runoff from the United States, for example, may be determined simply by totaling all stream flows into the oceans and adjacent countries. Some of this section is concerned with stream flows, and the term "runoff" is used to describe that type of water resource. One should recognize, however, that water flowing in streams originates only partly from discharge over the land surface; much of it can be attributed to the emergence of groundwater into streambeds. The "water yield" of a region includes the total of runoff and net underground outflow. The difference between water yield and rainfall on a drainage area provides an approximation of total evapotranspiration.

Some discussions in this section will be concerned only with flow actually passing over the ground surface, as contrasted with infiltration into the ground or evapotranspiration. To avoid possible confusion, the term "surface runoff" will be used to describe this phenomenon. The division of rainfall reaching the ground into surface runoff and infiltration depends on many factors, a few of which are discussed briefly in the following paragraphs.

Soil Characteristics

A key factor in determining what proportion of rainfall runs off over land surfaces directly into streams and lakes is soil perviousness—its ability to transmit water flow through it. This characteristic is very different from soil "porosity"—its ability to contain water. For example, both sand and clay have high porosity (that is, both can hold a lot of water), but only the sand also has high perviousness. Water can flow readily through sand deposits, but clay has a high resistance to its passage.

Other factors being equal, a soil with low perviousness will produce a higher proportion of surface runoff than a soil that allows water to infiltrate readily into the ground.

Topography

Shape and slope of ground surface also influence greatly the division of precipitation between surface runoff and infiltration. The passage of water into soil usually is relatively slow because of resistance to flow through small channels between the soil particles. Accordingly, any topographic characteristic that retards flow over land surface increases infiltration because it provides more time for water to pass into the ground.

That is why flat land areas produce less runoff than steep slopes for similar soils and rainfall patterns. Also, it explains the success of contour plowing in reducing surface runoff and erosion from farmlands; it creates more "depression storage" by producing ridges and furrows at right angles to the direction of water flow down the slopes. That detains water and increases the length of time available for infiltration.

Vegetation

The types and amounts of vegetation on a watershed can have major effects on the proportion of precipitation that flows directly into streams. The presence of vegetation decreases the amount of runoff by intercepting rainfall and increasing evaporation losses and by causing the transpiration of soil water through plants into the atmosphere. Also, plants tend to obstruct flow over the land surface, providing longer retention times and increased infiltration. However, some of the additional water that infiltrates into the ground may appear later as increased stream flows (runoff) during dry weather.

Clearly, the effects of vegetation on runoff may be complex and difficult or impossible to predict accurately. That accounts for much of the long-standing controversy concerning beneficial and harmful effects on water resources development by deforestation and reforestation projects.

Precipitation Intensity and Duration

The pattern in which precipitation occurs also may have major effects on surface runoff. In general, surface runoff is higher with increased precipitation intensity because water accumulates more rapidly on the ground surface, filling depression storage and giving less opportunity for infiltration before flowing into streams.

For a given rainfall intensity—inches per hour, for example—the proportion of surface runoff increases with rainfall duration because the soil structure becomes more saturated as precipitation continues and reduces the rate of further infiltration. For the same reason, the amount of surface runoff from a given storm often has been observed to depend heavily on elapsed time since the preceding precipitation.

Temperature

Temperature affects evaporation rates, as discussed earlier, as well as transpiration by plants whose metabolism depends heavily on this variable. Also, lower temperatures increase the viscosity of water, which increases its resistance to passage through the small pores in soil during infiltration. Because of these effects, surface runoff from a given rainfall tends to increase as temperature drops. The impact may be severe when precipitation falls on frozen ground; runoff may increase drastically over runoff from the same area when it is thawed.

The most radical temperature effects are observed when precipitation occurs as snow, producing little or no runoff because the water remains fixed in storage for weeks or months. Melting in the higher temperatures of spring thaw then can release water accumulated over extended periods within a relatively short time. That alters the timing of runoff relative to precipitation and increases the rate (intensity) at which water is applied to the soil. This phenomenon accounts for the disastrous floods that occur each spring in many areas of the United States.

Land Use Patterns

The factors just discussed indicate clearly that the division of precipitation between surface runoff and infiltration can be affected drastically by changes in land use patterns. This has been demonstrated often, for example, when forests and farmlands are converted to industrial and residential developments. Stripping of vegetation, changes in topography by grading, and reductions in perviousness by building roads, parking areas, and buildings can radically increase the surface runoff and raise downstream flood stages.

Calculation of Surface Runoff

Many techniques are available for estimating surface runoff, some of which are highly sophisticated and complex. However, because of space limitations, only one of the oldest and most straightforward methods will be outlined here. The "rational formula" has been commonly used for many years to calculate storm runoff from relatively small areas, especially in determining capacities of channels and sewers to carry the flow (Hammer and MacKichan, 1981; Langworthy, 1969; Viessman et al., 1977).

$$Q = CiA$$

where: Q = surface runoff, ft^3/sec (m^3/hr)
C = runoff coefficient (proportion of rainfall running off over surface)
i = rainfall rate, in./hr (m/hr)
A = drainage area, acres (m^2)

Statistical rainfall data, similar to those discussed earlier, are used to select an intensity that is appropriate for design purposes. The area that contributes drainage to the channel or pipeline is measured in acres, often from maps. The runoff coefficient is estimated by evaluating the land areas involved. Table 4.6 summarizes typical runoff coefficients for different types of contributing areas. The coefficient for a watershed may be determined by measuring the amount

TABLE 4.6
Values of Runoff Coefficient in the Rational Formula

Type of Drainage Area	Runoff Coefficient ("*C*")
Pavement	
Asphalt	0.70–0.95
Brick	0.70–0.85
Roofs	0.75–0.95
Lawns, sandy soil	
Flat slope (2%)	0.05–0.10
Average slope (2–7%)	0.10–0.15
Steep slope (>7%)	0.15–0.20
Lawns, heavy soil	
Flat slope (2%)	0.13–0.17
Average slope (2–7%)	0.18–0.22
Steep slope (>7%)	0.25–0.35
Business areas	
Downtown	0.70–0.95
Neighborhood	0.50–0.70
Residential areas	
Single family	0.30–0.50
Multiunits	0.40–0.75
Suburban	0.25–0.40
Apartments	0.50–0.70
Industrial areas	
Light industry	0.50–0.80
Heavy industry	0.60–0.90
Parks, cemeteries	0.10–0.25
Playgrounds	0.20–0.35
Railroad yard	0.20–0.35
Unimproved areas	0.10–0.30

Source: From *Design and Construction of Sanitary and Storm Sewers,* J. C. Lawler et al., copyright © 1969 by WPCF and ASCE. Used with permission of WPCF and ASCE.

of area in each category and calculating a weighted runoff coefficient for the total area. Techniques for doing this are described in detail in hydrology publications dealing with the design of storm runoff systems (Hammer and MacKichan, 1981; Langworthy, 1969; Viessman et al., 1977).

Accurate estimation of runoff from large areas is of interest in predicting flood flows and maximum water levels in streams as well as low flows for determining safe yields of water supply systems. Other more complex methods are used for those evaluations, for example, various types of direct statistical analyses of stream flow data (Chow, 1964; Fair et al., 1966; Hammer and MacKichan, 1981; Linsley et al., 1982; Riggs, 1968a, 1968b; Viessman et al., 1977).

Stream Flows

Global runoff, as measured by stream flows, averages about 35 to 40 percent of the precipitation on land. The U.S. runoff averages about 30 percent of precipitation (9.0 in./yr, or 0.23 m/yr). However, the runoff varies among different locations from less than 0.25 in./yr (0.006 m/yr) in some desert areas to over 80 in./yr (2.03 m/yr) in the Pacific Northwest.

Table 4.7 (Todd, 1970) summarizes the variability in U.S. annual runoff between 1931 and 1960. During those 30 years, the runoff varied from a low of 56 percent to a high of 129 percent of the average. The range of variability in runoff would be even greater for smaller, specific regions within the United States or periods shorter than a year. For example, some river flows range from near zero during dry periods to very high levels during flood stage. It is not uncommon for the maximum flow during a year to be over 500 times the minimum in that stream.

Table 4.8 summarizes annual stream flows from 21 major regions of the country, as adapted from a report of the U.S. Water Resources Council (USWRC, 1978). For each region, the average 1975 stream flow is given, as well as flows that were exceeded in 5 percent and 95 percent of the years, based on statistical analyses of past data. In New England, for example, the 1975 average flow was 78 billion gal/day (3415 m^3/sec). During 95 percent of the years, stream flows exceeded 48 billion gal/day (2100 m^3/sec) (62 percent of the average), and during 5 percent of the years they exceeded 108 billion gal/day (4730 m^3/sec) (138 percent of the average).

The most variable region is the Rio Grande, in which average 1975 flow was 1.2 billion gal/day (52.6 m^3/sec). The flow exceeded in 95 percent of the years was only 0.2 billion gal/day (8.8 m^3/sec) (17 percent of the average), while that exceeded in 5 percent of the years was 4.4 billion gal/day (193 m^3/sec) (367 percent of the average). Thus, the range between the 5 percent and 95

TABLE 4.7
Variability of Annual Runoff in the United States (Expressed as Percent of Mean Runoff)

Year	Percent of Mean	Year	Percent of Mean
1931	61	1946	101
1932	111	1947	110
1933	98	1948	106
1934	56	1949	112
1935	110	1950	117
1936	91	1951	111
1937	106	1952	127
1938	117	1953	87
1939	79	1954	73
1940	78	1955	73
1941	122	1956	83
1942	129	1957	105
1943	121	1958	122
1944	103	1959	80
1945	117	1960	100

Source: From *The Water Encyclopedia,* D. K. Todd, Copyright © 1970 by Water Information Center, Inc., Port Washingtion, N.Y., page 75. Reprinted by permission.

percent Rio Grande flows was 22/1 while, in contrast, that for New England was only 2.3/1.

Table 4.9 summarizes drainage areas and average discharges for several major U.S. rivers. The discharge for all of them averages 0.8 $ft^3/sec/mi^2$ of drainage area (8.73 liters/sec km^2) and ranges from a minimum of 0.28 for the Arkansas to 3.18 $ft^3/sec/mi^2$ for the Willamette.

GROUNDWATER

Factors affecting the division of rainfall between surface runoff and infiltration have been discussed in the preceding section. To review, infiltration is enhanced by an increase in soil porosity, flatter slopes of terrain, lower rainfall intensity, shorter duration, and land use practices that delay the overland passage of water into streams. It also is influenced by temperature, extent and types of vegetation, forestation practices, and several other factors, but the nature of these effects sometimes is less clear.

TABLE 4.8
Flow Variability in Selected U.S. Streams

Region	Average 1975 Flow (billion gal/day)	Flow Exceeded 5% of Years: billion gal/day	Flow Exceeded 5% of Years: % of average	Flow Exceeded 95% of Years: billion gal/day	Flow Exceeded 95% of Years: % of average
New England	78	108	138	48	62
Mid-Atlantic	79	115	146	48	61
South Atlantic, Gulf	228	357	157	122	54
Great Lakes	73	104	142	45	62
Ohio	178	254	143	105	59
Tennessee	41	58	141	31	76
Upper Mississippi	121	189	156	65	54
Lower Mississippi	433	757	175	202	47
Souris, Red, Rainy	6.0	11	183	1.8	30
Missouri	44	74	168	18	41
Arkansas, White, Red	63	121	192	22	35
Texas, Gulf	28	62	221	6.3	23
Rio Grande	1.2	4.4	367	0.2	17
Upper Colorado	10	16	160	3.9	39
Lower Colorado	1.6	1.7	106	1.2	75
Great Basin	2.6	4.7	181	1.2	46
Pacific Northwest	255	345	135	180	71
California	47	87	185	20	43
Total	1233	1957	159	675	55
Alaska	905	1030	114	705	78
Hawaii	6.7	10	149	3.8	57
Caribbean	4.9	7.1	145	1.6	33

Source: Adapted from U.S. Water Resources Council (1978).

Water infiltrates into the "aquifer," which is an underground stratum capable of containing and transmitting water. In its upper region, interstices between soil particles are not saturated, and the water moves downward under the influence of gravity. This part of the aquifer is referred to as the "zone of aeration" because the interstices are filled partly with water and partly with air.

The water passing downward finally reaches an aquifer zone in which the interstices are filled with water, making the soil saturated. This is referred to as the "zone of saturation," its surface is called the "groundwater table," and

TABLE 4.9
Discharges of Selected U.S. Rivers

River	Drainage Area (mi^2)	Average Discharge, 1941–70	
		ft^3/sec	$ft^3/sec/mi^2$
Mississippi	1,247,300	640,000	0.51
Columbia	258,000	262,000	1.02
Ohio	203,900	258,000	1.26
St. Lawrence	302,000	243,000	0.80
Yukon	327,600	240,000	0.73
Atchafalaya	95,100	183,000	1.92
Tennessee	40,900	64,000	1.56
Red	93,200	62,300	0.67
Mobile	43,800	61,400	1.40
Snake	109,000	50,000	0.46
Arkansas	160,600	45,100	0.28
Susquehanna	27,600	37,200	1.35
Willamette	11,200	35,700	3.18
Apalachicola	19,600	24,700	1.26
Hudson	13,400	19,500	1.46
Delaware	11,400	17,200	1.50
Totals	2,964,600	2,243,100	0.76

Source: Iseri and Langbein (1974).

its elevation may be located by noting the level at which water stands in a well penetrating the aquifer at that location. Movement of water in the saturated zone is predominantly horizontal because further migration downward is inhibited by the saturated soil and water below. The groundwater table, then, may be viewed as the surface of an underground stream or lake.

The behavior of water below the groundwater table is similar hydraulically to that in above-ground lakes and streams—it flows from higher elevations to lower ones, under the influence of gravity. Energy losses as water passes through soil cause the groundwater table to slope in the direction of flow, as does the surface of a river flowing from uplands to the ocean. Because resistance to flow through the small channels between soil particles is relatively high, water movement generally is much slower, and the slope of the groundwater table may be much steeper than normally observed with surface streams.

Figure 4.7 is a cross section through an aquifer, illustrating the locations of the zone of aeration, zone of saturation, groundwater table, and direction of water movement underground. Where the groundwater table intersects the sur-

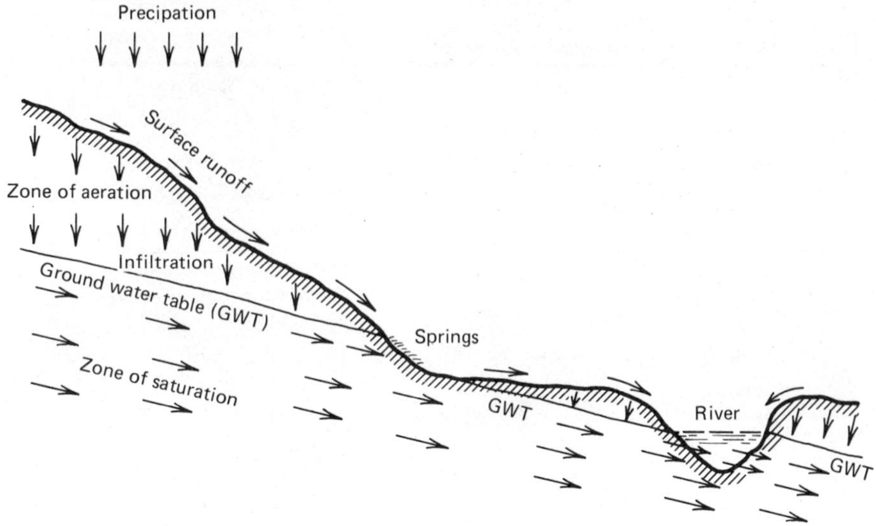

FIGURE 4.7
Section through a groundwater aquifer.

face, water discharges from the soil in springs or passes directly into a lake or stream, if the groundwater table slopes downward to that body of water, as shown where underground flow meets the river in Figure 4.7. Conversely, where the lake or stream water level is higher than the groundwater table, the flow may occur from surface water into the aquifer, as indicated near the right of Figure 4.7. This process is referred to as "recharge" of the groundwater system and is an important mechanism for augmenting underground water resources.

The relationships among groundwaters and surface waters are vitally important. Flows into and out of porous soils may greatly increase effective storage volumes by exchanges between a surface reservoir and an adjoining aquifer. Also, interchange between surface waters and groundwaters helps maintain relatively constant water levels in lakes during wet and dry seasons. Furthermore, transfers from aquifers into streams are responsbile for maintaining their flows during dry periods, as pointed out earlier.

The slow movement and long storage in groundwater systems makes them far more consistent in quantity and quality than surface water systems. Extended contact between water and soils makes chemical quality of a groundwater relatively constant with time, although there may be radical differences from place to place because of soil characteristics and chemicals that may dissolve in the water.

The amount of U.S. groundwater, as reported earlier, is not known accurately but is estimated at about 86 percent of the total freshwater inventory. Much of that is not available for practical supplies because of poor soil permeability that would restrict flow to wells or excessive lift necessary to bring the water to the surface. Also, much of it is poor in quality and would require extensive treatment. For example, many well waters in the U.S. Midwest and West contain salts far above levels acceptable for drinking, sometimes exceeding even the salinity of oceans.

The large inventory of water stored underground makes it possible and tempting sometimes to use it at rates substantially exceeding recharge into the aquifer. That situation has existed for several years in some U.S. areas, causing groundwater tables to drop many feet—sometimes even hundreds of feet. This use of water in excess of recharge is referred to as "mining" and can cause excessive pumping costs, ultimate exhaustion of the supply and, occasionally, subsidence of the ground surface above.

Groundwater supplies are difficult to locate and evaluate and present special problems in predicting available water quantities and qualities. Generally, they are higher in mineral content than surface waters and are more constant in quantity, quality, and temperature. Because of restricted movement, they respond more slowly to pollution, often requiring years for pollutants to travel through aquifers to points of use. An important corollary is that once pollution has reached the point of use, that same slow movement may require a very long time to clear the aquifer after cessation of pollution.

Because of difficulties in locating, evaluating, and developing large groundwater supplies, they are used relatively infrequently. On the other hand, great numbers of well supplies are used for individual homes and small communities. In that type of application, well supplies usually are desirable because they typically require little or no treatment other than the relatively long time of travel and filtration afforded by passage through the soil.

OVERALL SUMMARY OF U.S. FRESHWATER RESOURCES

Table 4.10 summarizes the U.S. dynamic freshwater resources, which originate in the hydrologic cycle (USWRC, 1978; Nace, 1976; U.S. Senate, 1961). The total precipitation averages 4200 billion gal/day (15.9 km^3/day). Seventy percent of that (2870 billion gal/day, or 10.9 km^3/day) is returned to the atmosphere through evapotranspiration, making it unavailable in streams, lakes, or groundwater systems for water supply. The remaining 1330 billion gal/day (5.0 km^3/day) are available as surface and underground supplies for potential use.

TABLE 4.10
Summary of U.S. Water Resources

Item	in./yr	m/yr	billion gal/day	km³/day
Total rainfall	30	0.76	4200	15.9
Evapotranspiration	21	0.53	2870	10.9
Net runoff				
Streams to Atlantic, Gulf			920	3.48
Streams to Pacific			300	1.14
Groundwater to oceans			100	0.38
To Canada and Mexico			10	0.04
Total	9	0.23	1330	5.03
Maximum sustained flow	9	0.23	1100	4.16
Maximum practical flow	5	0.13	650	2.46
1980 Fresh water used	3	0.08	380	1.44
Not used in 1980	6	0.15	950	3.60
1980 Evapotranspiration	1	0.03	100	0.38
1980 Return to resource	2	0.05	280	1.06
Total discharge to ocean	8	0.20	1230	4.66

Sources: U.S. Senate Select Committee (1961); U.S.W.R.C. (1978); Solley et al. (1983).

Construction of storage and other facilities to produce the maximum possible supply of fresh water could yield a sustained flow that has been estimated at, perhaps, 1100 billion gal/day (4.2 km³/day)—the remainder being lost through additional evapotranspiration incurred by the changes. However, it would not be feasible to develop all that flow. The best current professional judgment is that only about 650 billion gal/day (2.46 km³/day) should be viewed as the maximum practical freshwater resource of this country.

The withdrawals of fresh water for all uses in 1980 were 380 billion gal/day (1.44 km³/day) of the 450 billion gal/day (1.70 km³/day) total U.S. withdrawals (see Chapter 5; Solley et al., 1983). Of the fresh water withdrawn, about 100 billion gal/day (0.38 km³/day) was lost through further evapotranspiration associated with the uses, leaving 280 billion gal/day (1.06 km³/day) returned to the resource and available for reuse. Accordingly, the total U.S. discharge of fresh water to the oceans—used and unused—today is about 1230 billion gal/day (4.66 km³/day).

Figure 4.8 illustrates graphically the freshwater situation in terms of average inches of water per year, assuming uniform distribution over the entire United States (U.S. Senate, 1961). It shows that about 30 in./yr (0.76 m/yr) precipitates to produce our gross freshwater resource. Evapotranspiration

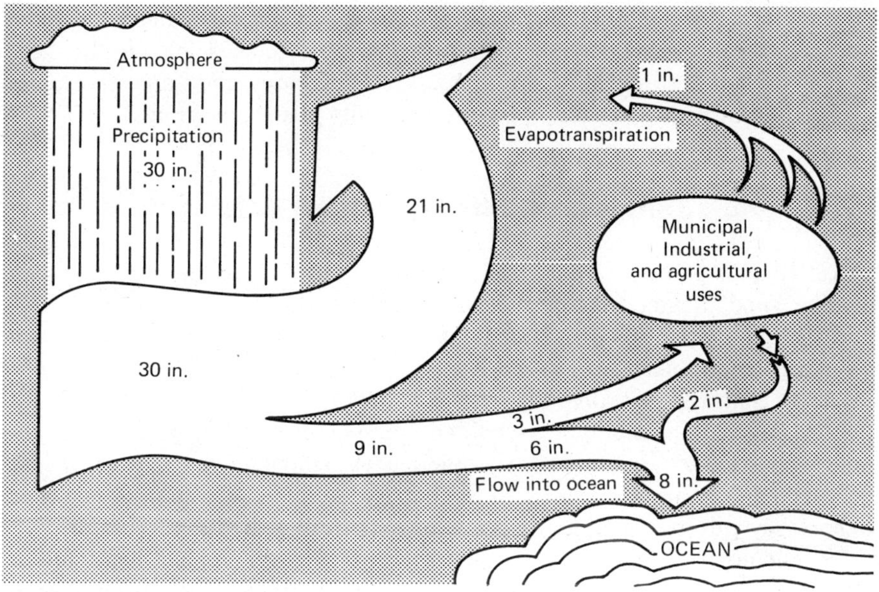

FIGURE 4.8
Summary of the distribution of the U.S. water resources, in inches per year.

returns 21 in./yr (0.53 m/yr) of that to the atmosphere, leaving 9 in./yr (0.23 m/yr) available for use as our potential supply.

Roughly one-third of that (3 in./yr) (0.076 m/yr) now is withdrawn from streams, lakes, and groundwater resources for use in municipalities, industries, and agricultural operations. About one-third of the water used (1 in./yr, or 0.025 m/yr) returns to the atmosphere through evapotranspiration. The remaining 2 in./yr (0.051 m/yr) passes to the oceans through streams, along with the 6 in./yr (0.15 m/yr) that is not withdrawn from the resource at all.

Information on water resources in various regions of the United States is available from several sources. The U.S. Weather Service, an arm of the National Oceanic and Atmospheric Administration, monitors, summarizes, and interprets data on precipitation. The U.S. Geological Survey is the agency with principal responsibility for compiling data on national water quantities. That organization conducts continuing monitoring programs of stream flows and inventories of surface and underground waters, as well as uses of those resources. Other agencies with banks of information on these subjects include the U.S. Corps of Engineers and the Department of Agriculture, Soil Conservation Service.

Data on water quality also are compiled by several agencies and usually can be accessed through computer facilities. Principal sources include the U.S. Geological Survey, the U.S. Environmental Protection Agency, and state agencies charged with responsibility for water quality control programs.

THE BASIC NATURE OF WATER AS A RESOURCE

Effective management of our freshwater resources requires an understanding of their fundamental characteristics. Perhaps the most important of these—and the one most often overlooked—is that most fresh waters are both transient and renewable.

That contrasts sharply with the character of many other important resources, including metal ores, coal, and petroleum reserves, which are available in fixed amounts and are not renewed at significant rates. They can be "conserved" in the classical sense of that term; that is, if they are not used today, they will be available tomorrow or for future generations.

On the other hand, fresh water flowing in a stream is replenished more or less continuously by the hydrologic cycle and is constantly moving to the oceans. If properly developed, the resource may have great utility. However, its value as fresh water usually is lost within a few hours or days when it reaches the ocean or a high-salinity estuary. Failure to use it for beneficial purposes before that happens could constitute the waste of a potentially valuable resource. Furthermore, when we bypass such values today they do not remain available for future use, as do the "fixed" resources previously referred to. The situation is similar for most lakes and groundwaters, although the time frame may be much longer.

One reasonably could view a flowing freshwater stream as similar in principle to other resources that are transient, such as fruits or vegetables. Fruit trees and vegetable gardens produce perishable products that have value only for limited periods. If not harvested within certain time limits, their values are lost irretrievably and usually little or no benefit accrues to anyone, now or later, from such misguided "conservation." True conservation of freshwater resources requires that they be protected and used intelligently to produce meaningful benefits among the wide range of uses to be discussed in Chapter 5. There is no virtue in deliberately bypassing their utility and allowing them to flow to the sea unused, discarded, and lost forever.

The complex considerations that must be faced in trying to optimize the use of freshwater resources include, on the one hand, concern that they should be developed to maximize benefits. On the other hand, in trying to generate benefits through some uses, the resources must not be allowed to degrade in quality or quantity to a degree that interferes unreasonably with other desired uses, either now or in the future.

Therein lies our dilemma in controlling water quality today—and also the source of most misunderstandings and controversies on the subject. The goal of this book is to provide the facts needed for an accurate understanding of the issues and resolution of the controversies.

STUDY QUESTIONS

1. Compare the quantities of water available globally in oceans, icecaps, groundwater, and freshwater lakes and rivers. Explain the significance of that distribution in the context of solving current U.S. water resource problems.
2. Draw a sketch showing the main features of the hydrologic cycle and explain it. Describe the roles of the cycle in meeting our freshwater needs.
3. Assuming a total ocean area of 139,000,000 mi^2 (360,000,000 km^2) and, using data given in this chapter, calculate the net global transfer of water from ocean to land areas in billions of gallons per day and cubic meters per second.
4. Using precipitation and evaporation data given in this chapter and a total land area of 52,000,000 mi^2 (135,000,000 km^2), calculate the net global runoff from land areas to the oceans in billions of gallons per day and cubic meters per second.
5. Distinguish clearly between evaporation and transpiration.
6. Identify six factors that affect evaporation rates and briefly explain the nature of each effect.
7. Explain the rationale behind the general type of equation presented in this chapter for estimating the rate of evaporation. Base your discussion on process concepts discussed in this chapter and in Chapter 2.
8. Should water returned to the atmosphere through evapotranspiration during its use be considered as a permanent "loss" from our freshwater resource? Explain the rationale for your answer.
9. Identify and explain three mechanisms that commonly are responsibile for cooling air masses and inducing precipitation.
10. Ouline and briefly explain the types of variations in precipitation that must be considered by those who plan for the development of water resources in the United States.
11. Indicate the circumstances under which water resource planners and designers would have special interest in *(a)* minimum precipitation rates, and *(b)* maximum precipitation rates at a specific location.

12. Using Figure 4.4 or 4.5, what is the average annual rainfall in Lincoln, Nebraska? If a facility in that community requires an annual rainfall of at least 400 mm/yr for success, how often should it be expected to fail in meeting its goal?
13. *(a)* What rainfall intensity in Boston (Figure 4.6) would be exceeded once in five years, with a duration of 30 minutes? *(b)* What duration would be exceeded once in 25 years with a rainfall intensity of 3.0 in./hr? *(c)* About how often would a rainfall of one-hour duration exceed an intensity of 2.0 in./hr?
14. Identify six factors that are important in determining the division of precipitation between surface runoff and infiltration. Briefly explain the types of effects for each factor.
15. The conterminous United States has an area of 3,020,000 mi^2 (7,820,000 km^2). Using data in this chapter for average annual rainfall and an average runoff of 1300 billion gal/day (4.92 km^3/day), calculate *(a)* the average annual precipitation, in billion gallons per day (cubic kilometers per day), and *(b)* the percent of rainfall that appears as runoff.
16. The storm runoff from an area of 1 mi^2 (2.59 km^2) is measured as 225 million gal/day (9.86 m^3/sec) during steady rainfall at a rate of 1.2 in./hr (30.5 mm/hr). What is the coefficient of runoff for that area? Explain just what that number means.
17. Make the proper conversion calculations in the rational formula to determine the percent of error inherent in it because of its set of inconsistent units.
18. Explain the difference between surface runoff and total river flow from a drainage basin. Which should be greater? Why?
19. Identify and explain the meanings of the following terms in groundwater hydrology: aquifer, zone of aeration, zone of saturation, and groundwater table.
20. Describe the relationships between groundwater and surface water and explain the significance of interchanges between the two.
21. Compare and contrast groundwater and surface water resources with respect to inventory on hand and their variability as sources of water supply.
22. What is meant by the term groundwater mining? Explain its occurrence and practical importance.
23. Identify, within 10 percent, the following data for U.S. water resources, in billions gallons per day: precipitation, evapotranspiration, net runoff, maximum sustained flow available, maximum practical flow available, and current total water withdrawals in the United States.

24. Explain what is meant by the statement that water resources are transient in character. What is the significance of that in planning for their utilization?

REFERENCES

Baldwin, H. L., and C. L. McGuinness, *A Primer on Ground Water*, U.S. Geological Survey, Government Printing Office, Washington, D.C. (1963).

Bernard, M., et al., *Hydrology Handbook,* Manual of Engineering Practice No. 28, American Society of Civil Engineers, New York (1949).

Chow, V. T., *Handbook of Applied Hydrology*, McGraw-Hill, New York (1964).

Fair, G. M., J. C. Geyer, and D. A. Okun, *Water and Wastewater Engineering,* Vol. 1, Wiley, New York (1966).

Feth, J. H., *Water Facts and Figures for Planners and Managers,* Circular No. 601-I, U.S. Geological Survey, Washington, D.C. (1973).

Foster, E. E., *Rainfall and Runoff,* Macmillan, New York (1948).

Gehm, H. W., and J. I. Bregman, *Handbook of Water Resources and Pollution Control,* Van Nostrand Reinhold, New York (1976).

Hammer, M. J., and K. A. MacKichan, *Hydrology and Quality of Water Resources,* Wiley, New York (1981).

Iseri, K. T., and W. B. Langbein, *Large Rivers of the United States,* Circular No. 686; U.S. Geological Survey, Reston, Va. (1974).

Koenig, L., "Economics of Groundwater Utilization," *Journal American Water Works Association,* Vol. 55, pp. 59–66 (1963).

Lawler, J. C., et al., *Design and Construction of Sanitary and Storm Sewers,* Manual of Practice No. 9, Water Pollution Control Federation, Washington, D.C. (1969).

Leopold, L. B., and W. B. Langbein, *A Primer on Water,* U.S. Geological Survey, Government Printing Office, Washington, D.C. (1960).

Linsley, R. K., M. A. Kohler, and J. L. Paulhus, *Hydrology for Engineers,* 3rd Edition, McGraw-Hill, New York (1982).

Nace, R. L., *Are We Running Out of Water,* Circular No. 536, U.S. Geological Survey, Washington, D.C. (1967).

Nace, R. L., "Hydrology," in *Handbook of Water Resources and Pollution*

Control, edited by H. W. Gehm, and J. I. Bregman, Van Nostrand Reinhold, New York (1976).

National Oceanic and Atmospheric Administration, *Comparative Climatic Data for the United States Through 1982,* National Climatic Data Center, Asheville, N.C. (1983).

National Oceanic and Atmospheric Administration, *The Hydrologic Cycle,* (pamphlet) Government Printing Office, Washington, D.C. (1979).

Riggs, H. C., *"Some Statistical Tools in Hydrology",* in *Hydrologic Analyses and Interpretation,* Chapter A-1, Book 4, U.S. Geological Survey, Government Printing Office, Washington D.C. (1968a).

Riggs, H. C., *"Frequency Curves",* in *Hydrologic Analyses and Interpretation,* Chapter A-2, Book 4, U.S. Geological Survey, Government Printing Office, Washington, D.C. (1968b).

Ruffner, J. A., and F. E. Bair, *The Weather Almanac,* Gale Research Company, Detroit (1981).

Solley, W. B., E. B. Chase, and W. B. Mann, *Estimated Use of Water in the United States in 1980,* Circular No. 1001, U.S. Geological Survey, Washington, D.C. (1983).

Tchobanoglous, G., *Wastewater Engineering: Collection and Pumping of Wastewater,* Metcalf and Eddy, Inc., McGraw-Hill, New York (1981).

Todd, D.. K., *The Water Encyclopedia,* Water Information Center, Port Washington, N.Y. (1970).

U.S. Congress, Senate Select Committee on National Water Resources, *Senate Report No. 29,* 87th Congress, Washington, D.C. (1959–61).

U.S. Geological Survey, *Ground Water,* (pamphlet), Government Printing Office, Washington, D.C. (1979).

U.S. Water Resources Council, *The Nation's Water Resources: 1975–2000, Second National Water Assessment,* Government Printing Office, Washington, D.C. (December 1978).

Viessman, W., J. W. Knapp, G. L. Lewis, and T. E. Harbaugh, *Introduction to Hydrology,* 2nd Edition, Harper & Row, New York (1977).

Wisler, C. O., and E. F. Brater, *Hydrology*, 2nd Edition, Wiley, New York (1959).

CHAPTER 5

Water Uses

The value of water as a resource lies in our ability to use it for a wide diversity of desirable purposes. This chapter describes several of those uses and discusses some basic concepts that must be understood to reach sound conclusions on the utilization of water resources. Also, data are presented on current and projected U.S. uses of water for various specific purposes.

BENEFICIAL USES OF WATER

Table 5.1 summarizes common water uses and classifies them as to whether they involve (1) its withdrawal from the watercourse, (2) its use in place without removal, or (3) the realization of benefits by using the flow characteristics of a stream. Entries in the table should be familiar because we now commonly use watercourses in all these ways to produce substantial benefits to individuals and organizations.

Withdrawal uses involve the removal of water from the resource, its utilization and, usually, the return of much of it to the watercourse. They include water supplies for domestic and other municipal purposes, industrial supplies to remove excess heat and satisfy manufacturing process needs, and water for irrigation (a major use in some areas, especially the West).

The propagation of fish, shellfish, and other marine life does not normally

TABLE 5.1
Major Beneficial Uses of Water

Type of Use	Withdrawal	In Place	Flow Use
Water supply			
Domestic and municipal	X		
Industrial	X		
Irrigation	X		
Stock and wildlife	X		
Propagation of marine life			
Fish		X	
Shellfish		X	
Other aquatic life		X	
Recreation			
Swimming		X	
Pleasure boating		X	
Sport fishing		X	
Other water sports		X	
Commercial navigation		X	
Aesthetic values		X	
Hydroelectric power production			X
Wastes disposal			X

require withdrawal of water from the resource, but is based on its utilization in place. That also is true for water-based recreation, commercial shipping, and the enjoyment of aesthetic values often associated with rivers and lakes.

Major beneficial uses that are based on the utilization of stream flow include power production in hydroelectric facilities and transportation and disposal of wastes.

Most uses in Table 5.1 are noncontroversial; that is, it is recognized generally that they are valid, reasonable, and beneficial to society under appropriate circumstances. However, some persons and groups question whether wastes disposal should be viewed as a legitimate beneficial use because of concern about stream pollution and the belief that wastes discharges to watercourses should be stopped altogether. That philosophy is reflected in 1972 Amendments to the Federal Water Pollution Control Act, which begins with the statement (U.S. Congress, 1972):

> . . . it is the national goal that the discharge of pollutants into the navigable waters be eliminated by 1985; . . .

There are some practical aspects of this question that require discussion before proceeding further.

WATER CARRIAGE OF WASTES

For many centuries we have depended on "water carriage" to dispose of certain wastes. By this we mean that water is used to collect and transport objectionable or harmful materials to some other location.

It may be surprising, but most of the water withdrawn from our resources by individuals, municipalities, and industries actually is used to collect and move wastes. For example, Table 5.2, from a study several years ago, shows that 41 percent of the water delivered to homes in Akron, Ohio, was used to flush toilets (U.S. Geological Survey, 1964). Another 37 percent was used for bathing, and most of the remainder was used for kitchen, laundry, and household cleaning. Other water use surveys have produced similar results, as illustrated by the data in Table 5.3. The numbers may vary among other communities and at different times, but the overall conclusion remains the same: most household water is used to collect and dispose of unwanted or potentially harmful materials—in other words, for water carriage of wastes.

We must not underestimate the importance of this. It is an established fact that the proper disposal of waste materials is an important aspect of our activities and often is vital in determining the quality of life, or even survival. For example, it is common knowledge that body wastes must be handled properly and carefully; otherwise, they may cause serious health hazards or nuisances, as discussed further in Chapters 8 to 10. Also, it is generally accepted that good sanitation practices in cooking, household cleaning, and personal hygiene are beneficial to the public health. All these activities require proper handling and disposal of unwanted, or waste, substances, and this usually is accomplished through water carriage. The availability of good water supplies and

TABLE 5.2
Typical Household Uses of Water[a]

Water Use	Percent of Total
Human consumption	5
Bathing	37
Flushing toilets	41
Kitchen	6
Laundry	4
Household cleaning	3
Lawn watering	3
Automobile washing	1
Total	100

[a]Based on data for Akron, Ohio.
Source: U.S. Geological Survey (1964).

TABLE 5.3
Estimated Per-Capita Daily Water Use by a Family of Four

Purpose	Average Use: gal per capita/day	Average Use: liters/day	Total Use (%)	Indoor Use (%)
Drinking and kitchen	2	7.6	2	3
Dishwasher	3.8	14	4	6
Toilet	24	91	28	41
Bathing	20	76	23	34
Laundry	8.5	32	10	15
Auto washing	2.5	9.5	3	—
Lawn watering	25	95	29	—
Garbage disposal	0.8	2.7	1	1
Total indoor use	59	223	68	100
Total use	86	327	100	

Source: Reprinted from *Urban Water Resources Research,* copyright © 1968 with permission of ASCE.

sanitary wastewater handling and disposal are responsible in large measure for the present high level of sanitation in this country. They have been key factors in improving our health statistics and increasing the human life span drastically during this century.

To see what could happen without them, we need only note the much higher incidence of diseases and shorter longevity in countries where water carriage is not used effectively. Currently, less developed countries of the world are striving to improve sanitation and quality of life in many ways, but especially by developing good public water supplies and sound systems for water carriage of wastes. That, in fact, is a major stated goal of the World Health Organization and many other agencies concerned with international development.

Most industrial uses also are for water carriage of wastes; that is, the removal of heat, impurities, or by-products from various types of processes. For example, about 48 percent of all water withdrawn in this country today is used to remove waste heat in steam power generation of electricity. That enhances the efficiency of the process, conserves energy, and reduces electricity bills. Almost all water used for that purpose, incidentally, is returned to the resource almost unchanged, except for temperature increase, and is available for reuse by others.

Large quantities of water also cool industrial manufacturing processes, making possible many important products and improving the economy of their production. Water removes heat from refining processes that produce gasoline and fuel oils. It is a necessary medium for many chemical reactions. It removes impurities during the manufacture of medicines and many other important chemcials. It is used in the food-processing industry for many operations, including equipment sanitation and cleaning the products.

These applications and others use water principally to remove and dispose of waste materials. They are vitally important and too deeply imbedded in our modern life-style to allow us to discard them lightly. The practical fact is that we cannot afford to discontinue water carriage of wastes without insuring that we have other efficient methods for coping with wastes-handling problems now solved in that fashion.

POTENTIAL FOR REDUCING WATER CARRIAGE

There are many ways in which water uses for the carriage of wastes could be reduced or eliminated. Household washing machines could be replaced by equipment using recoverable solvents to clean clothes, as practiced in dry-cleaning establishments, for example. However, that would be more complicated, less convenient, and more expensive and could produce a different set of environmental problems.

Water carriage of toilet wastes could be eliminated by using pit privies (outhouses) or composting devices similar to those installed in a few homes during the past decade or two. It appears unlikely that either approach will become popular in the United States because inconvenience and aesthetic problems make them unacceptable to most families. Furthermore, their widespread adoption could result in several environmental problems, some having serious health implications. Other types of toilets avoid the use of water by incinerating the wastes using gas or other fuel. These toilets cost more, consume fuels, and may cause air pollution.

Note, incidentally, that many homes use septic tanks and tile seepage fields for "onsite" treatment and disposal of household liquid wastes with highly satisfactory results; in fact, this techinque often represents by far the best approach for solving wastes problems of individual homes or small communities. It does not, however, avoid the use of water carriage, but merely results in wastes discharge into groundwaters instead of surface waters, which sometimes may be even less desirable. Also, the tile fields require substantial land areas to insure adequate operation, and that may be unattainable or prohibitively expensive in metropolitan areas.

The industrial use of water for cooling often could be eliminated by using equipment that discharges waste heat into the air instead of into watercourses. For example, heat removed by most home air conditioners, referigerators, and freezers is transferred to the air through cooling coils. Unfortunately, air-based cooling equipment for industrial systems is considerably larger, more expensive, and less efficient than that required in water cooling. Thus, this alternative usually would cost more and waste more of our limited energy resources.

Cooling towers now are used by many industries to reduce the heat content of discharges, but they substantially increase construction and operation costs. Also, the pumping needs of those systems consume large amounts of energy, and their higher water losses to the atmosphere through evaporation leave less water to return to the stream for reuse by others. At a large power plant, for example, water evaporated daily in cooling towers may be many millions of gallons. Furthermore, the chemicals added to control biological growths and corrosion in those systems often can have major environmental impacts on the watercourses into which they are discharged.

This is not intended to imply that these and other techniques that are available to reduce water carriage of wastes from industrial operations should be avoided. Frequently, their beneficial effects far outweigh the possible increased costs and undesirable environmental impacts. However, they must be implemented only after careful evaluation of available alternatives to insure that adoption would produce benefits to the environment and economy that outweigh potential adverse impacts. Also, note that there are instances in which viable alternatives simply are nonexistent in today's technology.

An inescapable conclusion is that municipal and industrial uses of water are directly related to, and largely based on, water carriage of wastes, and this is certain to continue for many years.

SOME PRACTICAL ENVIRONMENTAL IMPLICATIONS

Suitable substitutes are not available now for some of our present water carriage uses but undoubtedly could be developed with enough investment of time, effort, and money. The point is not really whether water carriage of wastes could be eliminated, but whether that would be our best move. Often, the benefits would not justify the expense. Worse, sometimes the environmental impacts of alternatives might be even less acceptable than properly designed and operated wastewater collection, treatment, and disposal systems, similar to those now in use.

Dr. Barry Commoner, in *The Closing Circle* (1971), stated his Second Law of Ecology: "Everything must go somewhere." We should recognize that the idea also must apply to wastes products and by-products from homes and industries. Consider, then, that only two broad choices are available to us for managing waste materials and by-products: (1) we can recover or reuse them, or (2) we can dispose of them.

Technology now is available for recovering or reusing some waste materials productively; hopefully, in time we will develop additional and improved ways to use materials now discarded as wastes. This is the ideal approach, when feasible, because it eliminates a problem on the one hand and generates a useful product on the other. However, practical considerations in homes and industries today often make it impossible or prohibitively expensive to reclaim or reuse most wastewater constituents because of their complex chemical characteristics and very dilute concentrations.

When recovery and reuse are not feasible, society must turn to the second choice: waste disposal. We have only three locations for doing that: (1) on land, (2) into the air, or (3) into watercourses. All of these currently present serious problems, and selection among them should be made reasonably on the basis of minimizing total adverse environmental impact. If we decide to eliminate disposal by the water route, then more must be released into the air or distributed on land. That raises a question as to whether we can afford to eliminate disposal into watercourses at the expense of compounding environmental problems that we already have with air pollution and land disposal.

Also, the discharge of wastes into the atmosphere usually does not prevent them from ultimately reaching the water and land anyway in precipitation, which removes constituents from the air. Anyone who doubts that need only consider the current "acid rain" problem that results from the release of sulfur and nitrogen oxides into the atmosphere during combustion, followed by their return later as damaging acids in rainfall. Also, many materials buried in landfills subsequently enter the aquatic environment by leaching into water passing through the deposit. There are many documented instances of serious groundwater and surface-water pollution by such leachates, and more are certain to be disclosed in the future.

In considering alternatives available to us for disposing of wastes, then, there is an important fundamental concern that must be recognized and understood clearly. If we decide arbitrarily to declare one area off limits for receiving wastes (for example, water resources, as suggested in the 1972 law described earlier), that places greater pressure on the others and ultimately could cause even more objectionable environmental effects. This observation is consistent with Dr. Commoner's First Law of Ecology—"Everything is connected to everything else"—although probably not in the sense in which he intended it (Commoner, 1971).

The fact is that there are few simple answers to wastes disposal problems; good solutions usually require complex decisions and actions. Suggestions that water carriage should be simply eliminated are simplistic and naive because they ignore many important aspects of the question. We are left only to conclude that water carriage of wastes often is, and must remain for a long time, a legitimate beneficial use of some of the water removed from watercourses.

Chapter 12 explains that current state-of-the-art wastewater treatment plants remove only some of selected constituents from wastewaters, leaving most of the waste materials carried in them to be released into streams. Accordingly, the use of water for municipal supply, mostly for carriage of wastes, clearly results in some of them being released into watercourses even after application of the most modern treatment processes used in the United States today and likely to be employed in the future. If all of the waste materials were removed from the wastewaters in the future, at a cost, that would result in an increase in discharges into the atmosphere and land disposal facilities, with some inevitable return of them to the water environment eventually.

None of this should be construed as implying that discharge of wastes into the water environment, or their movement to the sea through streams, must have prioroity over all other uses of those resources. A contrary position that water supply, for example, must have exclusive priority cannot be entirely correct, either; where water is used wastewaters are generated, and provision must be made for their disposal. *Both* are necessary in today's society. Fortunately, alternatives, such as treatment to alleviate quality problems, usually are available for one or the other or for both. The combination of alternatives that provides the best solution for the total of both problems is the solution that should be adopted.

COMPETITION IN USE OF WATER

Competition among water uses arises in many ways, which are divided here into three broad categories:

1. Competition based on water quantities.
2. Competition for water resource facilities.
3. Competition based on changes in water quality.

Competition based on quantity is easiest to visualize and occurs commonly. In 1980, for example, about 21 percent of all water withdrawn for municipal supplies in the United States was lost through evaporation or used in a way that precluded its subsequent return to the resource. This produced an identifiable reduction in the amount of water available for other purposes. Similarly, about 3 percent of the water used by industries was lost through evaporation and incorporation into manufactured products, and 55 percent of the water withdrawn for irrigation was "consumed" through evaporation and transpiration.

All of these uses could compete with other uses of water because of their potential effects on its availability. Also, the data cited show that quantitative impacts can vary widely among different uses. Irrigation, for example, produces about 20 times the consumptive water loss as that which would result from using the same amount of water for industrial purposes. This observation led Powell (1956) to suggest that national and regional goals sometimes might better be served by using our limited water resources in some regions for industrial and municipal development instead of encouraging their use for irrigation.

A less obvious example of quantitative consumption of water, as it might influence utility to downstream users, is the situation in which water is withdrawn from one watercourse and discharged into a different one after use. Another example is the intermittent interruption of stream flow below hydroelectric reservoirs, which may seriously change water flow patterns available to satisfy needs of downstream users.

The second type of competition—for facilities—may occur when they are developed to enhance the utilitzation of water resources or permit better control of them. Good sites for constructing impoundments, for example, are limited in number. That creates a competitive situation in making decisions about how to allocate their capacities and the techniques to be employed in operating them after construction.

The ideal method for operating a reservoir for flood control, for example, is to keep it empty most of the time to maximize storage available for containing high flows whenever they occur. On the other hand, optimum operation of the same facility for municipal or industrial supplies would require keeping it as full as possible to retain maximum stored water for use whenever needed. Decisions about good reservoir sites, then, must consider competition among potentially conflicting uses, both in construction and subsequent operation.

The third, and fundamentally different, type of competition exists when a use of water produces significant changes in quality. Although substantially all flow withdrawn may be returned to the resource, constituents added in wastewater discharges could depreciate downstream water quality seriously. That may reduce the value of the resource significiantly or even eliminate its utility for some subsequent uses, although neither quantity nor pattern of flow may be altered significantly. This situation may develop, for example, when wastewaters are discharged that diminish esthetic benefits, reduce or eliminate downstream fisheries, create less acceptable raw water quality for municipal or industrial supplies, or interfere unreasonably with the ability of the stream to assimilate potential pollutants from other dischargers.

The fact that wastewater discharges may interfere with other beneficial uses has been recognized by professionals and laypersons alike. It is less generally recognized that *any* use listed in Table 5.1 potentially could compete with the others. It is impossible to maximize utilization of a water resource for one purpose without diminishing its utility for others to some extent.

An example based on opposite extremes might be helpful to illustrate this point. Clearly, maximizing the use of a stream for wastewater disposal could diminish its suitability for municipal or industrial water supplies. Perhaps less clear, but equally true, a decision to maximize its value for a municipal water supply (i.e., to retain a section of it at the highest possible quality) must reduce its capacity and utility for wastewater carriage. It is insufficient and inaccurate to state merely that water supply is more important than wastewater disposal. As pointed out earlier, where water is used for municipal purposes, water carriage of wastes results, and subsequent wastewater disposal becomes a necessity!

Modern water and wastewater treatment technology always provide alternative approaches that planners and engineers can consider. Some of the choices in this example could be based on (1) more complete wastewater treatment, (2) using a different stream for wastewater disposal, (3) more complete water treatment, or (4) selecting a different source for the water supply. In some instances, only poor alternatives may exist for wastewater disposal, although there might be several good ones for satisfactory water supply. In others, the opposite may be true. The two scenarios in the example would place very different demands on watercourse quality, and it is likely that neither extreme would maximize its benefit to society.

In summary, a watercourse may be used for many beneficial purposes, but not simultaneously to the maximum possible extent for each. In practice, decisions must be reached—consciously or unconsciously—favoring some uses at the expense of others. Selection of the balance of uses to be maintained in a watercourse is the most important single decision in planning its development because all other actions follow from that step. This aspect of water quality management will be discussed in more detail in subsequent chapters.

COMPLEMENTARY ASPECTS OF WATER USE

Possibilities for competition notwithstanding, the several different uses of water resources are compatible under proper controls; furthermore, they often even may complement each other. Many multipurpose reservoirs, for example, successfully blend potentially competitive uses, such as municipal and industrial water supplies, hydroelectric power production, boating, fishing, wastewater disposal, flood control, and aesthetic benefits, all within the bounds of a single comprehensive facility. Those various uses not only coexist but may be mutually beneficial because they share in the economy of a large, single reservoir that enhances all of them simultaneously, whereas none might be economically justifiable alone.

For example, significant possibilities exist for serious competition when stream flow patterns are changed through hydroelectric releases from storage reservoirs. With good planning, however, there could be complementary effects if releases from the reservoir could be timed to meet downstream needs—perhaps to furnish increased flows for municipal and industrial supplies or to control water levels for navigation. Similarly, releases of stored floodwaters often can be timed to generate hydroelectric power to meet energy needs or to satisfy other downstream water demands.

A related approach involves the construction of combined wastewater storage and treatment facilities by dischargers to minimize environmental impact by permitting programmed releases in proportion to variable stream flows. This technique has been used very successfully by a large pulp and paper mill, for example, to optimize stream quality below its outfall (Okun et al., 1963).

Another example, based on quality control, is to regulate heated discharges into watercourses to enhance the production of aquatic life by accelerating growth and lengthening the growing season. The potential for using waste heat from power and other industries also has attracted interest in agriculture for similar reasons, but it appears less promising.

The concept of controlling wastewater discharges in ways that enhance the benefits of watercourses is a great, and mostly unanswered, challenge facing the water pollution control field. Only minimal attention has been directed so far to possibilities for using wastewater discharges to produce positive benefits. Past emphasis has been on building and operating treatment systems designed principally to destroy wastewater constituents, some of which could be valuable if they were handled more creatively.

Two areas that have been the subject of significant consideration for producing benefits include controlled releases of heated discharges and the use of wastewaters and their nutrient contents in irrigation. Little attention has been given, on the other hand, to the beneficial effects that could be realized by allowing carefully controlled discharges of some wastewater constitutents into natural watercourses. There, nutrients and trace elements could increase the productivity of fish and other aquatic life in ways analogous to enhanced growth of crops in land disposal systems—*if* properly controlled to minimize environmental impacts, of course.

Currently, some agencies require the systematic consideration of land disposal as part of the planning process for wastewater treatment and disposal. However, similar attention to possibilities for nutrient recovery in aquatic systems of receiving streams has been neither required nor encouraged. Instead of striving to develop techniques for that type of positive conservation, we have concentrated almost exclusively on systems that require major expenditures simply to remove or destroy potentially valuable nutrients before discharge.

These important questions deserve more careful attention from those entrusted with protecting and enhancing our environment and conserving our natural resources.

OPTIMIZATION OF WATER USES

The problem facing a water planner is not merely to establish which uses should be maintained or maximized but to develop a reasonable balance that produces optimum utilization of the resource. As indicated earlier, this requires careful evaluation of competitive and complementary aspects of uses within the context of societal needs in the region under consideration.

The balance that would be optimum varies widely from region to region, from stream to stream, and even along the length of any given stream. It depends on diverse factors, including the needs and desires of many persons and groups and availability of other water resources in the vicinity. One approach to the decision process is through the "stream classification" systems employed by many states; these systems identify specific water uses to be maintained in each section of each stream. How these complex and difficult matters are decided in practice is discussed in Chapter 15.

As will be discussed in more detail later, meaningful evaluation of water quality can be accomplished only within the context of purposes for which it is to be used; otherwise, the term quality can have little practical meaning. A corollary is that decisions about optimum uses must consider the quality of water available, as influenced both by natural phenomena and impacts from human activities. Attempts to establish an optimum balance in the uses of watercourses without careful attention to quality are doomed to failure because they ignore important questions about what is achievable in the real world.

Mathematical tools have been developed and promoted actively to assist in decision making for water use optimization. These are based on systems analysis concepts, with roots in the economic aspects of water resources, and offer the advantage of producing results systematically in ways that are more objective than most other approaches for optimizing these complex systems. Another advantage is that the use of economic models delivers answers in financial terms, which can be readily understood by the public.

A serious disadvantage, on the other hand, is that many parameters that must be used as raw data in the formulations may have little basis in fact. For example, it is impossible to establish objective dollar amounts for most aesthetic values of water, for the value of opportunities for pleasure fishing, or for the value of relaxation that accompanies boating or other recreation on or in clean waters. Accordingly, development of economic data for use in optimization models usually requires estimates that stray from the objective into areas that depend on subjective views of the modeler. That, in turn permits conscious or unconscious biasing of the analysis while giving the appearance of objective, mathematically derived conclusions—a potentially misleading and even dangerous situation.

CURRENT U.S. WATER USES

Table 5.4 summarizes population and total water use data for the United states at five-year intervals from 1950 through 1980 (Solley et al., 1983). During that

TABLE 5.4
U.S. Population and Water Used, 1950–80

Year	Population (millions)	Total Withdrawal: Billion gal/day	Total Withdrawal: gal/day/person	Hydroelectric (billion gal/ day)
1950	151	180	1200	1100
1955	164	240	1500	1500
1960	179	270	1500	2000
1965	194	310	1600	2300
1970	206	370	1800	2800
1975	216	420	1900	3300
1980	230	450	2000	3300

Source: Based on data from Solley et al. (1983).
Notes: Water uses are rounded to two significant figures.
km^3/day = billion gal/day × 3.79 × 10^{-3}.
liters/day/person = 3.785 × gal/day/person.

period, the national population increased by 52 percent and the total withdrawal of water for all purposes increased by 150 percent—from 180 to 450 billion gal/day (0.68 to 1.71 km^3/day). The data represent an increase of 67 percent in total withdrawals per person from 1200 to 2000 gal/person/day (4540 to 7570 liters/person/day), indicating that the increased water use cannot be attributed merely to population change.

Also shown in in Table 5.4 are estimated stream flows used to produce hydorelectric power, which tripled during the same 30 yr. Those data are not included in the total or per capita "offstream" use figures just discussed because water for hydropower is not withdrawn, but is used "instream."

Table 5.5 summarizes U.S. water withdrawals for various purposes from 1900 to 1980 (NWC, 1973; Solley et al., 1983). Although irrigation uses have increased substantially since the turn of the century, withdrawals for that purpose have increased more slowly, percentagewise, during the past two or three decades. This can be attributed partly to the limited availability of water resources in many areas where irrigation is practiced extensively. Also, it reflects a recognition that high rates of water "consumption"—through evaporation and transpiration—make it necessary to restrict irrigation withdrawals and to improve efficiency with which our limited water resources are used.

Withdrawals for public water supplies have increased about tenfold since 1900. As indicated earlier, only part of that increase can be attributed to population growth, although it should be noted that municipalities have grown more rapidly than the country as a whole. Some growth in public use can be attributed to the delivery of water service to steadily increasing proportions of

TABLE 5.5
Water Withdrawals for the United States, Including Puerto Rico (billion gallons/day)

Year	Total Withdrawal	Irrigation	Public Supplies	Rural Supplies	Industrial	Steam-Electric
1900	40	20	3	2.0	10	5
1910	66	39	5	2.2	14	6
1920	92	56	6	2.4	18	9
1930	110	60	8	2.9	21	18
1940	136	71	10	3.1	29	23
1950	180	89	14	3.6	37	40
1955	240	110	17	3.6	39	72
1960	270	110	21	3.6	38	100
1965	310	120	24	4.0	46	130
1970	370	130	27	4.5	47	170
1975	420	140	29	4.9	45	200
1980	450	150	34	5.6	45	215

Sources: 1900–40 data from National Water Commission (1973); 1950–80 data from Solley et al. (1983).

Notes: Totals may not check because of different sources of data and rounding of figures. $km^3/day = billion\ gal/day \times 3.79 \times 10^{-3}$.

municipal populations. Of great importance, however, is the substantial growth in household water uses because of increased reliance on flush toilets, washing machines, dishwashers, and garbage disposals as well as increased "luxury" uses such as swimming pools, lawn watering, and washing automobiles. Growth in rural water use can be attributed to the improved availability of individual water supplies to rural families, increased household demands, and growth in stock-watering.

Water uses by steam-electric generating plants have increased explosively in response to growth in demands for electricity. Water used to remove heat in those installations increased fivefold, from 40 billion gal/day (0.15km^3/day) in 1950 to 215 billion gal/day (0.81km^3/day) in 1980. Since 1975, the rate of increase in demand has been somewhat abated in response to the increased concern about energy supplies. Withdrawals by other industries that supply themselves—as contrasted with those that use water from public systems—grew from 37 to 45 billion gal/day (0.14 to 0.17 km^3/day) during the same period, or only 22 percent, in spite of a much more rapid increase in water actually needed and used within plants to satisfy the growth in manufacturing needs. That anomaly can be attributed to improved water conservation and increased reuse within industrial facilities, caused partly by installation of cooling towers, for example.

Table 5.6 presents more detail on various water uses and consumption (losses) in 1980. The withdrawal for public supplies was 34 billion gal/day (0.13 km^3/day), which served about 186 million people, or 81 percent of the U.S. population. This produces an average municipal water use of 183 gal/person/day (693 liters/person/day). About two-thirds of the water used in public supplies is for domestic and public purposes, and the remaining one-third satisfies the needs of industries served by the municipal systems; this is in addition to self-supplied industrial uses, reported separately in the table. Table 5.6 also shows that about 21 percent of the water used in public systems is "consumed"; that is, it is not returned to the resource from which it was withdrawn. Much of that loss can be attributed to evaporation and transpiration.

Although the largest user of water is the thermal power industry, its consumption is only about 1.5 percent of the flow used. That figure will increase in the future, however, as more cooling towers are installed and higher evaporation losses are incurred in those units. Other industries exhibit much higher consumption, mostly because of evaporation in cooling towers. Some industrial losses can be attributed to steam formation, other evaporation, and incorporation of water into the products produced.

TABLE 5.6
Summary of U.S. Water Uses and Consumption, 1980

Type of Use	Used (billion gal/day)	Consumed[a] (billion gal/day)	Consumed (%)
Public supplies			
Domestic	22		
Industrial	12		
Total	34	7.1	21
Rural use			
Domestic	3.4	2.0	
Livestock	2.2	1.9	
Total	5.6	3.9	70
Irrigation, total	150	83	55
Industries			
Thermal	215	3.2	1.5
Other	45	5.0	11
Total	260	8.2	3.2
Total	450	100	22

Source: Solley et al. (1983).

Note: km^3/day = billion gal/day × 3.79×10^{-3}.

[a]"Consumed" water is water not returned to the resource.

Industrial uses of water exhibit little or no relationship to population because the flow used in manufacturing depends on the type of process and rate of production, not on national or local population or even on number of persons employed by the industry. However, water needs can be affected substantially by changes in process technology and water use practices within industrial operations.

A similar situation also prevails for irrigation. Water uses do not correlate with population trends but depend mostly on climate, crops grown, and irrigation techniques. Note that the consumption of water in irrigation reported in Table 5.6 exceeds that in all other uses combined; in fact, irrigation accounts for 81 percent of the total U.S. consumption. Of course, that can be attributed to high evaporation and transpiration losses to the atmosphere.

TABLE 5.7
Sources from Which Water Was Withdrawn, 1950–80 (billion gal/day)

Source	1950	1960	1970	1980
Fresh groundwater	34	50	68	88
Saline groundwater	—	0.4	1.0	0.9
Fresh surface water	140	190	250	290
Saline Surface Water	10	31	53	71
Reclaimed sewage	—	0.6	0.5	0.5

Source: Solley et al. (1983).

Note: km^3/day = billion gal/day $\times 3.79 \times 10^{-3}$.

TABLE 5.8
Water Withdrawals from Various Sources, 1980 (billion/gal/day)

	Groundwater		Surface Water		Reclaimed
Water Uses	Fresh	Saline	Fresh	Saline	Sewage
Public supply	12		22		
Rural					
Domestic	3.3		0.2		
Livestock	1.2		1.0		
Irrigation	60		90		0.3
Thermoelectric industry	1.6		148	65	
Other industry	10	0.9	29	5.4	0.2
Totals	88	0.9	290	71	0.5

Source: Solley et al. (1983).

Note: km^3/day = billion gal/day $\times 3.79 \times 10^{-3}$.

Table 5.7 summarizes water withdrawals from each of several sources between 1950 and 1980, and Table 5.8 shows sources for each major use category in 1980 (Solley et al., 1983). Fresh surface water is by far the largest supply, followed by fresh groundwater and saline surface water, in that order. About 84 percent of all water withdrawn in 1980 was fresh, but the sevenfold increase in the use of saline surface water since 1950 is noteworthy and reflects the increased difficulty and expense in obtaining large freshwater supplies in some coastal regions. Much of that saline water is used for cooling purposes.

Figure 5.1*a* summarizes the 1980 withdrawals of water in the United States (Solley et al., 1983). Industry was the largest user, accounting for 58 percent of the total, followed by irrigation, which withdrew 34 percent of the total. Only 7 percent of the 450 billion gal/day withdrawal was for public water supplies, and 1 percent was for rural supplies.

The consumption data in Table 5.6 indicate that the 1980 loss of fresh water was 100 billion gal/day (0.38 km^3/day), or 22 percent of the total withdrawal. Figure 5.1*b* summarizes the consumption data and shows that 81 percent of all consumption was caused by evaporation and transpiration in irrigation—about 55 percent of all water withdrawn for that purpose. The largest water user, industry, is responsible only for 8 percent of the total consumption

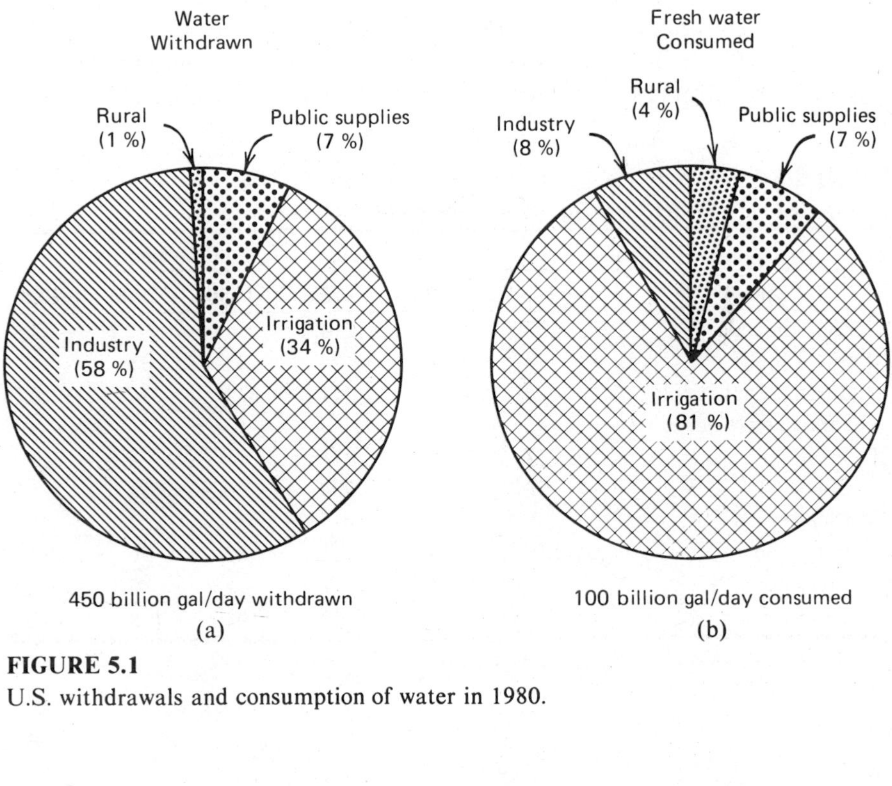

FIGURE 5.1
U.S. withdrawals and consumption of water in 1980.

because most industrial withdrawal is used for cooling and then returned to the resource, with little loss except when cooling towers are employed. Rural and public water supplies account for 4 percent and 7 percent of the total fresh-water consumption, respectively.

TRENDS AND PROJECTIONS

Table 5.9 summarizes estimated U.S. water withdrawals and consumption for various uses in 1980 and projects them to the years 2000 and 2020, as reported in 1973 by the National Water Commission (NWC). That report estimated that total withdrawal of water in 1980 would be about 440 billion gal/day (1.67 km^3/day), with steam-electric generating plants and agriculture remaining the largest users. Projections to the years 2000 and 2020 estimated total withdrawals at 800 billion gal/day (3.0 km^3/day) and 1350 billion gal/day (5.1 km^3/day), respectively. All water use categories were predicted to increase substantially, but industrial withdrawals greatly overshadowed the others, with steam-electric requirements estimated to grow about fivefold. A large fraction of the growth in demand for steam-electric generating plants was projected to be satisfied by the use of saline water. The predictions for water consumption seem to be consistent with the withdrawal trends.

Table 5.10 is a more recent projection of water uses and consumption, published in 1978 by the U.S. Water Resources Council (USWRC). A comparison

TABLE 5.9
Projected U.S. Water Uses for Various Purposes (billions/gal/day)

	Withdrawals			Consumptive Use		
Type of Use	**1980**	**2000**	**2020**	**1980**	**2000**	**2020**
Rural domestic	2.5	2.9	3.3	1.8	2.1	2.5
Public supplies	34	51	74	11	16	25
Industrial (self-supply)	75	127	211	6.1	10	16
Steam-electric power						
Fresh	134	259	411	1.7	4.6	8
Saline	59	211	504	0.5	2.0	5.2
Agriculture						
Irrigation	136	150	161	82	90	97
Livestock	2.4	3.4	4.7	2.2	3.1	4.2
Total	443	805	1368	104	128	157

Source: National Water Commission (1973).

Note: $km^3/day = billion\ gal/day \times 3.79 \times 10^{-3}$.

TABLE 5.10
Projected Water Uses, 1975–2000 (billion/gal/day)

	Withdrawals			Consumptive Use		
Type of Use	1975	1985	2000	1975	1985	2000
Rural domestic	2.1	2.3	2.4	1.3	1.4	1.4
Public and commercial	27	30	35	6.1	6.8	8.0
Industrial (self-supply)	58	32	31	8.2	12	18
Steam-electric power						
Fresh	89	95	79	1.4	4.1	10
Saline	60	91	119	—	—	—
Agricultural						
Irrigation	159	166	154	86	93	93
Livestock	1.9	2.2	2.6	1.9	2.2	2.6
Miscellaneous	1.9	2.2	2.5	1.2	12.5	1.7
Total	399	421	426	106	121	135

Source: U.S. Water Resources Council (1978).

Note: km^3/day = billion gal/day $\times$ 3.79 $\times 10^{-3}$.

of these data with Table 5.9 shows drastically different predictions. For example, Table 5.10 projects very little increase in total water withdrawal—from 399 billion gal/day (1.51 km^3/day) in 1975 to 426 billion gal/day (1.61 km^3/day) in 2000. That is less than the earlier prediction by the NWC, even for 1980, slightly less than 1980 figures recently compiled by Solley et al., (1983), and only about one-half of the NWC prediction for the year 2000.

The very large increases predicted in the NWC report for steam-electric generating plants were rejected in the USWRC report, which anticipated an actual *decrease* in freshwater withdrawal and a much smaller increase in saline water withdrawal. Public uses also were scaled down sharply in the 1978 report. The greatest disparity between the two is for self-supplied industrial withdrawals. The NWC predicted a 69 percent increase in this category by the year 2000. In contrast, the USWRC projected a *decrease* of 47 percent.

It is difficult to justify the differences between these two reports, both prepared in major federal studies only 5 yr apart. Although several circumstances changed during the 5-yr interval and different assumptions were made by the investigators, the final results still seem grossly out of line. Perhaps one should view the discrepancy as an extraordinary example of the uncertainty inherent in predicting future occurrences, similar to other experiences cited by Huff and Geis (1954). At least it gives clear warning to avoid placing too much faith in these types of projections.

OTHER CONSIDERATIONS

The amounts of water used for various purposes can vary considerably, depending on circumstances. This "resiliency" in demand is a function of several factors. Probably the most important one is the availability of water; uses generally are lower in water-short areas than in those where water is plentiful. Quality also is important in establishing the rate of water use, but the nature of that relationship is not so simple to determine. Under some circumstances, poor quality may correlate with lower uses, but under others the opposite effect may be observed.

Another important factor is the cost of the water, its treatment, and delivery; water uses tend to decrease with increase in cost. That relationship now is employed to influence water uses in some municipalities where supplies are limited. One approach is to change the basic rate structures for water, perhaps by eliminating the favorable rates often charged large users and substituting "inverted" schedules in which costs per gallon *rise* with an increase in the amount purchased.

A correlation between amount of water used and cost also may be noted in the increased attention to water conservation by companies that must build and operate expensive industrial wastewater treatment facilities; major water conservation programs frequently are undertaken to help minimize those costs. That could be responsible, to some extent, for the radical changes in projections for future industrial water withdrawals, discussed previously.

In summary, water withdrawals for various purposes are flexible. They are influenced by its availability, cost, quality, and many other factors. This makes water uses subject to control, within wide ranges.

STUDY QUESTIONS

1. Differentiate between "use" and "consumption" of water and outline examples of beneficial uses that exhibit widely different use-consumption relationships.
2. There are instances in which industries and municipalities located on streams have been cited as "consumptive users" of water, even though virtually no change in flow resulted from their withdrawals and uses of its water. Explain how this could happen and the conditions under which it could be a reasonable position for a regulatory agency to take.
3. List 10 beneficial uses of water. Identify at least two each in the withdrawal, on-site, and flow-use categories.
4. The 1972 amendments to the Federal Water Pollution Control Act

include a national goal of zero discharge of pollutants into navigable watercourses by 1985. Outline good and bad features of this legal position. Should it be retained as part of the law? Should it be enforced? Should it be repealed? Explain a rational basis for your position.

5. Outline arguments for and against including water carriage of wastes as a legitimate beneficial use of streams. Should water carriage be prohibited at some future date? Defend your position on this matter.
6. Differentiate between competition among water uses based on quantity considerations and competition based on quality. Illustrate your response with examples of each.
7. For any specific use of water listed in Table 5.1, explain how maximizing its utility for that purpose could interfere with other uses of the resource.
8. Describe two specific situations in which two or more water uses could complement each other. Explain the basis for that mutual benefit.
9. Describe and explain the relationship between population and total water use in the United States. Some persons have contended that water resources problems are principally the products of population growth. To what extent is this correct? Identify several other factors that have contributed to increased water use.
10. Of all water withdrawn for use in the United States, what proportion is used for municipal water supply? How much flow is used for that purpose? What proportion of that flow is consumed? Give examples of ways in which it is consumed.
11. Identify the principal uses of water in municipalities.
12. What is the average per capita water use in U.S. municipal systems? What is it, not including the industrial uses in municipalities? What is the total per capita water use in the United States?
13. Of all water withdrawn in the United States, what proportion is used for industrial water supply? How much total flow is used by industries? What proportion of that flow is furnished by municipal systems? What proportion of the total flow is used for cooling?
14. Assuming that 500 acres (202 hectares) of cropland is to be irrigated to the extent of 10 in. (254 mm) of water applied per year and using average demand figures for municipalities, what size community would that amount of water supply? Still using average figures, how many acres (hectares) of cropland could be irrigated to the same extent using effluent from the wastewater treatment plant of a community using that amount of water?
15. Of all water withdrawn in the United States, what proportion is used for irrigation? What is the average daily use for that purpose? What proportion and how many billion gallons per day of that water is consumed?
16. What municipal population could be served by the total flow consumed

in U.S. irrigation? What population would be required to produce the same amount of consumptive use in its municipal systems?

17. What percent of the total water withdrawal in the United States is supplied by reclaimed sewage? What is it used for?
18. What is the expected total water withdrawal in the United States by 2000 and what are the projected withdrawals for municipal systems, industries, and irrigation? Justify your figures.
19. Based on use projections and available water resources, will the U.S. water resources be exhausted by 2000? By 2020? Explain the basis for your answer.
20. Explain rationally why one might expect to find lower water uses in water-short areas. Explain why uses might sometimes be higher in those areas.

REFERENCES

American Society of Civil Engineers *Urban Water Resources Research,* Appendix G, p. G 34. Urban Hydrology Research Council, A.S.C.E., New York (1968).

Commoner, B., *The Closing Circle,* Knopf, New York (1971).

Davis, G. H., and L. A. Wood, *Water Demands For Expanding Energy Development,* Circular No. 703, U.S. Geological Survey, Washington, D.C. (1974).

Feth, J. H., *Water Facts and Figures for Planners and Managers,* Circular No. 601-I, U.S. Geological Survey, Washington, D.C. (1973).

Gehm, H. W., and J. I. Bregman, *Handbook of Water Resources and Pollution Control,* Van Nostrand Reinhold, New York (1976).

Huff, D., and I. Geis, *How to Lie With Statistics,* Norton, New York, (1954).

Leopold, L. B., and W. B. Langbein, *A Primer on Water,* U.S. Geological Survey, Government Printing Office, Washington, D.C. (1963).

National Water Commission, *Water Policies for the Future: Final Report to the President and to the Congress of the United States,* Government Printing Office, Washington, D.C. (1973).

Okun, D. A., J. C. Lamb, and C. C. Wells, "A Waste Control Program for a River with Highly Variable Flow," *Proceedings, 17th Purdue Industrial Waste Conference,* pp. 324–350 (1962), or *Journal Water Pollution Control Federation,* Vol. 35, pp. 1025–1043 (1963).

Powell, S. T., "Relative Economic Returns from Industrial and Agricultural

Water Uses," *Journal American Water Works Association,* Vol. 48, pp. 991–992 (1956).

Solley, W. B., E. B. Chase, and W. B. Mann, *Estimated Use of Water in the United States in 1980,* Circular No. 1001, U.S. Geological Survey, Washington, D.C. (1983).

U.S. Congress, "An Act to Amend the Federal Water Pollution Control Act," Public Law 92-500, 92nd Congress (1972).

U.S. Geological Survey, *Public Water Supplies of the 100 Largest Cities in the United States 1962,* U.S.G.S. Water Supply Paper 1812, Government Printing Office, Washington, D.C. (1964).

U.S. Water Resources Council, *The Nation's Water Resources 1975–2000,* Vol. 1, Summary, Government Printing Office, Washington, D.C. (1978).

CHAPTER 6

Methods of Evaluating Water Quality

Transfer of water quality information among individuals and groups requires the use of standardized techniques and terminology to assure accurate understanding by all concerned. Without a common "language," effective communication in planning and implementing quality control programs would be impossible. Water quality conditions could not be described in meaningful terms. Responsible agencies could not define clearly their goals or the requirements imposed on those being regulated. Enforcement would be severely hampered because it would be difficult or impossible to prove violations of laws and regulations. Most important, perhaps, it would be impossible to identify water quality trends, evaluate progress in pollution control programs, or learn from past successes and failures.

Therefore, those in the water quality field have found it necessary to establish comprehensive and complex systems of measurements and to standardize them as much as possible, consistent with maintaining both desirable accuracy and a reasonable work effort. The approach is based on using quantitative and reproducible parameters that are as descriptive and unambiguous as possible. It enhances the accuracy and ease of measuring water quality parameters and conveying that information to others.

This chapter reviews some conceptual aspects of water quality and approaches commonly used in practice to measure and specify it. Details of specific analytical test methods will not be discussed here—other available publications do that (ASTM, U.S. EPA, 1979; Water Pollution Control Federation, 1980; Brown et al., 1970; Slack et al., 1973; Sawyer and McCarty, 1978; U.S. Geological Survey, 1977).

THE CONCEPT OF WATER QUALITY

Often it is tempting to view quality in the context of a simple hierarchy based on purity, in which water with the lowest concentration of constituents is viewed as best. This may seem logical but, unfortunately, meaningful evaluation of water quality is not that simple. Many other factors must be taken into consideration.

The point may be estabished through a simple example. If one uses a scale based on purity, distilled water must rate very near the top, and seawater would be classified as relatively poor because it contains high concentrations of salts (35,000 mg/liter, or 3.5 percent total solids). It is well known that many types of fish, shellfish, and other aquatic life thrive in the ocean, specifically because they require salts, trace elements, and nutrients in concentrations approximating those actually there. Accordingly, ocean water is a desirable medium for growing those organisms and should be considered as high in quality for that purpose, in spite of (actually *because* of) the impurities. On the other hand, the same organisms would not survive long in distilled water because the necessary chemicals would be absent, making that an unacceptable environment for them. Thus, we must rate distilled water as poor in quality for that purpose.

Now consider a different example. Water added to high-pressure boilers evaporates and produces steam, leaving behind most salts that enter with the water. Excessive salt accumulations may cause boiler corrosion and scale formation problems. Water used for this purpose should be as low in salt content as possible, making distilled water excellent and seawater poor in quality for boiler makeup.

These two examples illustrate that water quality cannot be evaluated meaningfully without considering the use for which the water is being considered. Quality requirements for a stream that will be used as a municipal water supply are different from those that would make it desirable for trout fishing, and both may differ from the quality suitable for irrigation, swimming, or enjoyment of aesthetic aspects. An estuary with ideal quality for shellfish production may, nevertheless, be poor for many other uses.

Evaluation of the quality of a stream or lake must consider both (1) concentrations of various constituents in the water, and (2) uses that the resource will be called on to satisfy. Quality can be judged accurately only by comparing concentrations of various constituents in water with those that would be optimum for the intended use.

In short, *water quality is not absolute*. It has practical meaning only when related to some specific use of the water.

SYSTEMS FOR MEASURING WATER QUALITY

Water quality usually is dynamic—subject to major changes with or without human intervention. Also, many different parameters are used to describe and specify quality characteristics. Therefore, decisions on which parameters should be measured, how to collect samples, when to collect them, how to preserve them until they can be analyzed, and how to select appropriate analytical methods involve an extraordinarily complex matrix of variables.

Sometimes the best way to obtain the desired information is by measuring concentrations of one or more specific constituents. In other instances, it may be more appropriate to measure effects caused by them, whatever their identities and concentrations. Occasionally, time and analytical limitations may preclude reliable measurements of either, making it necessary to measure factors that may be unimportant themselves but are known to be associated with the constituents or their effects. When all else fails, it may be necessary to resort to qualitative descriptions.

For this discussion, it is convenient to divide the many types of analytical measurements into a few fairly broad categories. Although fundamentally different in some respects, as will be discussed shortly, they are not necessarily mutually exclusive; some analyses may fit into more than one category. The five groups to be considered are:

1. Analyses based on measuring concentrations of specific constituents of water, in which they are reported directly and quantitatively.
2. Analyses based on direct and quantitative measurements of quality characteristics. The parameters may be expressed on predetermined arbitrary scales, which actually may not relate to concentrations of any specific constituents.
3. Measurement of specific constituents or quality characteristics that are different from those of principal interest but are associated with them.
4. Measurement of the effects of constituents as contrasted with their concentrations in water.
5. Description or specification of water characteristics in qualitative terms.

Each category will be discussed further in the following sections.

DIRECT MEASUREMENTS OF CONSTITUENTS

Analytical techniques now are available that make it possible to measure concentrations of most elements in water, for example, calcium, magnesium, sodium, potassium, iron, manganese, mercury, and chromium, among others. Frequently, their concentrations must be known specifically because they are important in determining the quality of water for various purposes. For example, concentrations limits are defined for chromium, iron, mercury, and other metals in municipal waters because of potential health impacts or other undesirable effects.

In other instances, concentrations of certain chemical ions or compounds may be of special interest. For example, sulfates, sulfites, and sulfides all contain sulfur and all could be present in water, but these three chemical forms may have radically different effects on its quality. Accordingly, measurement of the concentration of elemental or total sulfur may convey inadequate information to someone interested in determining whether the water would be acceptable for a specific use; instead, it may be necessary to analyze for each chemical group. Examples of chemical groups having special significance in evaluating water quality include organic nitrogen, ammonia, nitrites, nitrates, sulfates, sulfides, and phenols.

Analytical techniques based on direct measurements of specific constituents also may be used in analyzing water to determine its microbiological content. Available procedures permit measurments of the numbers of specific types of bacteria, algae, protozoa, or other organisms. On the other hand, techniques also are available that permit measuring the total populations of certain groups that have common characteristics, including plate counts of bacteria, coliform organisms, fecal coliforms, algal content, fish populations, and others of special interest to those evaluating water quality.

Although different in details, methods in this category all have the common goal of permitting direct and quantitative measurements of selected constituents. Furthermore, they make it possible to report those measurements in terms that are understandable by others and reproducible in analyses conducted under similar circumstances. Accordingly, they are valuable tools for direct comparisons of water characteristics and for evaluating or defining quality needed for various purposes in which the specific constituents may be important.

DIRECT MEASUREMENTS ON ARBITRARY SCALES

Another category of methods for describing water quality is based on measuring characteristics that may be important in defining its quality but cannot be related directly to concentrations of one or more elements or chemical groups.

For example, "turbidity" refers to the scattering of light by particles in the water, which can cause a cloudy or milky appearance. This characteristic is especially important in determining aesthetic acceptability of drinking water, for example. Turbidity can be caused by many types of suspended materials, but the amount of turbidity produced by one may be radically different from that produced by the same concentration of another—because of different optical properties. Thus, if one wishes to measure and report the turbidity of a water in terms that convey a correct visual impression to others, it cannot be accomplished effectively simply by measuring the concentration of some specific constituent. Instead, it has been necessary to develop a quantitative analytical procedure that describes the extent to which the water exhibits a cloudy appearance.

Measurement of turbidity may be accomplished in practice by comparing the appearance of the sample with standards prepared to contain known concentrations of a specific suspended material that is similar to ones that cause turbidity in natural waters. The turbidity then is reported in "units," corresponding to the standard that most closely matches the water sample being examined. An alternative method is based on electronic measurement of light reflected from particles in the water, using an instrument developed for that purpose and calibrated in units equivalent to those for the standard suspensions.

This type of measurement provides a technique for quantitatively describing a quality characteristic of water without measuring concentrations of responsible constituents or, for that matter, without even identifying them. Analytical procedures based on comparisons with other arbitrary scales also have been developed for measuring important quality characteristics, including color, alkalinity, acidity, suspended solids concentrations, and dissolved solids. They all measure the parameters directly and quantitatively, based on specified and reproducible analytical techniques.

MEASUREMENTS OF ASSOCIATED FACTORS

Often, it is impossible, impractical, or too time consuming to measure the quality characteristic in which we are interested, but it may be feasible to measure some other parameter that is related to it. The measured value sometimes even may be used, by applying a previously determined correlation factor, to calculate an estimated concentration for the constituent or characteristic that is of principal interest. Two specific examples help illustrate the concept.

In stream pollution and wastewater treatment, the oxygen demand imposed by microorganisms is of special importance because of its impact on dissolved oxygen resources and aquatic life. The biochemical oxygen demand, (BOD) test is, perhaps, the most important single measure for evaluating performance of wastewater treatment plants and predicting impacts on receiving streams. The rationale and methodology for this test are discussed further in Chapters 8 and 10. Here it suffices to point out that, unfortunately, the BOD test usually requires incubating the sample for about five days before its results can be determined. This creates the undesirable situation in which a treatment plant operator only knows how the plant was performing five days earlier, but not its condition currently.

On the other hand, the chemical oxygen demand (COD) test uses a strong oxidizing chemical to react with organics in the wastewater and can be conducted in about two hours or less, but it produces a result that differs substantially from the BOD, both numerically and in physical meaning. Many treatment plant operators, however, have been able to establish empirical correlations between the COD and BOD that are specific to their systems, similar to the one shown in Figure 6.1. That makes it possible for them to conduct a relatively rapid COD test and apply the previously determined correlation to produce a predicted BOD, giving a tentative estimate of plant performance within about two hours instead of waiting the five days necessary to complete a BOD test. When feasible, that approach offers an obvious advantage to the operator who must keep up with system performance and needs an early warning of problems.

Another application of the approach is found in methods used to assess the bacteriological safety of drinking waters. Attempts to determine the types and numbers of pathogenic organisms, which are of great concern in identifying health risks, would be slow, difficult, and expensive. Furthermore, the results would indicate only whether they happened to be in the specific samples tested

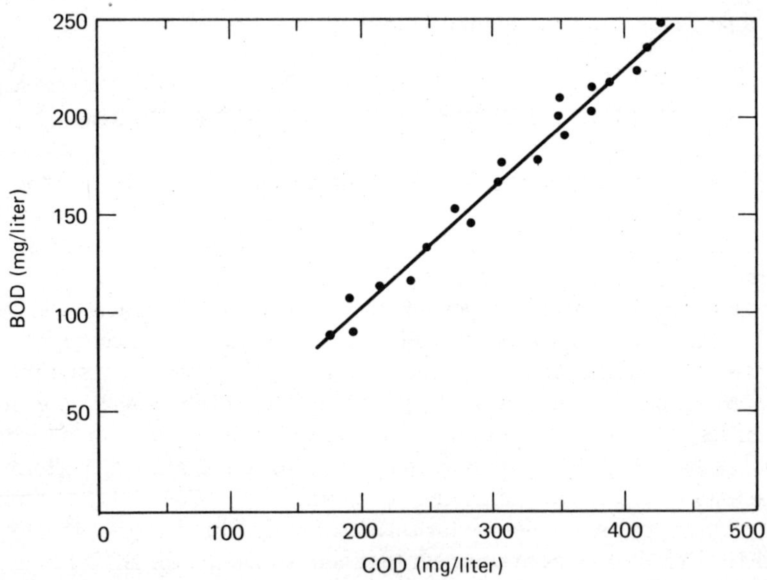

FIGURE 6.1
Relationship between BOD and COD for an industrial wastewater.

and would provide little information on the probability that pathogens might be in other portions of the water, either then or in the future.

The analytical procedure used in practice is based, instead, on measuring the number of "indicator organisms" in the water—specifically, the coliform group of bacteria. Coliforms are discharged from the intestinal tracts of warm-blooded animals, including humans, and their numbers may be used to estimate the extent to which a water has been contaminated by such wastes recently. The test, then, does not measure pathogens but the numbers of coliforms, which have been found to bear a clear association with the probability that pathogens could be encountered in that water, either then or at some future time. The coliform test is simple to conduct, relatively quick, economical, and produces information of great value for judging the risk of transmitting waterborne diseases. The approach will be discussed in more detail in Chapter 9.

The basic concept for this group of analyses is to measure a parameter that is associated with the quality characteristic of interest, but that can be determined more accurately, more easily, or more quickly. Other examples of tests based on associated factors include (1) measurement of volatile solids (loss upon ignition) to estimate organic content of waters and sludges, (2) measurement of conductivity to estimate dissolved solids in water, as illustrated by Figure 6.2 (Daniel et al., 1983), and (3) use of the total organic carbon (TOC) content to correlate with BOD in the manner outlined for COD.

MEASUREMENT OF EFFECTS OF CONSTITUENTS

Frequently, it is desirable or necessary to assess water quality by measuring the effects caused by constituents instead of trying to determine their identities and concentrations. Many procedures have been developed for that purpose.

For example, oxygen demand is vitally important in establishing impacts of wastewater discharges on fish and other aquatic life in streams and lakes. It is virtually impossible, in practice, to identify and measure concentrations of the many different chemicals that usually are present and to use those data to calculate oxygen demand that would be imposed on the receiving stream. It is more direct and accurate, as well as quicker and cheaper, to measure the BOD of samples directly.

That is why the BOD test persists as an important analytical measure of wastewater character, in spite of the many difficulties and frustrations that usually accompany its use, including complicated test setups, questionable accuracy, and poor reproducibility. The BOD test measures directly the amount of oxygen actually consumed by living organisms under conditions approximating those that might exist in a stream. Thus, it measures overall

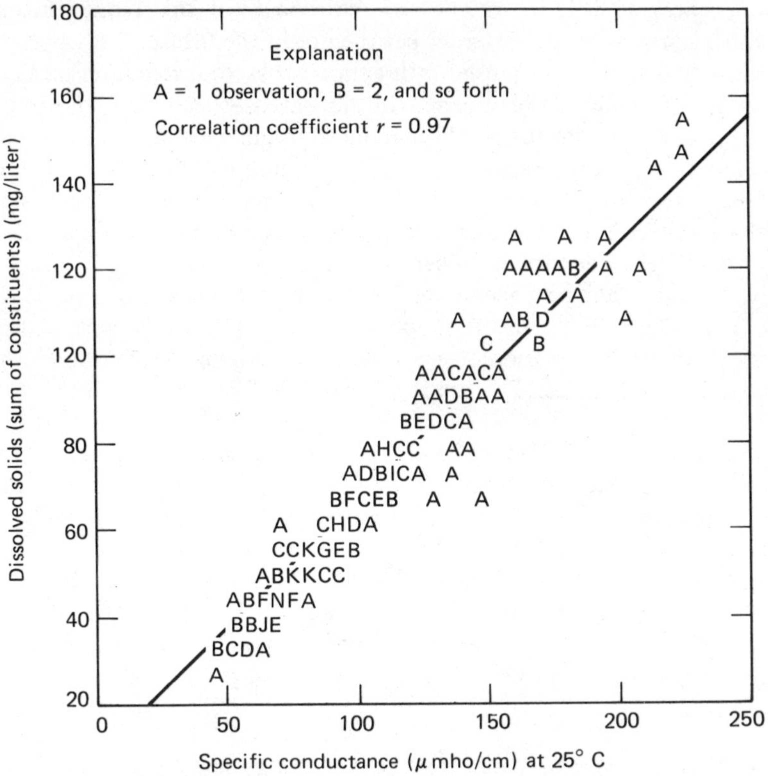

FIGURE 6.2
Relation between specific conductance and dissolved solids, French Broad River at Marshall, North Carolina, 1958–77 water years.

effect of the many wastewater constituents on oxygen resources without identifying directly any of those constituents. Because of its special importance in pollution control, the BOD test will be discussed in more detail in Chapter 10.

A similar rationale underlies fish bioassay testing. The current state of knowledge in fish toxicology does not permit accurate calculation of the toxic effects of complex chemical mixtures under different environmental conditions. Even with knowledge of all water constituents and their concentrations, one still could not estimate accurately their total impact on fish life in a stream.

Standard methods have been developed for conducting bioassays, using live fish in different concentrations of the chemical or wastewater in question. This procedure permits direct and quantitative evaluation of fish toxicity, regardless of the identity or character of the materials present, and can be used to evaluate impacts of waste streams—even those with unknown chemical content.

This can be very useful, for example, in tracing discharges that might be responsible for fish kills in streams.

Some other analytical tests based on measuring effects of constituents include (1) chlorine demand of water and wastewater, (2) threshold odor test, (3) hardness, (4) color, and (5) turbidity. Turbidity, for example, is an effect (cloudy appearance) caused by suspended matter in the water. However, it also is a quality characteristic that is measured on an arbitrary scale, as discussed earlier, making this an example of an analytical method that falls into two of the broad categories under discussion.

QUALITATIVE DESCRIPTIONS OF CHARACTERISTICS

As suggested earlier, it is preferable to use unambiguous, quantitative measures to describe water quality characteristics whenever possible, because they minimize misunderstandings. However, sometimes it is desirable or necessary to describe quality characteristics for which no quantitative measurement techniques are available. Under those circumstances, one must resort to qualitative descriptors.

For example, there is no standard test for measuring floating matter on streams or lakes, although it is an important quality characteristic. Regulatory agencies often must state merely that: "The stream shall be free of objectionable floating matter." Other examples often found in regulations include statements to the effect that water must be free of objectionable color, or that it must not contain toxic concentrations of chemicals.

Sometimes, regulatory agencies have issued standards stating that some specific chemical should be "none," or "absent," or another similar expression. These types of quality specifications should be viewed as necessary evils because they easily lead to misunderstandings. Although necessary occasionally because of limitations in analytical capability, they should be avoided as much as possible.

Also, use of the word "zero" should be avoided in water quality standards whenever feasible. To understand the ambiguity in that term, one need only consider the different implications of "zero" when expressed as: 0, 0.0, or 0.000. The word "zero" simply does not give enough information about the precision with which the measurement is made.

ON SELECTING SPECIFIC ANALYTICAL METHODS

Technical details of the many chemical and biological procedures used to measure various aspects of water quality are beyond the scope of this publication.

However, it is worthwhile to mention at least a few important points about how analytical methods are selected in practice, some facts about them that should be recognized, and sources of further information about them.

Analytical techniques used in water supply and wastewater disposal are based largely on standardized procedures, the most common ones being published in *Standard Methods For the Examination of Water and Wastewater* (Water Pollution Control Federation, 1980). That massive "cookbook" of chemical and biological methods is revised periodically in a joint effort by representatives of several associations that are involved in water quality evaluation and control.

It is important to understand that the procedures in *Standard Methods* are neither the simplest to conduct nor the most accurate of those available for measuring each quality characteristic. Each is selected among many alternatives by a subcommittee, after investigations and laboratory trials, based on judgment and compromise. Evaluation of the various available methods involves the consideration of several factors, including (1) abilities of analysts who will use the procedures in water and wastewater laboratories, (2) types of equipment that probably will be available in those laboratories, (3) time required to complete the analyses, (4) interfering substances likely to be present in municipal water or wastewater, and (5) overall accuracy that can be expected under conditions usually encountered in municipal practice. Thus, a method that is chosen and published should be recognized as a pragmatic "judgment call," not as a "referee" method that could be expected to produce accurate results even under conditions differing greatly from those assumed by the committee.

It is crucially important that those who produce and use data on water quality characteristics understand the advantages and limitations of "standard methods" and be prepared to adopt other techniques when necessary to insure accurate results. It is common in analyzing industrial wastewaters, for example, to find that some standard methods should not be used because of inaccuracies caused by wastewater constituents that interfere with them. In fact, the analysis of any water or wastewater differing substantially from that commonly encountered in municipal practice must be approached cautiously because of the potential for interferences. After all, most of the techniques in *Standard Methods* were selected predominantly to meet the needs in municipal water and wastewater practice.

A helpful book for understanding and interpreting standard analytical procedures, and the results obtained using them, is the one by Sawyer and McCarty (1978). Another important reference is published by the U.S. Environmental Protection Agency (1979). That one presents recommended or required procedures for collecting data in connection with wastewater discharge permits or for other purposes of concern to that agency. Some procedures specified by the EPA are simply incorporated from *Standard Methods* by reference. Others may differ substantially to cope with problems perceived by the EPA.

Other organizations concerned with water quality also publish collections of different "standard methods" for use by their personnel or others having similar interests and needs. They include the U.S. Geological Survey (Brown et al., 1970), the ASTM, agricultural chemists organizations, selected industrial groups, and others (U.S. Geological Survey, 1977). Caution must be exercised in using those procedures, however, or adopting data compiled by others who do, because the analytical methods sometimes may differ significantly from those commonly used in water and wastewater practice.

WATER QUALITY CRITERIA AND THEIR USE

The best efforts of knowledgeable persons and groups to define quality characteristics of water that are necessary or desirable for various specific uses are summarized periodically in the form of "water quality criteria." A comparison of data for the water in question with the appropriate published criteria provides a basis for judging probability that the resource would be satisfactory for beneficial uses that are under consideration and, if not, what changes in quality would be necessary to make it suitable for those purposes.

To illustrate this with an example, let us assume that the goal is for a stream to be used for irrigation of crops. For that to be feasible, the water should have certain quality characteristics, including limited boron content. The criteria in Table 6.1 were compiled by a panel of experts after careful study and consideration of past experience and show that sensitive crops will be protected adequately from the harmful effects of boron if its concentration in irrigation water is maintained at 750 μg/liter or less. Included in that figure is a safety margin judged by the experts to be reasonable and adequate to insure that boron problems should not occur, even under conditions differing significantly from those encountered in common irrigation practice. Thus, the quality criterion of 750 μg/liter boron can be used to judge, at least partly, the suitability of that water for irrigation.

Table 6.1 summarizes several water quality criteria published by the U.S. Environmental Protection Agency (1976) for various substances in waters to be used for selected purposes. It must be emphasized that they should not be viewed as valid and absolute limits, but only as guidelines that can be used for preliminary judgments. The effects of boron, for example, may vary widely, depending on other chemicals present in the water, synergistic action between those chemicals and boron, the type of crop involved, temperature, pH, and other environmental factors. With the safety margins included in the criteria,

TABLE 6.1
Summary of Quality Criteria for Water

Aesthetic qualities: All water free from substances attributable to wastewater discharges that settle to form objectionable deposits; float as debris, scum, oil, or other nuisances; produce objectionable color, odor, taste, or turbidity; injure or are toxic to humans, animals, or plants; and produce undesirable or nuisance aquatic life.
Alkalinity: should be >20 mg/liter for freshwater aquatic life, unless natural concentrations are less.
Ammonia: 0.02 mg/liter (as unionized ammonia) for freshwater aquatic life.
Arsenic: 50 μg/liter for domestic water supplies (health); 100 μg/liter for irrigation of crops.
Barium: 1 mg/liter for domestic water supply (health).
Beryllium: 11 μg/liter for aquatic life in soft fresh water;[a] 1100 μg/liter for aquatic life in hard fresh water;[a] 100 μg/liter for irrigation on all soils; except 500 μg/liter for irrigation on neutral to alkaline fine-textured soils.
Boron: 750 μg/liter for irrigation of sensitive crops.

Cadmium (μg/liter)

Domestic Use (Health)	Marine Life	Freshwater Life: Soft[a] Water	Freshwater Life: Hard[a] Water	
10	50	0.4	1.2	for sensitive fish
		4.0	12.0	for less sensitive aquatic life

Chlorine: Total residual 2.0 μg/liter for salmonid fish; 10.0 μg/liter for other organisms.
Chromium: 50 μg/liter for domestic water supply (health); 100 μg/liter for freshwater aquatic life.
Fecal coliform bacteria: For bathing waters, based on at least five samples taken over 30 days, should not exceed a log mean of 200/100 ml. Less than 10 percent may exceed 400/100 ml. For shellfish waters, median less than 14/100 ml. Less than 10 percent exceed 43/100 ml.
Color: No objectionable color (aesthetics); supply less than 75 units (domestic water); color and turbidity should not reduce depth of compensation point more than 10 percent.
Copper: 1.0 mg/liter for domestic water supplies (welfare). For aquatic life, 0.1 of 96-hour LC_{50}.[b]

[a]The concentration that will protect a given aquatic ecosystem can be determined by bioassays using ambient water and native fish and invertebrate species.
[b]Bioassays based on using sensitive resident species.

TABLE 6.1
(*Continued*)

Cyanide: 5.0 μg/liter for aquatic life and wildlife.
Gases, total dissolved: For aquatic life, dissolved gas concentrations less than 110 percent of saturation.
Iron: 0.3 mg/liter for domestic water supplies (welfare); 1.0 mg/liter for freshwater aquatic life.
Lead: 50 μg/liter for domestic water supply (health); 0.01 of 96-hour LC_{50} value for soluble lead for freshwater species.
Manganese: 50 μg/liter for domestic water supplies (welfare); 100 μg/liter for marine mollusks.
Mercury: 2.0 μg/liter for domestic water supply (health); 0.05 μg/liter for freshwater aquatic life and wildlife; 0.10 μg/liter for marine aquatic life.
Nickel: 0.01 of 96-hour LC_{50} for aquatic life.
Nitrates, nitrites: 10 mg/liter (as N) for domestic supply (health).
Oil and grease: For domestic supply, virtually free of oil and grease. For aquatic life, 0.01 of the lowest continuous flow 96-hour LC_{50}; no deleterious effects in sediment; and surface waters virtually free of floating oils.
Oxygen, dissolved: For aesthetics, sufficient dissolved oxygen to maintain aerobic conditions. For freshwater aquatic life, minimum dissolved oxygen of 5.0 mg/liter for fish.
Pesticides:

1. Aldrin-Dieldrin: 0.003 μg/liter for aquatic life.
2. Chlordane: 0.01 μg/liter for freshwater aquatic life; 0.004 μg/liter for marine aquatic life.
3. Chlorophenoxy herbicides: 2, 4-D—100 μg/liter for domestic water supply (health); 2,4, 5-TP—10 μg/liter for domestic water supply (health).
4. DDT: 0.001 μg/liter for aquatic life.
5. Demeton: 0.1 μg/liter for aquatic life.
6. Endosulfan: 0.003 μg/liter for freshwater aquatic life; 0.001 μg/liter for marine aquatic life.
7. Endrin: 0.2 μg/liter for domestic water supply (health); 0.004 μg/liter for aquatic life.
8. Guthion: 0.01 μg/liter for aquatic life.
9. Heptachlor: 0.001 μg/liter for aquatic life.
10. Lindane: 4.0 μg/liter for domestic water supply (health); 0.01 μg/liter for freshwater aquatic life; 0.004 μg/liter for marine aquatic life.
11. Malathion: 0.1 μg/liter for aquatic life.
12. Methoxychlor: 100 μg/liter for domestic water supply (health); 0.03 μg/liter for aquatic life.
13. Mirex: 0.001 μg/liter for aquatic life.
14. Parathion: 0.04 μg/liter for aquatic life.

TABLE 6.1
(*Continued*)

15. Toxaphene: 5 μg/liter for domestic water supply (health); 0.005 μg/liter for aquatic life.

pH: 5–9 for domestic water supplies (welfare); 6.5–9.0 for freshwater aquatic life; 6.5–8.5 for marine aquatic life.
Phenol: 1 μg/liter for domestic water supply (welfare), and to protect against fish flesh tainting.
Phosphorus: 0.10 μg/liter for marine or estuarine water.
Phthalate esters: 3 μg/liter for freshwater aquatic life.
Polychlorinated biphenyls: 0.001 μg/liter for aquatic life.
Selenium: 10 μg/liter for domestic water supply (health); 0.01 of 96-hour LC_{50}[b] for aquatic life.
Silver: 50 μg/liter for domestic water supply (health); 0.01 of 96-hour LC_{50}[b] for aquatic life.
Solids (dissolved) and salinity: 250 mg/liter for chlorides and sulfates in domestic water supplies (welfare).
Solids (suspended, settleable) and turbidity: For freshwater aquatic life, suspended solids should not reduce depth of compensation point more than 10 percent.
Sulfide, hydrogen sulfide: 2 μg/liter undissociated H_2S for aquatic ife.
Tainting substances: Materials should not be present in concentrations that produce undesirable flavors detectable in organoleptic tests on aquatic organisms.
Temperature: See explanation in *Quality Criteria for Water (U.S. EPA, 1976)*
Zinc: 5000 μg/liter for domestic water supplies (welfare); 0.01 of 96-hour LC_{50}[b] for freshwater aquatic life

Source: Based on data from *Quality Criteria for Water*, U.S. EPA, 1976.

the values in Table 6.1 may be viewed as concentrations that usually promise safety across a wide range of circumstances found in practice. Lower concentrations would be even safer. Higher concentrations might not preclude the specified use of the water but might only represent a reduction in the margin of safety.

The criteria summarized in Table 6.1 should be used with caution because they have been abstracted from a document of several hundred pages which, in turn, is a summary of many thousands of pages describing conditions under which the data were compiled. There always is a risk inherent in using such data without full understanding of how, when, and by whom they were compiled.

Information is being developed continuously in this area, and each criterion is subject to change as more is learned about the effects of water constituents on beneficial uses. No published criteria ever should be considered as final; they are valuable principally as guidelines. Unfortunately, that limitation is not always recognized by regulatory agencies and others.

STUDY QUESTIONS

1. Explain the rationale behind using a standardized system of water quality parameters.
2. Why is it necessary to consider the uses to be made of a water resource in judging its quality?
3. Identify several uses of water in which high purity is either necessary or very desirable. Briefly explain why for each one.
4. Identify several uses of water in which high purity is neither necessary nor especially helpful. Briefly explain why for each one.
5. Outline and briefly differentiate between the five broad categories of water quality measurements described in this chapter.
6. Do you feel that any of those types of measurements are inherently superior to the others? If so, identify which and explain why you feel they are better. If not, explain your rationale.
7. Describe the rationale for using water quality parameters that can be measured directly. Under what circumstances are these methods preferable to others? When would they be inappropriate for use?
8. Under what circumstances would the use of water quality parameters based on arbitrary scales of measurement be most appropriate? Explain why that is so, using examples as might be appropriate.
9. Explain a reasonable rationale behind measuring a factor associated with an important water quality characteristic instead of measuring the characteristic itself.
10. List three water quality parameters that are based on measuring the effects of constituents. Explain why each might be preferable for practical use instead of other types of measurements.
11. Explain the difference in implications between standards written to limit concentration of a water constituent to (*a*) "none," (*b*) 0 mg/liter, and (*c*) 0.000 mg/liter.
12. Explain the limitations of qualitative water quality measurements as regulatory tools and justify their uses in practice.
13. How could strongly different views of an industry and a regulatory agency about interpretation and enforcement of a specific qualitative type of standard (for example, "objectionable color") be reconciled in practice?
14. Explain the rationale for using "standard methods" for measuring water quality characteristics.

15. Describe the limitations of "standard methods" for measuring water quality characteristics. How could you assess the validity of such a method?
16. Specifically, where would you look for an analytical procedure appropriate for measuring some water quality characteristic? Assuming doubts about that procedure, where would you look for further information?
17. What are water quality criteria? How are they established? Where would you look to find criteria for some specific constituent of water?
18. What are the reasons for establishing and publishing water quality criteria? What are their limitations?

REFERENCES

American Society for Testing Materials, *Book of ASTM Standards, Part 31,* Philadelphia (published annually).

Brown, E., M. W. Skougstad, and M. J. Fishman, "Methods for Collection and Analysis of Water Samples for Dissolved Minerals and Gases," in *Techniques of Water Resources Investigations of the U.S. Geological Survey,* Book 5, Chapter A-1, Government Printing Office, Washington, D.C. (1970).

Daniel, C. C., H. B. Wilder, and M. S. Weiner, "Water Quality of the French Broad River, North Carolina—An Analysis of Data Collected at Marshall, 1958–77," in *Water Supply Paper 2185A-D,* U.S. Geological Survey, Government Printing Office, Washington, D.C. (1982).

McKee, J. E., and H. W. Wolf, *Water Quality Criteria,* Publication No. 3-A, State Water Quality Control Board, Sacramento, Calif. (1963).

National Academy of Sciences, National Academy of Engineering, *Water Quality Criteria, 1972,* Government Printing Office, Washington, D.C. (1974).

National Technical Advisory Committee, Federal Water Pollution Control Administration, *Water Quality Criteria,* Government Printing Office, Washington, D.C. (1968).

Sawyer, C. N., and P. L. McCarty, *Chemistry for Environmental Engineering,* 3rd Edition, McGraw-Hill, New York (1978).

Slack, K. V., R. C. Averett, P. E. Greeson, and R. G. Lipscomb, "Methods for Collection and Analysis of Aquatic Biological and Microbiological Samples," in *Techniques of Water Resources Investigations of the United States Geological Survey,* Book 5, Chapter A-4, Government Printing Office, Washington, D.C. (1973).

U.S. Environmental Protection Agency, *Handbook for Analytical Control in Water and Wastewater Laboratories,* EPA, Cincinnati (1979).

U.S. Environmental Protection Agency, *Methods for Chemical Analysis of Water and Wastes,* EPA, Cincinnati (1979).

U.S. Environmental Protection Agency, *Quality Criteria for Water,* EPA Report No. 440/9-76-023, Washington, D.C. (June 1976).

U.S. Geological Survey, *National Handbook of Recommended Methods for Water-Data Aquisition,* Government Printing Office, Washington, D.C. (1977, with periodic revisions).

Water Pollution Control Federation, *Standard Methods for the Examination of Water and Wastewater,* 15th Edition, WPCF, Washington, D.C. (1980).

CHAPTER 7

Constituents in Natural Waters

This chapter discusses some characteristics of "natural" waters—that is, waters free of gross pollution or other major impacts initiated by society. Note, however, that no waters really are free of such influences, because even rainfall in remote arctic regions contains constituents that were discharged into the atmosphere from processes installed and operated by humans. The distinction is more a matter of degree. Most of the water characteristics outlined here can be attributed predominantly to natural phenomena, with only occasional or relatively minor or diffuse effects caused by human activities. Major impacts from point-sources and other discharges will be discussed in subsequent chapters.

RAIN AND OTHER PRECIPITATION

At the instant of formation, precipitation is very pure because it has its origin in a massive solar distillation process—the hydrologic cycle, discussed in Chapter 4. However, purity of the water deteriorates rapidly as the falling rain or snow accumulates dissolved chemicals and particulates from the atmosphere. This "scrubbing" action is the major mechanism by which our atmosphere is purged of materials that otherwise could accumulate to deadly concentrations. Many impurities enter the air through natural phenomena, but others originate in industrial operations, automobile exhausts, smoke and gases from home heating systems, increased dust from developed land areas, and many other sources.

Examples of particulates picked up by precipitation include ash from volcanic eruptions, fires, power generation, and other industrial operations as well as dust swept into the atmosphere by wind action and aerosols produced in ocean surf, manufacturing processes, and wastewater treatment plants. Some serve as nuclei for the formation of raindrops, and others are entrained by the rain or snow during its fall. Their removal from the atmosphere accounts for the strikingly improved visibility often observed after heavy rain or snow.

Dissolved gases in rainwater include some of each gas present in the atmosphere, including oxygen, nitrogen, carbon dioxide, sulfur dioxide, and nitrogen oxides. The last two are of special interest today because they are primarily responsible for the phenomenon of "acid rain."

Increased combustion of fossil fuels (for example, in steam-powered electric generating plants) has resulted in sharp increases in amounts of sulfur dioxide and nitrogen oxides released to the atmosphere in many areas. Subsequent reactions between those gases and water increase the acidity of rainfall significantly and can result in chemical attack on buildings and statues exposed to the acid rain (U.S. EPA, 1979). Also, harmful effects on vegetation have been observed (Likens et al., 1974; Marx, 1975; U.S. EPA, 1979). In and downwind from highly industrialized areas of the United States, acid rain has become a serious water pollution problem. For example, many lakes in the northeast United States and Canada have suffered major impacts on fish life because their waters lack the buffer capacity needed to absorb the additional acid load without destructive changes in chemical characteristics. Current increase in the occurrence of acid rain represents a serious threat to the environment on a broad scale, and its effects now are being studied extensively (U.S. EPA 1979, 1980).

Chemical concentrations in precipitation typically are low when compared with streams and groundwaters, but the total amounts can be very large. Chemical contributions by precipitation are illustrated by data on characteristics of rainfall in eastern North Carolina and southeastern Virginia, collected by the U.S. Geological Survey (Fisher, 1968). Table 7.1 summarizes concentrations of selected constituents in rainfall over the two-year study, the total contribution of each constituent per square mile of area, and the estimated proportion of that constituent in local stream flow that could be attributed to contributions in rainfall. Table 7.2 presents similar information collected during a study in the northeastern United States (Pearson and Fisher, 1971). As might be expected, data for the industrial northeast differ somewhat from those collected in the more rural area represented by Table 7.1.

Data in the two tables show that constituents in precipitation may account for a substantial fraction of impurities in a stream. Precipitation must be rec-

TABLE 7.1 Chemical Composition of Rainfall in North Carolina and Virgina

Constituent	Concentration Range (mg/liter)	Approximate Contribution (tons/mi²/yr)	Approximate Percent of Stream Load
Chloride	0.19–1.70	2	25
Sodium	0.33–1.19	2	25
Calcium	0.39–1.33	2	50
Magnesium	—	0.5	30
Potassium	—	0.3	20
Sulfate	—	7	100
Nitrate	0.17–1.00	2	120
Ammonium	0.1	0.4	—

Source: Fisher (1968).
Note: Tons (metric)/km^2/yr = tons/mi^2/yr × 0.35.

TABLE 7.2 Chemical Composition of Rainfall in the Northeastern United States

Constituent	Concentration Range (mg/liter)	Approximate Contribution (tons/mi²/yr)	Approximate Percent of Stream Load
Chloride	0.5–2.1	1–7	30–80
Sodium	0.3–1.1	1–4	40
Calcium	0.6	2	25
Magnesium	0.1	0.4	25
Potassium	0.2	0.5	80
Sulfate	4	14	90
Nitrate	0.3	1	130 (Nitrate and Ammonium combined)
Ammonium	0.2	0.7	
Phosphate	0.04	1.4	—

Source: Pearson and Fisher (1971).

ognized as a significant, and variable, source of those materials—one that can exert a profound influence on the composition of many surface waters and groundwaters.

In some instances, the concentrations of chemicals in rain may be high, even when compared with levels sometimes considered unacceptable in the context of pollution control activities. For example, mercury often has been reported in rainfall at concentrations at or above 0.2 μg/liter and occasionally exceeds 0.5 μg/liter in some areas. Those concentrations are higher than some criteria or standards (0.05 to 2.0 μg/liter) adopted by regulatory agencies for the chemical (see Table 6.1). Results of a recent U.S. Geological Survey study of chem-

TABLE 7.3 Summary of Dissolved Chemicals in Bulk Precipitation in the North-Central and Northeastern United States (1980–81)

Constituent	Concentration: Median	Concentration: Range
	(μg/liter, unless otherwise noted)	
Ammonium, as N (mg/liter)	0.34	<0.01–3.7
Arsenic	<0.1	<0.1–4
Barium	2	<2–30
Beryllium	<0.5	<0.5–0.6
Cadmium	<1	<1–191
Calcium	0.6	0.05–41
Chloride	0.44	0.07–3
Cobalt	<3	<3.00–4
Copper	<10	<10–120
Fluoride (mg/liter)	0.07	0.01–0.64
Iron	25	<3–128
Lead	5	<1–78
Lithium	12	<4–59
Magnesium (mg/liter)	0.1	<0.14–3.4
Manganese	7	<1–91
Mercury	<0.1	<0.1–0.3
Molybdenum	<10	<10
Nickel	1	<1–34
Nitrate, as N (mg/liter)	0.7	0.1–2.5
Potassium (mg/liter)	0.1	<0.1–2.8
Selenium	<0.5	<0.5
Silica (mg/liter)	0.17	<0.009–2.6
Sodium (mg/liter)	0.2	0.2–1.6
Specific conductance (μmho/cm)[a]	23	0.3–250
Strontium	3	<0.5–81
Sulfate (mg/liter)	2.9	0.23–47
Total inorganic carbon (mg/liter)	0.4	0.1–8.6
Vanadium	<6	<6–7
Zinc	13	<3–592

Source: Peters and Bonelli (1982).

[a] μmho/cm = micromhos per centimeter.

icals in precipitation in the Northeast are summarized in Table 7.3 (Peters and Bonelli, 1982). This shows mercury contents ranging from less than 0.1 μg/liter to 0.3 μg/liter. Examination of data in the report reveal that the concentration was 0.1 μg/liter or higher in 45 samples (about 15 percent of the total). It is worth noting, incidentally, that news stories about the Great Lakes mercury "crisis" a few years ago failed to report that the contribution of mercury to Lake Erie by rain falling directly into it was approximately equal to that contributed by industrial wastewaters.

Another example of the importance of constituents in precipitation was reported by Murphy et al. (1975), who measured concentrations of phosphorus in the precipitation around Lake Michigan. They found that about 1 million kg/per yr—18 percent of the total lake budget for that nutrient—is contributed from the atmosphere. Currently, a program is under way to remove phosphorus from municipal and industrial wastewaters discharged into the Lake. After its implementation is complete, the phosphorus input contributed in precipitation will be about 30 percent of the total from all sources.

In summary, precipitation can contribute significant amounts of chemicals to our water resources. In a few instances, for example acid rain, the chemical content of precipitation reasonably could cause it to be viewed as a major source of pollution because of adverse effects on aquatic life or other serious problems. More commonly rainfall contributions alone do not cause problems but, nevertheless, should be considered as among the significant sources of some important chemicals.

RIVER WATERS

River waters are composed of surface runoff and groundwater flows into streams. The constituents in stream water include (1) those contributed in precipitation, discussed earlier, supplemented by (2) materials added by land erosion and solution of chemicals during travel over and through the soil, and diminished by (3) removal of chemicals by plants or reactions with soil constituents. Of course, constituents also are added through human activities. These may originate in point-source discharges of wastewaters or increased contributions from nonpoint-sources because of changes in land use patterns.

The quality characteristics of stream water may change drastically from time to time. Much of that variability can be attributed to changes in proportions of surface runoff and groundwater in the stream—those two sources usually being very different in physical and chemical characteristics. Immediately after heavy precipitation, the flow may be predominantly surface runoff, but most of the time it is composed entirely of water entering the streambed from underground sources. Also, the characteristics of input from either source may change with time because of impacts of rainfall intensity on runoff quality or effects of hydraulic changes on groundwater flows from different aquifers into the riverbed along its length.

This is illustrated by Figure 7.1, showing changes in dissolved solids content of a North Carolina stream as a function of its flow (Harned, 1982). Figure 7.2 shows short-term changes in flow and suspended matter in another stream, as influenced by local precipitation, demonstrating a close relationship between those variables (Daniels et al., 1982). Figure 7.3 shows that the lead and manganese content of the Neuse River vary in association with the suspended solids content of that stream, making the concentrations of those constituents also a function of stream flow (Wilder and Simmons, 1982).

Table 7.4 summarizes average concentrations of several constituents in riv-

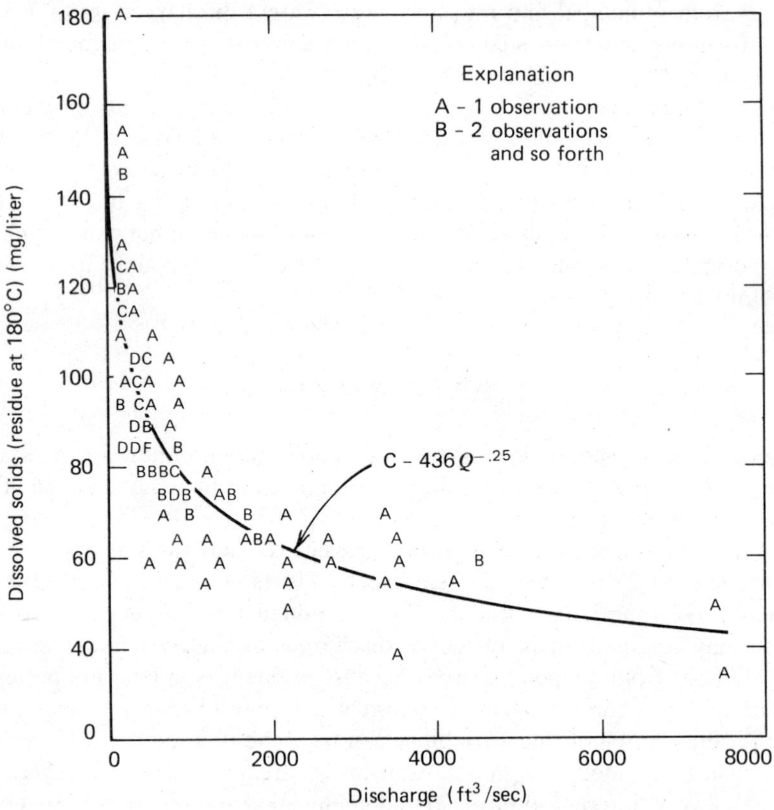

FIGURE 7.1
Relationship between river flow and dissolved solids concentration for the Neuse River near Clayton, North Carolina (1956–67).

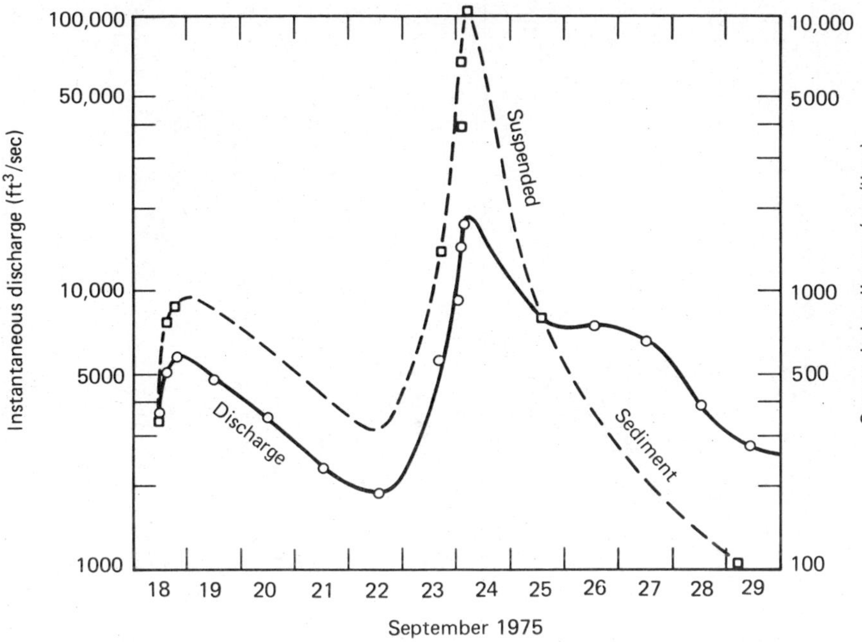

FIGURE 7.2
Changes in flow and suspended solids for the French Broad River, Marshall, North Carolina (September 18–29, 1957).

ers around the world, as reported by Livingstone (1963). Those data, and the preceding discussions, show that concentrations of constituents vary widely from stream to stream and time to time, making the information in Table 7.4 valuable only to illustrate the order of magnitude of concentrations that might be encountered in river waters. The variability also is illustrated in Table 7.5, which presents median concentrations of common constituents of U.S. rivers and concentration ranges that are estimated to include about 90 percent of all U.S. stream flows (Warren, 1971).

Dissolved solids represent an important category of constituents in stream waters and include organic compounds, gases, and inorganic chemicals, some of which are listed in Tables 7.4 and 7.5. These constituents enter river waters in precipitation, through solution of soil consituents, in wastewater discharges from point-sources, and in non-point-source discharges.

Leifeste (1974) analyzed dissolved solids discharges into the oceans from 54 U.S. river basins and found them to total 264 million tons/yr, as summarized in Table 7.6. The yields of dissolved solids, calculated in tons per square mile per year, ranged from 26 to 231 for the different areas studied, averaging about 100 for the entire country. Average concentrations of total dissolved solids

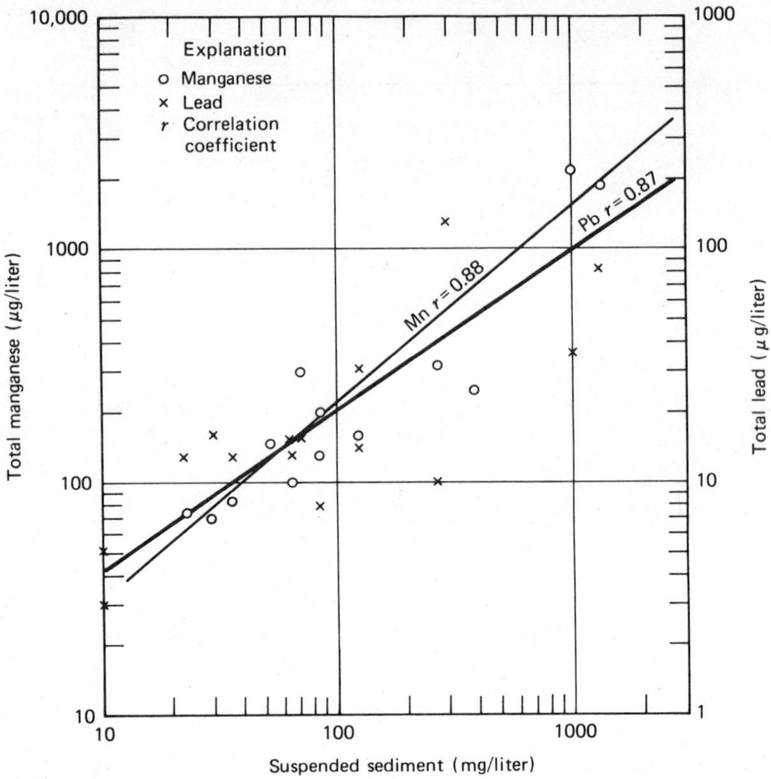

FIGURE 7.3
Effect of suspended solids concentration on manganese and lead content of the Neuse River at Clayton, North Carolina.

TABLE 7.4 Weighted Mean Composition of World Rivers (parts per million)

Constituent	North America	South America	Europe	Asia	Africa	Australia	World
Bicarbonate	68	31	95	79	43	32	58
Sulfate	20	4.8	24	8.4	14	2.6	11
Chloride	8	4.9	6.9	8.7	12	10	7.8
Nitrate	1	0.7	3.7	0.7	0.8	0.1	1
Calcium	21	7.2	31	18	12	3.9	15
Magnesium	5	1.5	5.6	5.6	3.8	2.7	4.1
Sodium	9	4	5.4 }	9.3	11	2.9	6.3
Potassium	1.4	2	1.7 }		—	1.4	2.3
Iron	0.2	1.4	0.8	0.0	1.3	0.3	0.7
Silica (SiO_2)	9	12	7.5	12	23	3.9	13
Total	142	69	182	142	121	59	120

Source: Livingstone (1963).
Note: Totals do not check because they include other constituents of the waters, as well.

TABLE 7.5 Major Constituents of U.S. Rivers (mg/liter)

Constituent	Median	Range[a]
Total dissolved solids	169	72–400
Bicarbonate (HCO_3)	90	40–180
Sulfate (SO_4)	32	11–90
Chloride (Cl)	9	3–70
Calcium (Ca)	28	15–52
Magnesium (Mg)	7	4–14
Sodium and potassium (Na and K)	10	6–85

Source: W. B. Hart, P. Doudoroff, and J. Greenback, *The Evaluation of the Toxicity of Industrial Wastes, Chemicals and Other Substances to Fresh Water Fishes,* Atlantic Refining Company, Philadelphia, 1945. Reprinted with permission of the Atlantic Richfield Company.

[a] Range estimated to include about 90 percent of total U.S. stream flow.

TABLE 7.6 U.S. Dissolved Solids Discharge Data

Characteristic	Atlantic Ocean	Gulf of Mexico	Pacific Ocean	Total United States
Drainage area				
mi^2	285,900	1,750,500	632,500	2,667,900
% of total	10.7	65.6	23.7	100
Water discharge				
ft^3/sec	359,400	887,400	499,060	1,745,860
% of total	20.6	50.8	28.6	100
Dissolved solids				
1000 tons/yr	37,500	183,000	43,400	264,000
% of total	14.2	69.3	16.5	100
Dissolved solids/mi^2/yr				
Maximum tons	231	159	219	
Minimum tons	40	30	26	
Average tons	131	104	69	100
Concentration, mg/liter				
Maximum	904	951	704	
Minimum	23	20	38	
Average	105	210	85	153

Source: Leifeste (1974).

Note: Maximum and minimum figures refer to average data for various river basins in each region.

ranged from 20 to 951 mg/liter for different streams and averaged 153 mg/liter for the United States as a whole.

Suspended solids also are important constituents of stream flows. They enter water as particulates removed from the atmosphere during precipitation, from erosion of land during surface runoff, through scouring of channels during high stream flows, and in wastewater discharges. Curtis et al. (1973) studied U.S. sediment discharges to the oceans and estimated that the total approximated 491 million tons/yr, distributed as shown in Table 7.7. Sediment yield for the country is about 185 tons/mi^2/yr, producing an average concentration of 286 mg/liter and a range of 15 to 1634 mg/liter.

Average figures are relatively meaningless except to illustrate ranges encountered because of the extraordinarily wide differences between areas and changes within any given area at different times. An example described by Curtis et al. (1973) illustrates the point—improved land use practices and the construction of reservoirs reduced sediment discharges in the Colorado River near Yuma, Arizona from 966 tons/mi^2/yr in 1911 to 1916 to 0.63 tons/mi^2/yr in the period 1965 to 1967.

TABLE 7.7 U.S. Suspended Solids Discharge Data

Characteristic	Atlantic Ocean	Gulf of Mexico	Pacific Ocean	Total United States
Drainage area				
mi^2	287,166	1,739,200	632,410	2,658,776
% of total	10.7	65.6	23.7	100
Water discharge				
ft^3/sec	359,350	887,400	499,065	1,745,815
% of total	20.6	50.8	28.6	100
Suspended solids				
1000 tons/yr	14,204	378,179	99,067	491,450
% of total	2.9	77.0	20.1	100
Suspended solids/mi^2/yr				
Maximum tons	111	398	9426	
Minimum tons	7.7	6.2	0.04	
Average tons	50	217	157	185
Concentration, mg/liter				
Maximum	58	820	1634	
Minimum	15	15	49	
Average	40	433	201	286

Source: Curtis et al. (1973).

Note: Maximum and minimum figures refer to average data for various river basins in each region.

LAKES

Inputs to lakes include streams, surface runoff, groundwater inflows, and precipitation that enters the lake directly. Many chemical characteristics of lake waters are similar to those of the streams feeding them, but additional factors may influence the water quality significantly. An important difference is that reduced velocity and turbulence allow much of the particulate matter entering with river waters to settle, producing lower suspended solids and turbidity in the lake. Also, the longer holding time and internal mixing reduce the range of quality variations with time far below those observed in streams.

Reduced turbidity and relatively long retention times in lakes encourage the growth of some types of aquatic organisms. For example, greater penetration of sunlight facilitates photosynthetic activity and the growth of algae, increasing productivity of the water and changing the ecological balance radically from that in feeder streams, as outlined in Chapter 3. That causes changes in the organic and inorganic content of the water and may profoundly influence its suitability for various purposes.

A phenomenon that often has great impact on water quality in lakes is thermal stratification, caused by the lower density of warm surface waters in the summer. As discussed in Chapter 3, the temperature differential may isolate the lower strata of the lake, prevent reaeration of that water, and produce low or zero dissolved oxygen concentrations below depths of about 20 ft. That can impact severely on water quality for supporting fish and other aquatic life that require dissolved oxygen. It also can produce conditions that favor the growth of organisms that develop undesirable tastes and odors in the water. The anaerobic conditions also can cause solubilization of iron and manganese in bottom deposits, producing concentrations in the water that make it unacceptable for municipal supplies unless appropriate treatments are applied before use.

GROUNDWATERS

Groundwaters are derived originally from precipitation and surface waters, so there often are many similarities in chemical content between groundwater and surface water. However, passage through the ground and contact with soils for long periods—often for many years—sometimes can alter the physical, chemical, and biological characteristics radically. For example, filtration through soil over long distances usually removes virtually all suspended materials, including turbidity and bacteria.

On the other hand, the slow movement increases opportunities for contacting and dissolving chemicals present in the soils, and biological or physical-chemical reactions may either reduce or increase the concentrations of various constituents. For example, ion-exchange reactions with solid particles may remove some constituents from the water, replacing them with other ions from the soil. This may have good or bad effects on water quality, depending on what ions are removed or added in the process.

An example illustrating differences between surface water and groundwater is found in the solution of limestone deposits. Calcium carbonate (limestone) is only slightly soluble at pH values normally encountered in surface waters. However, as shown in Figure 7.4, when precipitation moves down through the topsoil it may acquire substantial quantities of carbon dioxide produced in biological decomposition of organic matter. The solution of that carbon dioxide in water produces weak carbonic acid, which can react with calcium carbonate deposits to produce calcium bicarbonate—a chemical that is much more soluble in water than calcium carbonate. This mechanism increases the calcium

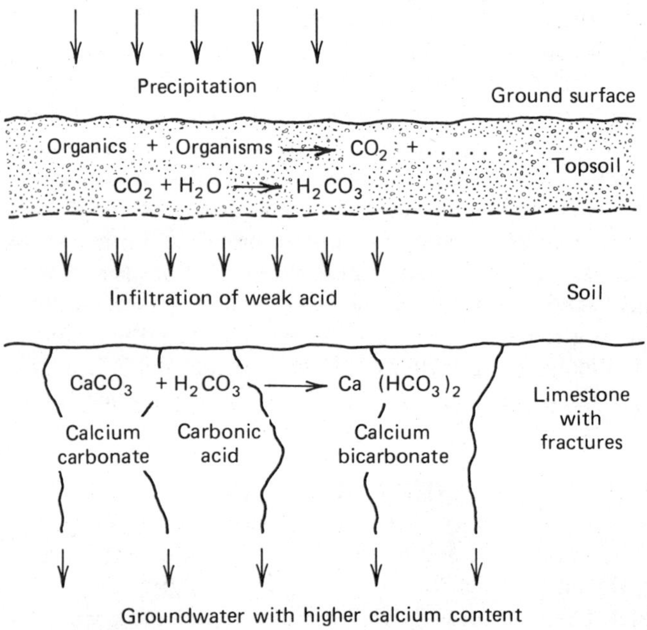

FIGURE 7.4

Illustration of a mechanism by which limestone is dissolved. Adapted from *Chemistry for Environmental Engineering,* Third Edition, C. N. Sawyer and P. L. McCarty, Copyright © 1978 by McGraw-Hill Book Company, New York, Page 380. Reprinted by permission.

content of groundwater to a level much higher than would be possible at the original pH, even over very long periods of contact. Therefore, groundwaters are much more likely to contain high concentrations of calcium than are surface waters. This, of course, also is the mechanism by which caverns and "sinkholes" may be formed in limestone areas.

Because of large storage volumes and long retention times, groundwaters exhibit less variability in quality with time than do surface waters, even lakes. They tend to be much more uniform in temperature, chemical characteristics, and biological content. Filtration through soil usually removes essentially all of the suspended matter, including organisms, but the mineral content of groundwater usually is higher than that of surface water in the same region.

OCEANS

The oceans receive inputs from precipitation, stream flows, groundwaters, discharges from point-sources of wastewaters, and flows from nonpoint-sources. Also, there are direct inputs of constituents from marine volcanic action and other geologic unheavals.

Table 7.8 summarizes several constituents of seawater, which is relatively constant in composition around the world except for local influences. The total dissolved solids content is about 35,000 mg/liter, or 3.5 percent by weight. Roughly 99 percent of the total dissolved solids can be attributed to chlorides,

TABLE 7.8 Selected Constituents of Seawater

Element	Parts per Million
Chlorine	18,980
Sodium	10,560
Sulfates	2,560
Magnesium	1,270
Sulfur	885
Calcium	400
Potassium	380
Bromine	65
Carbon	28
Strontium	13
Boron	4.6
Silica	3.0
Fluoride	1.0

sodium, magnesium, sulfate, calcium, and potassium. The predominant characteristic of seawater is high salinity, which is based on its content of chloride, sodium, and sulfate.

The high concentrations of constituents in seawater can be attributed partly to volcanic action over geologic ages and partly to continued concentration effects by the hydrologic cycle. Constituents of rivers flowing into the ocean are mostly nonvolatile. When water evaporates and returns to the land through the hydrologic cycle, virtually all of the constituents remain behind in the ocean, accumulate with time, and slowly increase in concentration.

However, the situation really is more complicated than that. Some suspended solids settle, removing them from the water environment. Some colloidal and dissolved solids precipitate through chemical reactions that occur after they arrive in the ocean and also settle to become part of the bottom deposits. Still other constituents may be used in the food chains of the thousands of living organisms in the ocean. Ultimately, they may be incorporated into cell materials, converted to by-products of biochemical reactions, or added to bottom deposits after death of the organisms. Also, biological action in bottom deposits may return compounds to the water above, providing interchanges in both directions between deposits and water. Similarly, there is an interchange of volatile and other constituents between the ocean and the atmosphere. All of these phenomena may be modified substantially on a local basis by inputs of streams, wastes, runoff, or other sources of constituents. Ocean water, then, is a dynamic system with many inputs and outputs and defies simple explanations of trends in concentrations of its constituents.

OTHER WATERS

Some waters not discussed in the preceding sections may be encountered less frequently but have quality characteristics that are worthy of note.

Some groundwaters contain very high concentrations of dissolved constituents—often too high for the water to be of practical value to humans in everyday activities. Saline underground waters are encountered, especially in some areas of the Midwest and West. They include slightly saline supplies, which sometimes are used after treatment to reduce salt concentrations, and extremely saline ones—often greatly exceeding the salt content even of seawater. Table 7.9 presents a descriptive classification system developed by the U.S. Geological Survey and often used to describe salinity of various waters (Swenson and Baldwin, 1965).

Great Salt Lake, in Utah, is the largest inland lake in the Western Hemisphere without an outlet to the sea. Its salinity varies with flows of tributary

TABLE 7.9 U.S. Geological Survey Classification of Saline Waters

Dissolved Solids (parts per million)	Classification
1000–3000	Slightly saline
3000–10,000	Moderately saline
10,000–35,000	Very saline
>35,000	Brine

Source: Swenson and Baldwin (1965).

streams over a range from about 150,000 ppm (15 percent) to a high approximating 280,000 ppm (28 percent). The Dead Sea contains dissolved solids of about 250,000 ppm (25 percent) (Swenson and Baldwin, 1965).

SUMMARY

All waters that occur in nature contain impurities. The types of constituents and their concentrations vary greatly, depending on natural phenomena that may be involved and on human activities. The data in Table 7.10 illustrate typical dissolved solids concentrations that might be encountered in waters from various sources. The concentrations vary widely from place to place and

TABLE 7.10 Examples of Dissolved Solids in Water

Source of Water	Dissolved Solids (parts per million)
Distilled	0
Rain	10
Lake Tahoe	70
Suwannee River	150
Lake Michigan	170
Missouri River	360
Pecos River	2,600
Ocean	35,000
Brine well	125,000
Dead Sea	250,000

Source: Swenson and Baldwin (1965).

time to time. It must be recognized that only limited generalizations can be made about water quality in nature.

STUDY QUESTIONS

1. List several chemical constituents of rainfall and identify a reasonable concentration range that might be expected for each.
2. What is "acid rain?" Explain the causes for it and the types of problems that result from it.
3. Identify several chemicals that have been reported to be present in precipitation at concentrations exceeding the recommended water quality criteria for various uses.
4. Explain why the chemical characteristics of river waters often vary widely with changes in stream flows.
5. Compare, contrast, and explain the relationships often observed between stream flow and concentrations of (*a*) dissolved solids, and (*b*) suspended solids.
6. Outline the reasons for the wide concentration ranges reported for various chemicals in U.S. rivers.
7. Assuming that municipal wastewater flow averages about 80 percent of the water intake and that the average content of suspended solids in the treated effluent is about 30 mg/liter, calculate the percentage of total suspended solids discharged from the United States that can be attributed to that source.
8. Explain the differences that may be expected between quality characteristics of water in a lake and that in feeder streams entering it.
9. Contrast the differences in quality that might be expected between groundwater and surface water in an area and briefly explain the reasons for those differences.
10. Describe the natural processes by which limestone (calcium carbonate) may be dissolved by waters passing through it to produce concentrations of calcium in the water greater than that expected based on reported solubility of calcium carbonate.
11. Identify and explain the factors that have been responsible for the accumulation of many constituents (chlorides, for example) in seawater over the ages.
12. Outline the reasons why other materials may not accumulate in ocean water to the degrees that might be expected based on their inputs.
13. Explain why the salinity of the Great Salt Lake in Utah is so high.
14. Compare and contrast the quality characteristics that might be encountered in raw water supplies taken from precipitation, surface streams, lakes, and groundwater sources.

REFERENCES

Biesecker, J, E., and D. K. Leifeste, *Water Quality of Hydrologic Bench Marks—An Indicator of Water Quality in the Natural Environment* Circular No. 460-E, U.S. Geological Survey, Washington, D.C. (1975).

Curtis, W. F., J. K. Culbertson, and E. B. Chase, *Fluvial Sediment Discharge to the Oceans from the Conterminous United States,* Circular No. 670, U.S. Geological Survey, Washington, D.C. (1973).

Daniel, C. C., H. B. Wilder, and M. S. Weiner, "Water Quality of the French Broad River, North Carolina—An Analaysis of Data Collected At Marshall, 1958–77," in *Water Supply Paper 2185A-D,* U.S. Geological Survey, Government Printing Office, Washington, D.C. (1982).

Durum, W. H., J. D. Hem, and S. G. Heidel, *Reconnaissance of Selected Minor Elements in Surface Waters of the United States,* Circular No. 643, U. S. Geological Survey, Washington, D.C. (1971).

Eisenreich, S. J., *Atmospheric Pollutants in Natural Waters,* Ann Arbor Science Publishers, Ann Arbor, Mich. (1981).

Fisher, D. W., *Annual Variations in Chemical Composition of Atmospheric Precipitation Eastern North Carolina and Southeastern Virginia,* Water Supply Paper 1535M, U.S. Geological Survey, Government Printing Office, Washington, D.C. (1968).

Fleischer, M., *Recent Estimates of the Abundances of the Elements in the Earth's Crust,* Circular No. 285, U.S. Geological Survey, Washington, D.C. (1953).

Harned, D. A., "Water Quality of the Neuse River, North Carolina—Variability, Pollution Loads, and Long-Term Trends," in *Water Supply Paper 2185A-D,* U.S. Geological Survey, Government Printing Office, Washington, D.C. (1982).

Hart, W. B., P. Doudoroff, and J. Greenbank, *The Evaluation of the Toxicity of Industrial Wastes, Chemicals and Other Substances to Fresh Water Fishes,* The Atlantic Refining Company, Philadelphia, (1945).

Leifeste, D. K., *Dissolved Solids Discharge to the Oceans from the Conterminous United States,* Circular No. 685, U.S. Geological Survey, Washington, D.C. (1974).

Likens, G. E., and F. H. Bormann, "Acid Rain: A Serious Environmental Problem," *Science* Vol. 184, pp 1176–1179 (1974).

Livingstone, D. A., "Chemical Composition of Rivers and Lakes," (professional paper no. 440-G), in *Data of Geochemistry,* 6th Edition, U.S. Geological Survey, Government Printing Office, Washington, D.C. (1963).

Marx, J. L.,, "Air Pollution: Effects on Plants," *Science,* Vol. 187, pp. 731–733 (February 28, 1975).

McGauhey, P. H., *Engineering Management of Water Quality,* McGraw-Hill, New York (1968).

Murphy, T. J., and P. V. Doskey, *Inputs of Phosphorus from Precipitation to Lake Michigan,* Report EPA-600/3-75-005, U.S. Environmental Protection Agency, Duluth, Minn. (1975).

Pearson, F. J., and D. W. Fisher, *Chemical Composition of Atmospheric Precipitation in the Northeastern United States,* Water Supply Paper No. 1535p, U.S. Geological Survey, Government Printing Office, Washington, D.C. (1971).

Peters, N. E., and J. E. Bonelli, *Chemical Composition of Bulk Precipitation in the North-Central and Northeastern United States, December, 1980 Through February 1981,* Circular 874, U.S. Geological Survey, Washington, D.C. (1982).

Reuss, J. O., *Chemical/Biological Relationships Relevant to Ecological Effects of Acid Rainfall,* Report EPA-660/3-75-032, U.S. Environmental Protection Agency, Corvallis, Ore. (1975).

Sawyer, C. N., and P. L. McCarty, *Chemistry for Environmental Engineering,* 3rd Edition, McGraw-Hill, New York (1978).

Sverdrup, H. U., M. W. Johnson, and R. H. Fleming, *The Oceans,* Prentice-Hall, New York (1942).

Swenson, H. A., and H. L. Baldwin, *A Primer on Water Quality,* U.S. Geological Survey, Government Printing Office, Washington, D.C. (1965).

U.S. Environmental Protection Agency, *Acid Rain,* Report EPA-600/9-79-036, EPA, Washington, D.C. (1980).

U.S. Environmental Protection Agency, *Research Summary, Acid Rain,* Report EPA-600/8-79-028, EPA, Washington, D.C. (1979).

Warren, C. E., *Biology and Water Pollution Control,* W. B. Saunders, Philadelphia, (1971).

White, D. E., J. D. Hem, and G. A. Waring, "Chemical Composition of Subsurface Waters" (professional paper no. 440-F), in *Data of Geochemistry*, 6th Edition, U.S. Geological Survey, Government Printing Office, Washington, D.C. (1963).

Wilder, H. B., and C. E. Simmons, "Program for Evaluating Stream Quality in North Carolina," in *Water Supply Paper 2185A-D,* U.S. Geological Survey, Government Printing Office, Washington, D.C. (1982).

CHAPTER 8

Types and Causes of Water Quality Problems

Water quality problems have intensified through the ages in response to the increased growth and concentration of populations and industrial centers. Often, the problems have been viewed as inevitable consequences of community development and, sometimes, even have been accepted as evidence of affluence and progress. That syndrome is similar to the old ironic statement that air pollution " . . . smells like money." Only after recognizing that serious health impacts could result from uncontrolled wastewater discharges did we finally initiate systematic water pollution control activities.

As discussed in Chapter 5, most municipal water is used for carriage of wastes. Accordingly, increased municipal uses require the construction of more and larger sewerage systems to remove the correspondingly greater wastewater flows. Those discharges, in turn, can serve as vehicles that facilitate the transmission of diseases to downstream water users. The obvious irony is that extension of public water supplies can improve the quality of life through better sanitation but, simultaneously, may increase health risks to others.

For many years, a prime motivation in controlling stream pollution was to protect the public from waterborne diseases. Although health is still an important consideration today, views of water pollution have broadened to recognize that its impacts on us are far more complex than health alone. In fact, most pollution control programs now are based principally on reasons only remotely, if at all, related to health. Examples illustrating that point in subsequent sections of this chapter demonstrate that most current regulations are not implemented to protect human health but to meet entirely different goals.

NATURE OF WATER POLLUTION

There are major differences of opinion about what constitutes water "pollution" and the rationale that should be employed in establishing and implementing pollution control programs. Some can be attributed to fundamental differences in philosophy and perceptions about needs for using water resources. Others are based simply on semantics—failures to identify accurately and clearly just what is meant by the term "pollution."

The word pollution implies undesirable quality, but it may be interpreted in various ways by different individuals. This may be attributed partly to emotional reactions but often involves substantive differences about our goals in using water resources and quality characteristics desired or needed to meet those goals. The presence of constituents that may be viewed as pollution by someone seeking a municipal supply could be acceptable or desirable to someone else who wishes to use the resource for swimming or fishing. Conversely, quality characteristics that could be satisfactory in a municipal water supply source might be unacceptable for swimming or in waters used to support or propagate fish life. Those wishing to use a watercourse in ways that produce benefits principally from its aesthetic qualities (for example, for boating or scenic enjoyment) may be interested in entirely different characteristics. Of course, those divergent views of what constitutes "pollution" merely amount to different statements of the concept outlined in Chapter 6: water quality cannot be evaluated meaningfully without considering specific uses to which it will be put.

Based on that foundation, it is possible to define pollution as:

> The presence of materials in water that interfere unreasonably with one or more beneficial uses of it.

This definition relates pollution to specific quality problems in using the watercourse. If a municipal or industrial wastewater discharge, either alone or in combination with others, actually interferes with a beneficial use that otherwise would be desirable and reasonable, then the watercourse is "polluted." On the other hand, if that discharge would not create or threaten an identifiable water quality problem, it would not constitute pollution. This definition is at odds with that favored by some individuals and groups who feel that *any* change in water quality from the natural state should be considered as pollution. However, it does not exclude aesthetic qualities of water resources or maintenance of a balance in aquatic life in any degree because those have been recognized earlier, in Table 5.1, as included among legitimate beneficial uses and should be considered on a par with all of the others.

An important point to note about the proposed definition is that it is rooted in the idea of reasonableness. This recognizes that sometimes local needs and conditions make it unrealistic or virtually impossible to require that *all* beneficial uses be protected all of the time. For example, some streams are fundamentally incapable of supporting certain types of fish or shellfish, even without receiving wastewater discharges, because their natural characteristics are unsuitable for propagation of those organisms. In that situation, an attempt to justify standards to protect these types of marine life must fail, and money or other resources expended specifically to meet that goal would be wasted.

Furthermore, the practical needs of society could dictate that even though the stream or estuary might be capable of supporting that type of aquatic life, it still could be unreasonable to invest large sums of money and other resources in facilities to maintain the water quality necessary for that purpose. Frequently, the same investments diverted to programs to solve other problems facing society could produce much greater returns.

Similarly, some streams (ditches, for example) may have little utility for aesthetic purposes, making the benefits expected from them in that regard minimal, although very substantial benefits may accrue through other beneficial uses of those watercourses. The effort and expense justified in maintaining the aesthetic quality of the water in such streams could be considerably less than that for a beautiful stream or lake.

Those examples raise the question of benefit-cost relationships in pollution control programs. In a world with finite limits, society must carefully consider the alternatives in expending its efforts and resources, selecting among them in ways that maximize the benefits received for each unit of investment. Examples of some alternatives could be decisions as to whether limited community resources should be spent for improved water quality in a stream, food for the poor, improved medical care, or upgraded education.

In summary, the concept proposed here is that pollution should be viewed in practice as based on *unreasonable* interference with some beneficial use of a watercourse. A corollary is that without unreasonable interference the situation does not involve pollution, even though substantial amounts of materials capable of causing quality changes may be discharged into the watercourse. Under this approach, it would be overly simplistic to view the mere existence of municipal or industrial wastes discharges as proving, necessarily, that a stream is polluted. That is true only when they cause or threaten water quality problems downstream and, furthermore, only when those problems are unreasonable, considering local conditions, desires, and needs.

POTENTIAL POLLUTANTS

The effect of a water constituent on use of the resource depends on its concentration. If it is sufficiently dilute, it would have no adverse impact on utility of the water for any desired purpose. In fact, many constituents of water that could be objectionable at high concentrations actually may be beneficial, or even necessary, in small amounts for some uses. As the concentration of a constituent increases, the water quality may be impacted adversely for some beneficial uses and, ultimately, may become unsuitable for virtually all desirable uses.

Although *any* constituent of water, if sufficiently high in concentration, could interfere with the use of the resource, it could be misleading to refer to it as a pollutant. Often, a chemical or other constituent does not cause pollution of water in the sense outlined earlier, because its concentration is below that which would cause objectionable quality or because the change in quality does not interfere with a use that is reasonable for that watercourse.

A rational approach is to consider every constituent of water as suspect until its concentration is demonstrated to be below a level that creates an unreasonable quality problem by impacting beneficial uses that should be protected. Accordingly, it is preferable to refer to water constituents as "potential pollutants" (that is, as having the potential for causing water quality problems, but not necessarily always doing so).

Table 8.1 lists eight categories into which it is convenient to divide potential pollutants. Each describes a class of constituents having characteristics or effects with special significance in determining the suitability of water quality. Because of their vital importance in water quality control programs, these categories are discussed briefly in the following sections. Two of them—infectious and toxic agents, and oxygen-demanding substances—are of such fundamental importance in today's water supply and pollution control programs that they are developed in more detail in Chapters 9 and 10.

TABLE 8.1
Types of Potential Pollutants

Infectious and toxic agents
Oxygen-demanding substances
Persistent organic chemicals
Plant nutrients
Minerals and chemicals causing specific problems
Suspended matter
Radioactive substances
Heat

INFECTIOUS AND TOXIC AGENTS

This category includes a wide variety of substances demonstrated or suspected to have adverse impacts on human health when consumed in drinking water or otherwise dependent on water as a vehicle for transmitting the harmful agents.

The first agents identified in this category, about a century ago, were bacteria capable of causing the classical waterborne diseases—typhoid fever, cholera, paratyphoid fever, and dysentery. They were responsible for massive epidemics that decimated many urban populations and spread terror wherever they occurred. Control of those diseases was the original battleground for improved sanitation and protection of the public health. Identification of their waterborne modes of transmission and development of methods for controlling their spread through improved water treatment and distribution remain as outstanding examples of public health practice, even to this day.

Improvements in water supply, wastewater treatment, and general sanitation during the past century have combined to cause those diseases to decrease almost to the vanishing point in many countries. Nevertheless, in spite of extensive knowledge about how they could be controlled, they remain endemic today in many less-developed countries. That tragic fact can be classified only as international negligence because the unnecessary human suffering is so great and methods for solving the problem are so clear.

It is widely recognized that water is a key vehicle in direct transmission of the infectious agents of several diseases, including the partial list in Table 8.2. They include bacteria, viruses, and other types of pathogenic organisms known to be transmitted through the water route. Water also is vitally important in the routes of many diseases in which it plays indirect but necessary roles, including some of the leading killers of people: malaria, yellow fever, filiariasis, and schistosomiasis. Finally, there are many other diseases for which water seems to be an important, but as yet undefined, part of the transmission mode.

In many countries today, including the United States, the greatest public health concern about water is directed to the ever-increasing numbers of toxic substances, including arsenic, lead, mercury, cadmium, other metals, and a wide variety of organic chemicals. A few of the more important ones are listed in Table 8.3. The effects of some are known, or may be reasonably estimated, but the impacts of most of them on humans at the low concentrations commonly encountered in waters are uncertain.

Their possible impacts include (1) acute, or short-term, health effects (2) chronic effects from accumulations over long periods of time and (3) genetic influences on future generations. We know surprisingly little about the acute effects of most of them, even less about the possible chronic effects from consuming water containing low concentrations over many years, and almost noth-

TABLE 8.2
Some Diseases in Which Water Plays an Important Role

As a Direct Vehicle of Transmission
Cholera
Typhoid fever
Salmonella infections
Infectious hepatitis
Diarrheal diseases
Schistosomiasis
Amebic dysentery
Paratyphoid fever
As an Indirect Vehicle (Vector Habitat)
Malaria
Yellow fever
Dengue
Encephalitis
Filiariasis
Onchocerciasis
Schistosomiasis

TABLE 8.3
Some Toxic Agents Transmitted by Water

Metals
Lead
Mercury
Cadmium
Selenium
Arsenic
Chromium
Organic Chemicals
Pesticides
Chloroform
Benzene
Carbon tetrachloride
Vinyl chloride
Polychlorinated biphenyls (PCB)
Trihalomethanes
Other Toxic Constituents
Cyanide
Nitrate
Fluoride
Radioactive substances

ing about the genetic damage that could result from the concentrations encountered in drinking waters.

Although toxic constituents often enter the water environment in municipal and industrial wastewaters, they also may be introduced in general land runoff or through various natural phenomena. A source of special concern that recently has been identified is their formation as by-products of chemical disinfection during the treatment of water for municipal use.

The relationship of water quality to the public health, and agents that are of special interest in that regard, will be discussed in more detail in Chapter 9.

OXYGEN-DEMANDING SUBSTANCES

Many organic chemicals may be used by microorganisms in watercourses as sources of energy and chemicals necessary for growth. Metabolic processes in those transformations cause the breakdown of organic constituents to simpler compounds and, often, ultimately to inorganic ash and gases. The biochemical reactions result in utilization of oxygen dissolved in the water, imposing a BOD (biochemical oxygen demand) on the limited oxygen resources available in watercourses. The amount of BOD exerted depends on the types and amounts of organic chemicals present, numbers and types of organisms in the water, temperature, pH, presence of nutrients and trace elements necessary for growth, and many other environmental factors. Oxygen used in the biochemical reactions can be replenished through transfer of that gas from the atmosphere into the water and by photosynthetic action of algae and other aquatic green plants.

A stream or lake can tolerate the imposition of some BOD without serious upset of its dissolved oxygen balance or of organisms that depend on that element for survival. However, when dissolved oxygen (DO) is used faster than it can be replenished, its concentration in the water decreases. Reduction of DO to less than perhaps 3 to 5 mg/liter can cause an adverse impact on fish that require a relatively high oxygen concentration to meet their metabolic needs. A further increase in oxygen demand would result in an even lower DO and progressively worse conditions for fish and other aquatic life. The addition of enough oxygen-demanding materials to the watercourse could cause the total depletion of DO and the death of all fish. Furthermore, the absence of DO could result in the growth of microorganisms that produce by-products that cause foul odors in the water and its surroundings.

For many years the depletion of DO in streams has received priority attention from regulatory agencies, making the control of oxygen-demanding materials (BOD) in wastewater discharges the backbone of their efforts. For this reason, the BOD test (discussed in Chapters 6 and 10) remains a key measure for evaluating characteristics of wastewater discharges, in spite of severe analytical limitations. This test is the only practical and direct method for measuring oxygen consumption during biochemical stabilization of organic matter.

It is important to understand that attention devoted to BOD and DO in streams is not motivated by concern about public health impacts of discharges. It has never been demonstrated that there is any relationship between the dissolved oxygen content of water and the health of persons drinking or otherwise contacting it. Since most stream pollution control activities in the United States are based principally on controlling BOD loadings and stream DO, they should not be viewed as public health measures. On the contrary, they usually are based simply on maintaining DO levels that are adequate to support the desired types of fish life, protect recreational interests, and insure desirable aesthetic characteristics of the watercourse. In short, the health of aquatic life—not of humans—is the principal focus of today's pollution control activities.

Because of the importance of oxygen-demanding substances in today's pollution control programs, the concepts of oxygen balance in streams will be developed further in Chapter 10.

PERSISTENT ORGANIC CHEMICALS

As just described, some organic chemicals are of concern in stream pollution control programs because they biodegrade and impose oxygen demand on watercourses. In contrast, other types of organic chemicals cause concern for the opposite reason; they are not decomposed through biological action and may persist for long periods, or indefinitely. Because there is no mechanism by which nature can cleanse itself of these constituents, they accumulate in the environment and ultimately may reach objectionable concentrations in the water or in aquatic life. That is why the growing number of nonbiodegradable synthetic organic chemicals being produced in industrialized countries causes great concern among environmentalists.

Several years ago, some types of nonbiodegradable synthetic detergents received great publicity because the foaming that they caused provided the public with visual, and sometimes dramatic, evidence of the problem. Those chemicals generated concern principally because of aesthetics and other problems caused by the foam, occasional difficulties in water and wastewater treatment, and possible impacts on aquatic life. They generally were not viewed as increasing the risk of serious public health impacts on persons drinking the water.

On the other hand, several pesticides that are highly resistant to biochemical attack do cause concern because of suspected or demonstrated impacts on health. Some chlorinated hydrocarbons, for example, are highly resistant and may have acute or chronic health effects on humans. Because they are so persistent and are strongly adsorbed by cell materials, they often accumulate in microorganisms to concentrations several times as high as that in the water. The consumption of those organisms by others higher in the food chain and the subsequent repetition of that process produce ever-higher concentrations at each step and, ultimately, may result in accumulations to levels hundreds or even thousands of times that in the water. This sometimes produces concentrations in fish high enough to make them unacceptable as food, although concentration of the chemical in water itself may not be high enough to cause concern.

DDT presents an interesting case history in this respect. It is a chlorinated hydrocarbon that has been used extensively throughout the world for many purposes but principally in mosquito control. Its use has caused dramatic reductions in malaria and other mosquito-borne diseases in less developed countries, making it one of the most effective and economical weapons for battling diseases that cause millions of illnesses and deaths annually. It also has been used with great effect in controlling pests capable of massive destruction of crops in countries where hunger is the number one human problem. While producing those substantial public health benefits, there has been no accumulation of solid evidence that DDT has caused problems in the health of populations exposed to it through food or water consumption.

Because DDT is extremely resistant to biodegradation, virtually every human on earth now carries a small accumulation of that chemical in their fatty tissues. Although no harmful impacts by those accumulations have been demonstrated, they cause great concern because of uncertainty about the long-range effects that might appear later. Accordingly, the continued use of DDT is highly controversial—one side expressing concern about its possible unknown effects, while the other points to known benefits through preventing illnesses and saving lives by mosquito control and improved crop production. The history of DDT provides a concise picture of some dilemmas and complexities faced by those who try to evaluate the benefits and impacts of various chemicals.

Some of the thousands of persistent organics in existence today are objectionable because of known harmful effects on humans, fish, or wildlife. Others are considered objectionable because of adverse effects that are suspected but whose nature and extent are unclear now. This problem is very different in character and dimensions from the relatively simple concerns about oxygen demand and dissolved oxygen levels in streams, which represent the principal target of stream pollution control programs today.

PLANT NUTRIENTS

Biological growth requires favorable environmental conditions, including access by the organisms to chemicals needed to manufacture new cells. The necessary elements include carbon, oxygen, hydrogen, nitrogen, phosphorus, sulfur, and several others that must be present, at least in trace amounts (see Chapter 3). Growth can continue only so long as all of those essential constituents are available to the organisms. When one is depleted from a biological system, growth ceases and can resume only to the extent that the supply of that material is supplemented. Under this situation, the chemical that restricts growth is referred to as the "limiting nutrient."

The concept of the limiting nutrient may be illustrated by examining the role of phosphorus as a specific example. The composition of algal cells may be represented crudely by a chemical formula that summarizes elements contained in them. One of several that have been proposed is $C_{106}H_{180}O_{45}N_{15}P$. Calculations based on relationships in this formula indicate that 1 lb of cells contains about 0.013 lb of phosphorus. A corollary is that 0.013 lb of phosphorus must be available to permit the biological system to produce 1 lb of cells. Expressed another way, each pound of phosphorus available in the system has the potential for allowing production of up to 77 lb of cells, assuming proper environmental conditions and the presence of all other required constituents in adequate amounts.

Algal "productivity" of many freshwater lakes actually is limited by low concentrations of phosphorus in the water. When phosphorus is added to such a lake in land runoff, wastewater discharges, or from other sources, the restraint is removed and the extent of growth that can occur is directly related to the amount of phosphorus added. Note that the phosphorus does not cause the growth but allows it to occur by removing a constraint that had existed before its addition. This is why phosphorus is viewed with such interest in evaluating discharges into lakes and provides the basic rationale for removing it from many wastewater discharges. Where phosphorus is the limiting nutrient in a body of water, preventing its addition in wastewaters can provide a practical method for controlling some aquatic growths.

On the other hand, where phosphorus is not the limiting nutrient and productivity is controlled by some other chemical—nitrogen, for example—a more practical method for controlling aquatic growths could be to limit the supply of that constituent instead. Nitrogen and phosphorus are especially important "nutrients" because phosphorus often limits growths in lakes and nitrogen sometimes may be limiting in estuaries.

The concern about this type of potential pollutant also is based on factors other than human health—specifically, aesthetic qualities of the watercourse, character of the aquatic habitat, tastes and odors in municipal water supplies, and other problems caused by excessive accumulations of algal growths.

MINERALS AND CHEMICALS CAUSING SPECIFIC PROBLEMS

This category of potential pollutants includes a wide variety of organic and inorganic chemicals that may have undesirable effects on water uses. Some may cause aesthetic problems in drinking water supplies; for example, phenol reacts with chlorine used in disinfecting water to produce chlorophenols and cause highly objectionable tastes and odors. In other instances, chemicals may accumulate in fish flesh, producing taste and odor problems and adversely impacting commercial and recreational fishing in streams and lakes.

Many minerals and chemicals may be harmful to aquatic life either through acute toxicity or chronic effects from accumulations in the organism over extended periods. Chemicals that bioaccumulate in food chains may create problems, even though their concentrations in the water are well below objectionable levels. Information on the identities of several important constituents and their concentrations that may cause problems are found in water quality criteria, and some are summarized in Table 6.1.

Some chemicals act indirectly by inducing secondary reactions in the watercourse. For example, sugars in effluents from paper mills sometimes have been observed to stimulate growths of *Sphaerotilus,* a nuisance filamentous organism, in watercourses. The chemicals themselves do not impart objectionable characteristics to the water, but organisms growing in response to their presence may create difficulties in downstream water supplies, diminish recreational values, interfere with commercial fishing by clogging nets, or be responsible for causing undesirable appearance of the water.

Table 8.4 summarizes a few specific chemicals that are of special interest in evaluating water quality, with brief descriptions of problems that they may cause. Note especially the differences in effects on water quality caused by various chemical forms of the same element, for example, sulfur. This illustrates the importance, sometimes, of assaying specific ions or molecules instead of merely total content of the element itself, as discussed in Chapter 6.

SUSPENDED MATTER

Suspended matter is material with a particle size larger than dissolved molecules or ions. It may be viewed as falling into two broad and rather loosely defined categories: suspended solids and colloidal particles.

TABLE 8.4
Potential Water Quality Problems That May Be Caused by a Few Selected Chemicals

Chemical	Potential Problems
Arsenic	Toxicity to humans Toxicity to aquatic life
Chlorine	Organic reactions form trihalomethanes Toxicity to fish and other aquatic life
Calcium	Causes "hardness" in water May result in scale formation in pipes
Iron	Causes stains in laundry and on fixtures May kill fish by clogging their gills
Nitrogen: ammonia	May accelerate eutrophication in lakes May improve productivity of the water May be toxic to aquatic life
Nitrogen: nitrates	May be toxic to babies May accelerate eutrophication in lakes May improve productivity of the water
Oxygen, dissolved	Low concentrations harmful to fish Low concentrations may cause odor production High concentrations accelerate metal corrosion Low or zero concentration may allow sulfide formation and concrete corrosion
Phenolics	Tastes and odors in drinking water Can cause tainting of fish flesh May be toxic to aquatic life
Sulfur: sulfides	Objectionable odors in and near water May be toxic to aquatic life May corrode concrete through acid formation Oxidation of sulfide to sulfate exerts an oxygen demand
Sulfur: sulfites	React with DO and exert oxygen demand
Sulfur: sulfates	Increase water corrosiveness to metals Decompose anaerobically to form sulfides Salty taste and laxative effects

Suspended solids are empirically determined as those particles large enough to be removed from water during its passage through a specific type of laboratory filter. They usually are large enough to settle in quiescent water to become part of the bottom deposits of streams and lakes or to rise to the surface as scum if they are lighter than water. Colloidal particles are larger than dissolved molecules and ions but too small to settle at significant rates. In many instances, they may remain in suspension indefinitely and usually will pass through ordinary filters.

Suspended matter may have many undesirable effects on water quality. For example, fine silt or chemical precipitates may increase turbidity, making the water less attractive for most purposes. Also, the particles may interfere with penetration of sunlight, causing a substantial impact on the aquatic organisms that depend on light for survival and growth. This may have a profound influence on the ecological balance in the watercourse.

Suspended matter lighter than water may float and form scum that is unsightly or interferes with passage of light and oxygen through the surface. Heavier suspended solids may settle and accumulate in deposits referred to as sludge banks, which can obstruct channels and require more frequent dredging. If they are high in organic content, decomposition of sludge banks may give off foul odors, especially when exposed to the atmosphere at low water levels. More direct effects include reducing the capacity of water reservoirs because of the accumulations, killing fish by clogging their gills, and destroying bottom life by blanketing it under accumulated sediment.

In many instances, the suspended matter in streams and lakes originates in runoff from farmlands and urban areas, or it may be discharged in municipal and industrial wastewaters. Sometimes, precipitates may form after the discharge of dissolved materials into a stream and their reaction with other constituents in the water. For example, iron compounds in effluents from steel mills or abandoned coal mines may react with the alkalinity and oxygen in streams to form precipitates that impact on fish life and cause several other problems. Dissolved or colloidal constituents may precipitate and coagulate after mixing with the saline waters of estuaries or the ocean. This important phenomenon is largely responsible for the formation of deltas near the mouths of large rivers.

RADIOACTIVE SUBSTANCES

Radioactive materials in drinking waters have caused substantial concern in instances where their levels have approached or exceeded those specified in drinking water standards. However, such occurrences have been rare exceptions because of the unique care and attention that have been devoted to disposal of radioactive wastes. No other potential pollutant has been subjected to such intense scrutiny, control, and monitoring.

Generally, there has been more justifiable concern about the levels of radioactivity sometimes attained by accumulation in aquatic food chains. This "biomagnification" may cause objectionable radioactivity in aquatic life, although its level in the water may be low enough to be entirely acceptable from all other points of view. The concentration effects that can be realized depend on many factors, including the types of organisms involved, their feeding habits, the type of radioactive chemical, its role in the food chain, and its level of radioactivity.

The presence of radioactivity in water sometimes has caused serious problems in manufacturing and developing X-ray and photographic films and papers by spotting them, even though the radioactivity level in the water was exceedingly low. This problem usually has been solved by subjecting the water to extensive treatment to insure the removal of radioactive particles.

HEAT

Heat provides an excellent example of the complexity of water quality questions because it can have many effects on water uses, either harmful or beneficial, depending on the circumstances. Perhaps the most obvious impact of heat added to a stream is that it reduces the value of that resource for subsequent use as a coolant by others downstream. Also, higher temperatures often intensify taste and odor problems in drinking water supplies.

The impact of BOD discharged into a stream usually is increased substantially by temperature rise, for several reasons. First, the rates of biochemical reactions in the stream are accelerated with the rise in temperature, increasing the rate of exertion of BOD and its impact on DO resources. Also, the solubility of DO in water decreases with the rise in temperature, which reduces the DO available in the system. Furthermore, higher temperature usually decreases the rate at which oxygen is transferred from the atmosphere into the water to replenish the oxygen consumed by biochemical reactions. This compounds the adverse impact of temperature rise on DO resources in streams.

As stream temperature rises, the death rates of some fish are increased significantly. This effect begins with the most sensitive types of fish, but a sufficient temperature increase may eliminate all types. In addition to deaths caused by exceeding the thermal death points for various species, rapid rises or falls in temperature may cause lethal thermal shock.

Also, there may be many more subtle adverse effects on fish populations. For example, mature fish may not be impacted seriously, but reproduction may be halted because of unfavorable temperatures for spawning, or fish food organisms may be impacted. Even relatively small changes in temperature can alter the balance among different forms of aquatic life and, ultimately, cause significant changes in the ecological balance. Interactions have been observed between temperature and toxicities of certain chemicals to fish, with rises in temperature usually increasing the potential for harmful effects.

On the other hand, elevated temperatures sometimes may have beneficial effects on certain uses of watercourses. Of special interest, it is possible to increase the growth rates of fish and shellfish and lengthen the growing season by carefully using waste heat to raise water temperature. Another possibility that has attracted substantial attention, but whose benefits are less promising economically, is using heated waters to irrigate crops, enhancing growth rates and providing longer growing seasons.

Although heated discharges may originate from several sources, those of principal importance are cooling waters from large industries, especially steam power electrical generating plants. That industry uses exceedingly large quantities of cooling water to remove massive amounts of waste heat—usually approximating 60 to 70 percent of the total energy consumed in fuel supplied to the plant. Where a once-through cooling system is employed, most of that heat is discharged to the receiving stream. When cooling towers are used, some water is evaporated to reduce the heat released to the stream. This reduces thermal impact on the water resource, but at the expense of increasing the water lost through evaporation

SOURCES OF POLLUTIONAL MATERIALS

Table 8.5 summarizes several sources from which potential pollutants may enter watercourses. Sometimes water pollution problems may be caused principally by substances originating from one source but, more commonly, they are the result of accumulated contributions from several sources. It is important to recognize the several sources and know something about the types of materials that may be contributed by each.

It is helpful to keep in mind two very broad categories into which the many sources may be divided: (1) point-sources, and (2) nonpoint-sources. Point-sources include municipal and industrial wastewaters, storm sewer outfalls, and any other sources for which specific points of entry can be identified. The nonpoint-sources include general land runoff and other diffuse contributions that cannot be narrowed to specific points of addition.

This method of classification is used mostly for convenience in considering regulatory questions. Usually, flows from point-sources can be treated and controlled before discharge. On the other hand, non-point-sources are more difficult to manage and must be approached in different ways.

TABLE 8.5
Sources of Potential Pollutants

Natural Sources
Atmospheric
Dissolved minerals
Decay of vegetation
Aquatic growths
Storm runoff
Agricultural Sources
Increased erosion
Animal wastes
Fertilizers
Pesticides
Irrigation return flows
Wastewaters
Municipal sanitary sewage
Industrial wastewaters
Municipal storm runoff
Wastewaters from boats
Water treatment by-products
Impoundments
Leaching from bottom deposits
Aquatic growths
Miscellaneous Sources
Construction activities
Mines
Dumps and landfills
Groundwater

Natural Sources

Many potential pollutants enter watercourses from "natural" sources, discussed in Chapter 7. Some of them have little or no relationship to human activities, while others may. For example, rain falling through the atmosphere may pick up substantial amounts of acid constituents, particulates, mercury, other metals, nitrogen, phosphorus, and many other potential pollutants that originate in volcanic action, wind erosion, or other purely natural phenomena. Also, however, the concentrations of these and other constituents may be increased substantially by air pollution from industrial operations—often at locations remote from the watercourse in question. Several examples of these were discussed in Chapter 7.

Many other natural sources of constituents also contribute heavily to streams and lakes. Groundwaters may add high concentrations of dissolved chemicals leached from soils contacted during extended underground passage. The organic content of swamp streams may be high, even without receiving wastewater discharges, because of dissolved decomposition products released by decaying vegetation. Those materials often produce an objectionable yellow-brown color in the water or cause low or zero DO by exerting BOD through decomposition of the organic chemicals of natural origin. This water quality contrasts sharply with that in clear, sparkling mountain streams that have had little opportunity for contact with decaying organic materials.

An important constituent of water that often is present in large quantities through natural phenomena is suspended matter. Much of that originates in the erosion of land by storm runoff, even without human influences through clearing and tilling the soil.

Aquatic growths occur in steams and lakes as natural phenomena, even without the addition of nutrients in wastewaters. Eutrophication, or aging of lakes, occurred long before the appearance of humans on earth. The result of human contributions has been to accelerate that process because of increased releases of nutrients into watercourses.

Agricultural Sources

The amount of suspended matter entering watercourses in land runoff may be increased substantially when the water passes over land developed for agriculture, because of clearing and cultivating the soil. This is especially true where land management practices are poor. On the other hand, contour plowing, provision of buffer strips between tilled soil and streams, proper grading, and other methods of control may be very effective in reducing erosion losses in agricultural runoff.

Some of the fertilizer added to agricultural lands is transported into streams by runoff. This can have a substantial impact on the nutrient concentrations in watercourses, sometimes representing as much as 80 percent of the total contributions of nitrogen and phosphorus. In the Lake Erie drainage area, for example, it has been estimated that about 50 percent of the phosphorus entering the lake originates in runoff, much of that from agricultural lands. The significance of that in pollution control activities is apparent when one considers the major efforts now under way to remove phosphorus from municipal and industrial discharges into the lake to slow its eutrophication rate. Even complete success in removing nutrients from point-sources still would leave 50 percent of the present total contribution intact because it is contributed in land runoff. This makes it clear that to retard the eutrophication sharply, it will be necessary to impose effective controls on land runoff as well as on point-sources.

As pointed out earlier, pesticides are causing concern with respect to potential health impacts on humans. These chemicals often are carried into streams in runoff from agricultural lands and in irrigation return flows. Their control must be attained largely through judicious chemical selection, careful application to crops, and improved land management practices to minimize their transport in runoff.

Irrigation return flows may contain high concentrations of many chemicals leached from the soil, nutrients, and suspended solids. All of them usually are at concentrations much higher than those in waters applied to the land because evapotranspiration removed about 50 percent of the water. That roughly doubles concentrations of chemicals in the irrigation raw water, in addition to constituents added during flow over and through the soil.

Runoff from farms also frequently contains animal wastes, which may add substantial quantities of BOD and suspended matter to the stream. These originate primarily from dairy farms, animal feedlots, and operations in which large numbers of chickens, turkeys, and ducks are raised.

Wastewaters

Municipal and industrial wastewaters are especially important point-sources of potential pollutants; in fact, they frequently are viewed by much of the public as being responsible for most or all water pollution problems. Because of their special importance, their characteristics and control will be discussed in Chapters 12 and 13. For now, it is sufficient to indicate that they frequently contribute major quantities of oxygen-demanding materials (BOD), suspended matter, bacteria, metals, and many specific chemicals capable of causing a wide variety of problems in watercourses or downstream uses.

Less generally recognized as a major source of potential pollutants is urban storm runoff, which can contribute large amounts of suspended solids, oxygen-demanding materials, metals, bacteria, nutrients, and many other potential pollutants. In fact, urban storm runoff sometimes may contribute potential pollutants in quantities even higher than those in untreated sanitary sewage from the same community. Where the sanitary wastes of a community are effectively treated, storm runoff becomes a main source of several potential pollutants and may have a significant impact on stream quality.

Wastewaters from ships and pleasure boats have received special attention in recent years, sometimes with little justification and often without sound judgment. Sanitary wastes from vessels contain oxygen-demanding materials and suspended solids, usually in quantities too small to cause significant problems in the receiving stream, although on rare occasions they conceivably might be capable of causing difficulties in crowded harbors. The principal concern about these wastes has been the potential health risk of transmitting diseases through body contact with water receiving them, through drinking waters, or through the consumption of shellfish and other aquatic life. It has not yet been documented that waterborne diseases actually have been transmitted in this fashion, because of discharges from boats, although the possible route of transmission obviously is there.

Control systems for sanitary and oily discharges from vessels have been implemented at considerable cost. However, they offer special regulatory problems because the thousands of individual boats are widely scattered and frequently on the move, making adequate surveillance through inspection virtually impossible.

Other wastewaters that only recently have been recognized as important in the United States are the by-products originating in the treatment of water for municipal and industrial supplies. These wastes, mostly ignored for many years, include sludge accumulations in settling basins and wastewater flows from backwashing sand filters. They contain most of the suspended matter and microorganisms in the intake water as well as the chemicals added to facilitate their removal. Some water treatment plants also discharge large quantities of dissolved solids from ion-exchange processes or from other treatment systems needed to produce the required final water quality. Today, U.S. discharges from water treatment plants are subject to the same regulatory controls as wastewaters from other industrial operations.

Impoundments

Reservoirs used for water supply storage, hydroelectric power production, and other purposes do not result in the direct addition of potential pollutants to the water. Nevertheless, they may cause deterioration of water quality through several mechanisms.

The change from a flowing stream to an impoundment (lake) with long retention and low flow velocity increases the likelihood for developing certain types of aquatic growths, especially algae, that may be objectionable for many reasons. Accordingly, impoundment construction for water supply or hydroelectric power may cause water quality problems that would not have existed had the stream remained in its natural, free-flowing state.

A phenomenon that can cause serious problems is the stratification of deep impoundments during warm weather, as discussed in Chapter 3. That can cause the production of odorous compounds through anaerobic biochemical reactions in the lower strata. Also, the anaerobic conditions are favorable for dissolving iron, manganese, and other chemicals from the bottom deposits of the reservoir, which may have undesirable impacts on subsequent uses of the water.

The discharge of water from lower levels of a stratified reservoir can cause serious reduction in downstream DO—far below that which would have been present in the former free-flowing stream. That can reduce the capacity of the system for assimilating oxygen-demanding substances (BOD) discharged by municipalities and industries below the reservoir, reducing the value of the water resource to them. The change in stream characteristics must be compensated by downstream users by installing and operating more efficient—and more expensive—wastewater treatment plants if the same stream quality conditions are to be maintained.

Miscellaneous Sources

There are many other sources of potential pollutants that could be of great importance in determining stream quality. For example, large quantities of suspended matter may originate from highway and building construction sites where extensive areas are graded. Without proper controls, the effects of those discharges on water quality could be disastrous.

In many areas of the United States, drainage from operating or abandoned mines contributes heavy loads of acids and metals, sometimes rendering the rivers below them largely devoid of life. This situation is found in many streams of Pennsylvania and West Virginia that are polluted by "acid mine drainage" from abandoned coal mines.

As indicated previously, even relatively clean flows of groundwater into streams may add large quantities of dissolved chemicals, making the water less desirable for some purposes. Seepage of organics, minerals, and toxic substances from refuse dumps can cause serious impacts on water quality. Currently, special attention is being directed to the question of groundwater pollution by leachate from dumps and landfills. Chemicals that are of great concern are metals and toxic organics that can be carried by percolating rainfall to reach and pollute groundwater aquifers used for drinking water supplies. Methods for preventing such migration are being pursued by implementing regulations that require the adequate handling and treatment of hazardous wastes and the construction of safer disposal sites designed to prevent contact between materials deposited in them and percolating water. This is an area of environmental protection that is developing and will receive concentrated attention over the next several years.

EVALUATION OF STREAM POLLUTION

Meaningful evaluation of stream pollution requires that water quality be viewed in the context of uses that are practical and feasible for the watercourse.

Some watercourses are unacceptable in quality for certain uses. Sometimes that situation may arise because of the impact of one or more specific discharges from municipalities or industries. In other instances, the quality may be unsuitable even without such discharges because quality characteristics depend on many factors, only some of which are determined by wastewater discharges.

Placing the matter into accurate perspective requires the consideration of an exceedingly complex matrix of factors, including (1) the wide variety of uses that may be made of a water resource, (2) quality characteristics necessary for those uses, (3) potential pollutants present in the stream, (4) sources of those materials, (5) methods that are available for correcting quality deficiencies, and (6) other alternatives available to satisfy societal needs and desires.

This is complicated further by the fact that the impacts of potential pollutants may be viewed from entirely different perspectives, depending on the situation. For example, the addition of heat to a stream commonly is viewed as potentially harmful, for reasons discussed earlier. However, under some circumstances and with proper controls, it could be beneficial for some uses. This creates a clear conflict in perception of the role of that potential pollutant—perhaps harmful and beneficial simultaneously for different uses of the same resource.

Similarly, nutrient additions to a watercourse could cause serious quality problems by promoting excessive growths of algae. On the other hand, their addition to a watercourse deficient in nutrients could increase the growth of fish food organisms, enhancing productivity of the resource. These potential pollutants must be viewed under some circumstances as harmful—if the goal is to maintain maximum clarity of the water for scenic and recreational purposes. Under other circumstances, when the goal is to maximize fish productivity, their addition could be beneficial. Many similar situations can be described for other potential pollutants including, for example, even the biodegradable organic chemicals responsible for BOD of wastewater discharges.

To evaluate water quality and determine the nature and extent of pollution, we must begin with a careful consideration of which uses society wishes to make of the resource, each of which requires specific quality characteristics. Constituents present at concentrations high enough to interfere unreasonably with those uses should be classified as "pollutants." Materials not present at levels causing or threatening problems should not be viewed as pollutants, even though they are added to the watercourse in discharges from municipalities or industries.

When the uses that we wish to make of streams or lakes seem to be incompatible with their water quality, we must evaluate carefully the costs of correcting the deficiencies and balance these costs against the benefits that we could realize by doing so.

STUDY QUESTIONS

1. Identify the eight categories of potential pollutants discussed in this chapter.
2. For each of the categories of potential pollutants, describe the types of water quality problems that could be expected to result from excessive concentrations in a watercourse.
3. What is the justification for statements to the effect that today's pollution control programs are based principally on reasons other than human health?
4. What is the main goal of water pollution control programs that are based largely on controlling BOD, suspended solids, and pH of wastewaters from municipalities and industries? Justify your answer.
5. Take a position agreeing or disagreeing with the definition of pollution proposed in this chapter and justify your position.
6. Take a position for or against the concept that virtually any change in water quality represents pollution of that resource and justify your position.
7. What is the justification for referring to the constituents of a watercourse as *potential* pollutants?
8. Identify several diseases that may be transmitted directly through the water route.
9. Describe the role of water in the transmission of diseases in which it is an indirect vehicle (for example malaria or yellow fever).
10. Identify several types of toxic agents that are of special concern today and are transmitted by water.
11. Evaluate the significance of oxygen-demanding substances (BOD) with respect to human health impacts.
12. Why are oxygen-demanding substances (BOD) of such major interest to pollution control agencies?
13. What are persistent organic chemicals? Why are they important in water pollution control activities?
14. Identify two "nutrients" that are of prime interest in water pollution control activities. Explain why they are important.
15. What is meant by the term "limiting nutrient" in a watercourse? Explain its significance and how an understanding of the concept and identity of the limiting nutrient in a watercourse can be used to control certain types of pollution problems.
16. What is "bioaccumulation," how does it occur, and why is it important?
17. Identify and briefly explain undesirable effects that can be caused by suspended matter in a watercourse.

18. Outline potential harmful impacts and benefits that can result from the discharge of heat into watercourses.
19. Identify five major categories of sources of potential pollutants and give examples of the types of materials that might be discharged from each.
20. Differentiate between point-sources and nonpoint-sources of potential pollutants.
21. Explain why an impoundment may be important as a source of water quality problems, even though nothing may be added to water flowing through it.
22. Outline the factors that should be considered in evaluating water pollution problems.

REFERENCES

American Water Works Association, "Organic Contaminants in Water Supplies," Committee Report in *Journal American Water Works Association*, Vol. 67 pp. 418–424 (1975).

American Water Works Association, *Water Quality and Treatment*, Chapters 1, 11, 12, 13, 19, McGraw-Hill, New York (1971).

Craun, G. F., And L. J. McCabe, "Problems Associated with Metals in Drinking Water," *Journal American Water Works Association,* Vol. 67, p. 593 (1975).

Curtis, W. F., J. K. Culbertson, and E. B. Chase, *Fluvial Sediment Discharge to the Oceans from the Conterminous United States*, Circular No. 670, U.S. Geological Survey, Washington, D.C. (1973).

Harris, R. H., and E. M. Brecher, "Is the Water Safe to Drink?", *Consumer Reports* Vol. 39 Part I, pp. 436–443 (June 1974); Part II, pp. 538–542 (July 1974); Part III, pp. 626–627 (August 1974).

Hoak, R. D. "Physical and Chemical Behavior of Suspended Solids," *Sewage and Industrial Wastes*, Vol. 31 pp. 1401–1408 (1959).

McGauhey, P. H. *Engineering Management of Water Quality*, McGraw Hill, New York (1968).

National Academy of Sciences, *Drinking Water and Health*, Safe Drinking Water Committee, Washington, D.C. (1977).

National Academy of Sciences, *Drinking Water and Health,* Vol. 2, Safe Drinking Water Committee, Washington, D.C. (1980a).

National Academy of Sciences, *Drinking Water and Health,* Vol. 3, Safe Drinking Water Committee, Washington, D.C. (1980b).

National Academy of Sciences, *Drinking Water and Health*, Vol. 4, Safe Drinking Water Committee, Washington, D.C. (1982).

National Academy of Sciences, *Drinking Water and Health,* Vol. 5, Safe Drinking Water Committee, Washington, D.C. (1983).

Sawyer, C. N., and P. L. McCarty, *Chemistry for Environmental Engineering*, 3rd Edition, McGraw-Hill, New York (1978).

Sobsey, M. D., "Enteric Viruses and Drinking Water Supplies," *Journal American Water Works Association*, Vol. 67, pp. 414–418 (1975).

Stockinger, H. E., "Toxicology and Drinking Water Contaminants," *Journal American Water Works Association*, Vol. 69, pp. 399–402 (1977).

CHAPTER 9

Water and Health

The relationship between water quality and health occupies a niche of special importance in the public's perception. There is a strong and widespread belief that water pollution problems must be closely related to human health effects. Sometimes they are, but not nearly as often as imagined.

There are sound historical reasons for that public feeling because only a relatively few years ago water was a principal vehicle for transmitting some of the most dreaded diseases of humanity—especially cholera and typhoid fever. Also, the public intuition is reinforced by recognition today of a relationship between drinking water and a variety of intestinal diseases encountered by travelers in many other countries. Furthermore, the concern, or even fear, is compounded by frequent and highly publicized pronouncements that yet another health "crisis" can be attributed to water quality defects—often based on little or no sound evidence.

Actually, relationships between water quality and health are more complicated than realized by most individuals and include both negative and positive effects—a fact often overlooked. The roles of most water constituents are not yet clear enough to give reasonable assurance that we even know whether they have any impact on human health. That is true, for example, for most of the potential carcinogens that are of such great concern currently but for which epidemiological evidence is sparse and inconclusive. Only a few relatively simple water-related problems have been investigated sufficiently to permit accurate evaluation of dose-response relationships. The question of which constituents may be hazardous in drinking waters at concentrations encountered in practice remains highly speculative and controversial.

The relationships between water and health are divided into four categories in the discussions presented in subsequent sections of this chapter:

1. Water often plays an important role as a vehicle for the direct transmission of microorganisms that cause infectious diseases.
2. In other instances, water may have an indirect role by serving as the habitat for mosquitoes and other vectors that transmit disease-producing organisms.
3. Water also can be an important vehicle for transmitting many types of chemical toxicants.
4. Water may have important beneficial effects on health, either because of its use in general sanitation or because of its chemical constituents.

WATER AS A DIRECT VEHICLE FOR DISEASE TRANSMISSION

A recent book by McJunkin (1982) presents a wealth of information on water-related diseases and effects of control methods on their incidence.

A publication of the U.S. Geological Survey (Greeson, 1981) lists 36 infectious diseases that probably can be transmitted directly by water. They are listed in Table 9.1 with the type of causative agent for each, and include 12 diseases that are caused by bacteria, 4 by viruses, 19 by protozoa or other parasites, and one that may be caused by several different infectious agents. A few diseases that are of special interest are discussed briefly in the following subsections.

Typhoid Fever

This is a classical waterborne disease that was a worldwide scourge many years ago and remains of major importance in the less developed countries today. Because its former impact may be difficult for some people to visualize now, consider that in the 1880s the death rate from this disease in the United States averaged about 100 persons annually per 100,000 population. In Pittsburgh, in the late 1880s, the annual typhoid deaths exceeded 160 persons per 100,000 population. In 1910, typhoid was number 10 among all causes of death in Pennsylvania.

A single 1885 outbreak in Plymouth, Pennsylvania produced 1000 cases of typhoid fever among a population of only 8000 and caused 114 deaths. Major outbreaks occurred in Lawrence, Massachusetts, in 1890, in Lowell, Massa-

TABLE 9.1
Waterborne Infectious Diseases

Disease	Type of Organism
Amebiasis	Protozoan
Angiostrongyliasis	Nematode
Anisakiasis	Nematode
Balantidiasis	Protozoan
Cholera	Bacterium
Clonorchiasis	Trematode
Diphyllobothriasis	Cestode
Dracontiasis	Nematode
Echinostomiasis	Trematode
Enteropathogenic diarrhea	Bacterium
Epidemic viral gastroenteritis	Virus
Fascioliasis	Trematode
Fasciolopsiasis	Trematode
Giardiasis	Protozoan
Gnathostomiasis	Nematode
Hepatic capillariasis	Nematode
Hepatitis A	Virus
Heterophyidiasis	Trematode
Hydatidosis	Cestode
Inclusion conjunctivitis	Bacterium
Intestinal capillariasis	Nematode
Leptospirosis	Bacterium
Melioidosis	Bacterium
Paragonimiasis	Trematode
Paratyphoid fever	Bacterium
Pharyngoconjunctival fever	Virus
Poliomyelitis	Virus
Primary amebic meningoencephalitis	Protozoan
Salmonellosis	Bacterium
Schistosomiasis	Trematode
Shigellosis	Bacterium
Traveler's diarrhea	Virus or bacterium
Tularemia	Bacterium
Typhoid fever	Bacterium
Vibrio parahaemolyticus food poisoning	Bacterium
Yersiniosis	Bacterium

Source: Greeson (1981).

chusetts, in 1891, in Chicago in 1892, and in Pittsburgh in 1900. The impact by typhoid fever in England in the middle 1800s, as well as the groping efforts to identify its causes, were recorded vividly by Budd (1874). It is significant, incidentally, that thorough epidemiological investigations of outbreaks made it possible to determine that this disease was waterborne several years before the discovery that the disease is caused by bacteria. Experience with typhoid fever demonstrates that the mode of transmission and methods for controlling public health problems often can be identified and implemented without fully understanding those problems.

Table 9.2 shows the combined death rates for typhoid and paratyphoid fever in Massachusetts and the United States during the past century. The trend has been consistently and dramatically downward to today's rate of less than 0.05 deaths annually per 100,000 people. Much of that reduction can be attributed to improvements in water treatment and delivery systems. Figure 9.1 sometimes is used to illustrate this point by showing the parallel declines in typhoid fever death rate in Massachusetts and percentage of the population in that state not served by public water systems (Fair et al., 1966). The clear impli-

TABLE 9.2
Trends in Combined Typhoid and Paratyphoid Death Rates (1865–1970)

Period Ending	Massachusetts Deaths/yr/100,000	United States Deaths/yr/100,000	United States Cases/yr/100,000
1865	105		
1870	79		
1875	82		
1880	45		
1885	50		
1890	42		
1895	32		
1900	24	31	
1905	18	25	
1910	12	25	
1915	6.7	16	80
1920	2.4	11	54
1925	1.8	7.5	36
1930	0.9	5.1	26
1935	0.3	3.6	19
1940	0.2	1.8	11
1945	0.1	0.5	4.6
1950	<0.05	0.2	2..5
1955	<0.05	0.1	1.4
1960	<0.05	<0.05	0.7
1965	<0.05	<0.05	0.3
1970	<0.05	<0.05	0.2

Source: U.S. Bureau of the Census (1975)

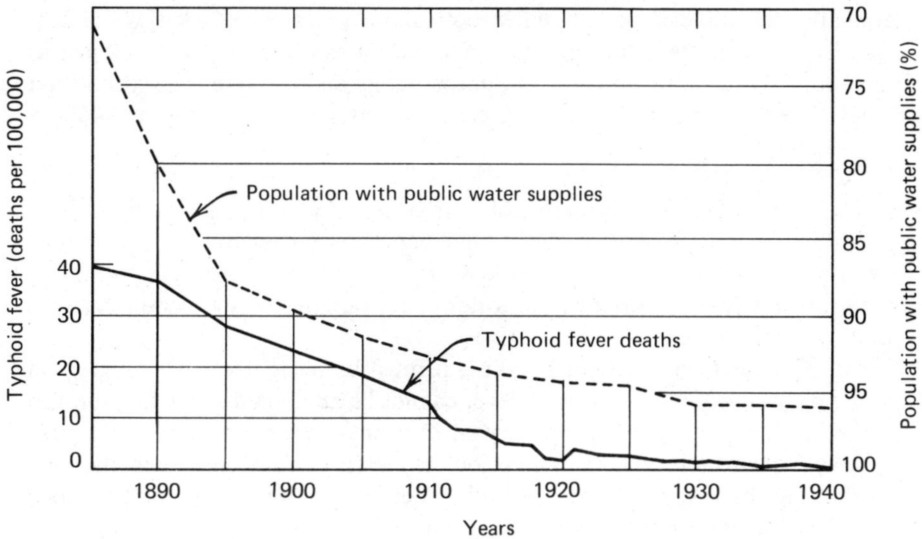

FIGURE 9.1

Typhoid deaths and public water supplies in Massachusetts (1885–1940). From *Water and Wastewater Engineering,* Volume 1, G.M. Fair, J.C. Geyer and D.A. Okun, Copyright © 1966 by John Wiley & Sons, Inc., New York, Page 1–16. Reprinted by permission.

cation is that the reduced death rate was caused by the movement to public water supplies and the resulting improvement in bacteriological safety of water consumed by the public.

Table 9.3 illustrates the same idea, perhaps more directly and validly. The data show typhoid fever death rates in 26 cities along the Ohio River that did not have water treatment plants in 1906. The 10 cities that did not install treatment facilities between 1906 and 1914 had rates averaging about 75 deaths

TABLE 9.3
Effects of Water Treatment on Typhoid Death Rates in 26 Ohio River Cities

	Typhoid Deaths/yr/100,000	
Conditions	**1906**	**1914**
Ten cities that did not install treatment	76.8	74.5
Sixteen cities that did install treatment in period 1906–14	90.5	15.3

Source: After McJunkin (1982).
Note: The water sources were not changed for any of the 26 cities between 1906 and 1914.

annually per 100,000 population in both those years. The 16 cities that did install treatment plants during the period had rates averaging about 90 deaths per 100,000 population in 1906, but dramatically decreased to one-sixth of that level in 1914. None of the 26 cities changed water sources between 1906 and 1914 (Frost, 1941).

Although the death rate due to typhoid fever is now at an extremely low level in the United States, there still is an occasional outbreak (Pfeiffer, 1973, for example). They usually occur in small communities and are traced to defective water supply systems or errors in their operation. The enemy is still there, lying in wait for the careless; the water route remains a key mode for transmitting this disease.

All of the evidence notwithstanding, it must be recognized that progress in reducing the incidence of typhoid fever cannot be attributed solely to improved control of water quality. During the past 75 years, there have been many simultaneous advances in general sanitation, especially improvements in the processing and handling of food and milk. These also have had major beneficial effects on typhoid fever rates and must not be overlooked.

The typhoid fever situation in less developed countries is very different from that described for the United States. Although the situation is generally better than it was a few years ago because of improved water supplies and other sanitation advances, there still are some areas in which typhoid is widespread and substantial outbreaks occur from time to time. For example, from 1967 to 1974 the number of cases annually per 100,000 population averaged 40 in Sri Lanka (WHO, 1976a). Between 1970 and 1975, the incidence of cases in Singapore varied from 7.0 to 23 annually per 100,000 population (WHO, 1976b). The disease may be under excellent control in some countries, but globally it still represents a significant problem. Perhaps the most tragic aspect of the situation is that we know how to control it but have not yet managed to do so.

Cholera

This disease has been responsible for some truly catastrophic waterborne outbreaks and remains a major global problem today. It is caused by a bacterium—*Vibrio cholerae*—and has produced fatalities in epidemics ranging from 5 percent to as high as 75 percent of those contracting the disease. Cholera is endemic* in many Asian countries today, and epidemics† often occur in those areas. The most recent pandemic‡ occurred during the 1960s and 1970s and covered most of Africa and Asia as well as some of southern Europe.

---

*Endemic—constantly present in a particular country.

†Epidemic—prevalent among many individuals in a community at the same time.

‡Pandemic—epidemic over a large region.

A classic example of water epidemiology is the case of the Broad Street pump, in London. This involved a careful investigation in 1854, by Snow, of over 500 deaths from cholera, most of which were traced to polluted water supplied by a single well and hand pump. Snow was credited with stopping the outbreak simply by removing the handle from the pump, preventing its further use and blocking the spread of the disease organisms. A highly significant aspect of this experience was that the problem was identified, traced, and solved about 30 years before the bacteria responsible for causing cholera were discovered by Koch, in 1883. As indicated earlier in discussing typhoid fever, this also demonstrates that sometimes it is possible to solve a public health problem without even knowing its exact cause.

Another outbreak of great historical significance is the Hamburg-Altona cholera epidemic of 1892 (Miller, 1962). Hamburg withdrew its drinking water, without treatment, from the Elbe River below an immigrant camp in which there was a cholera outbreak. Hamburg suffered 17,000 cases of the disease, with 8065 deaths, in a population of 640,000 over a period of 2 months. Altona also used the Elbe for its drinking water source, but treated its water by sand filtration. In spite of being downstream from both the immigrant camp and Hamburg, Altona had a total of only 328 deaths in a population of 143,000. Furthermore, many of the Altona victims worked in Hamburg and presumably contracted the disease organisms by drinking water from that system. The important lesson here was that sand filtration of water could have a highly beneficial effect in protecting consumers from cholera.

Cholera has not been a major disease in the United States during this century. There have been about 20 cases recorded in the past decade, most of which were imported, according to the WHO weekly epidemiology reports. However, cholera remains a serious disease globally and is endemic in many countries, especially in Asia and Africa. Since 1950, the number of cases reported worldwide has varied from 25,000 to 250,000 per year. This situation represents a constant threat to all areas of the world because people are highly mobile and the disease may be contracted and transferred to other countries in only hours or days. That is illustrated by outbreaks that have occurred periodically in Europe and by the recent U.S. cases.

Infectious Hepatitis

Hepatitis A is an important waterborne viral disease that was implicated in 66 U.S. outbreaks between 1946 and 1974 according to Craun (1976). A problem in accurately assessing the extent and trends of waterborne infectious hepatitis is that the disease also is transmitted through several other routes, including milk and food, and especially through the consumption of shellfish polluted by sewage. About 40,000 cases of hepatitis A were reported in the United States

TABLE 9.4
Numbers of Waterborne Outbreaks of Hepatitis A, by Sources of Contamination (1946–74)

Cause of Outbreak	Municipal Systems	Semipublic and Individual Systems
Untreated surface water	1	10
Untreated groundwater	4	26
Inadequate or interrupted disinfection	3	5
Distribution system contamination	11	1
Insufficient data to classify	3	2
Total	22	44

Source: Journal AWWA, Vol. 68, No. 8. Reprinted by permission. Copyright © 1976, the American Water Works Association.

in 1974, but only a small fraction of those cases were transmitted directly by water (NAS, 1977). Batik et al. (1980) state that waterborne outbreaks in the United States have caused 35 to 120 cases of hepatitis annually and are responsible for less than 1 percent of the total reported cases of that disease.

Another problem is that only a relatively small fraction of infectious hepatitis cases actually are reported to public health authorities—less than 10 percent as estimated by some investigators. The cases being reported annually may be only the tip of an iceberg approximating, perhaps, 500,000 total cases per year in the United States.

Reported sources of outbreaks are summarized in Table 9.4 and may be attributed mostly to failures in treatment or distribution of water (Craun, 1976, 1979, 1981a). There is no evidence that the disease is transmitted through properly operated water systems, in spite of occasional concern about the efficacy of conventional water treatment processes for removing or destroying its virus.

The largest recorded outbreak of hepatitis occurred in Delhi, India in 1955. A combination of low stream flow, which allowed sewage to enter the city water intake, and failure to maintain adequate disinfection at the water treatment plant produced over 30,000 cases (Dennis, 1959; NAS, 1977).

Other Waterborne Diseases

Many other waterborne diseases listed in Table 9.1 have caused serious problems at various times; space permits only the briefest mention of some.

Paratyphoid fever is similar in symptoms to typhoid, although milder, and may be caused by any of several salmonella bacteria. It occurs principally in

Europe and Asia, but seldom in the United States (Greeson, 1981). Many types of salmonella can cause other severe intestinal upsets. A major outbreak of salmonellosis (15,000 cases) occurred in Riverside, California in 1967 and was caused by sewage contamination of unchlorinated well water.

Shigellosis—bacillary dysentery—may be caused by many types of shigella. These organisms are commonly identified in the United States during outbreaks of diarrheal diseases.

Amebic dysentery epidemics occur occasionally in the United States, especially when there is gross contamination of drinking water by sewage. Another protozoan that causes an important waterborne disease is *Giardia lamblia*—giardiasis now is becoming a major waterborne disease in the United States. There is special concern because the organism may not always be removed adequately by conventional water treatment systems and because the possibility of its presence may not be reliably detected through the use of tests for coliform bacteria (Craun, 1981).

About 50 percent of the 99 waterborne outbreaks reported by Craun (1976) for 1971 to 1974, and summarized in Table 9.5, were classified as "... acute gastrointestinal illness of unknown etiology." The largest single disease outbreak was of giardiasis, with about 4800 cases in Rome, N.Y.

Schistosomiasis—also known widely as bilharzia—usually is transmitted through direct contact with water containing the cercarium larvae, which can penetrate the skin (McJunkin, 1970, 1982). This is an exceedingly important and widespread waterborne disease globally—with a total incidence of, perhaps, 200 million persons—but it is of little public health significance in the United States.

A disease that has received special attention recently is Legionnaire's disease. After long and difficult epidemiological investigations, it has been estab-

TABLE 9.5
Etiology of Waterborne Disease Outbreaks and Cases (1971–74)

Disease	Outbreaks	Cases
Gastroenteritis	46	7992
Hepatitis A	13	351
Shigellosis	13	2747
Giardiasis	12	5127
Chemical poisoning	9	474
Typhoid	4	222
Salmonellosis	2	37
Total	99	16,950

Source: Journal AWWA, Vol. 68, No. 8. Reprinted by permission.

lished that the disease is waterborne, in spite of its being a respiratory ailment. It may be contracted by inhaling the organisms in aerosols generated in air-conditioning cooling towers or even by shower heads (Dufour, 1982).

In summary, between 1971 and 1978, 224 waterborne outbreaks were reported in the United States (Craun, 1981a), causing two deaths and 48,193 reported illnesses. The few deaths should not be interpreted as indicating that waterborne diseases are only minor in importance but as a tribute to the waterworks and medical professions. The occasional outbreak that does occur reminds us that the risk always is present and waiting to trap the careless.

EVALUATING RISKS FOR WATERBORNE INFECTIOUS DISEASES

The probability of becoming a victim of a waterborne disease is a function of many variables, including (1) the type and number of pathogens ingested, (2) virulence of the cells, (3) age of the potential victim, (4) general health of the individual, (5) extent of acquired immunity, and (6) other factors that are less clear, such as the presence of certain types of chemicals or synergistic relationships with other organisms. The risk cannot be evaluted simply by determining whether or not pathogens are present in the water because it is not a simple yes or no situation. The leading method for evaluating the risk of transmitting waterborne diseases, based on tests for coliform organisms, was outlined briefly in Chapter 6. This section will discuss in more detail the rationale and methodology of that and other techniques used for assessing the risks.

Enumeration of Pathogens

Techniques are available for identifying and enumerating many types of pathogens commonly transmitted by water. However, the examination of samples to determine the number of each type of pathogen present, if any, would be long, complex, and expensive. Furthermore, that approach is severely limited in value if the goal truly is to assay the probability of acquiring a disease through drinking or otherwise contacting water from that source.

The chain of events involved in transmitting a waterborne disease usually includes the contamination of water by pathogens in human wastes, the transport of those organisms to the potential victim, contact between that individual and the water, and the transfer of pathogens into the body. The probability of contracting the disease through this route depends on several factors, two of the most important ones being the initial contamination of water by human wastes and the presence of pathogens in those wastes. The absence of pathogens in a water sample may be either because it did not receive human wastes or because wastes discharged into it originated from persons who were not carriers of the disease. Thus, the fact that examination of a sample fails to detect pathogens does not necessarily demonstrate that there is no risk; the pathogens may appear without warning at some future time when the contributing population gains a carrier of the disease in question.

This situation, compounded by the difficulty, time, and expense of the analyses, explains why enumeration of pathogens is not ordinarily used to evaluate the microbiological quality of water. An even more important reason is that simpler, more convenient, and quicker methods are available that can assess the risk more meaningfully.

The Concept of Indicator Organisms

Techniques usually used to evaluate the microbiological safety of water are based on determining its content of "indicator organisms." These are organisms that occur extensively in human wastes, whether or not the person suffers from a waterborne disease. Accordingly, they do not necessarily relate directly to the populations of pathogens but provide an estimate of the extent to which the water has been contaminated recently by human wastes. The principal indicator organism used in evaluating water quality is the coliform group of bacteria, to be discussed shortly.

Desirable characteristics of indicator organisms have been discussed by many authors (NAS, 1977; McFeters, 1978; U.S. EPA 1976a; Allen, 1978) and are summarized here.

1. They should be harmless to humans.
2. They should be present in polluted waters when pathogens are, or might be, present.
3. The number of indicator organisms in polluted water should be correlated with the probability that pathogens are present.
4. They should be present in polluted waters in numbers higher than those of pathogens.
5. They should be easy and quick to identify through relatively simple laboratory tests.
6. They should be easy to enumerate in those tests.
7. They should not multiply under conditions where pathogens do not multiply.
8. They should survive unfavorable environmental conditions longer than pathogens do. This insures that waters that have been treated to produce low or zero populations of indicator organisms are safe.

The Coliform Group of Bacteria

For the past century it has been known that large populations of certain bacteria *(Escherichia coli)* grow in the intestinal tracts of humans and are excreted in fecal wastes. The "coliform group" of bacteria includes *E. coli* and some different types that originate from other sources as well as in human fecal discharges (Allen, 1978; Newsletter, 1966; U.S. EPA, 1976a).

The coliform group is used to evaluate the microbiological quality of waters; the number present is interpreted as an indicator of the extent to which that water has been contaminated recently by human fecal discharges. Actually, the presence of coliforms does not establish that pathogens necessarily are there. Some coliforms are capable of causing disease, but most are not viewed as pathogenic; however, the number present does provide a measure of the probability that waterborne pathogens might be there.

Although the coliform group meets some of the criteria just listed, it does have certain limitations. For example, many bacteria in the group do not originate in the intestinal tracts of humans and, therefore, really bear little or no relationship to the opportunity for pathogens to be present (Newsletter, 1966). Also, coliforms often can multiply in treatment facilities or watercourses. This can produce high populations without corresponding sanitary significance, because pathogens are far less likely to multiply under those same environmental conditions.

Coliforms usually outlast most of the common pathogenic bacteria during natural die-off in streams or in treatment facilities, but they do not always survive viruses and several other organisms. There is concern that the reduction of coliforms to low levels, or even their elimination, may not always be accurate in indicating that the water is safe. For example, there is evidence that *Giardia lamblia* often may be present in unacceptable numbers in drinking waters, even with negative or low coliform tests (Lippy, 1978; Craun, 1981b).

Its shortcomings notwithstanding, use of the coliform test to evaluate the risks of transmitting waterborne diseases has served us well. It has been used for over 70 years as a key part of many successful programs around the world for controlling the spread of waterborne diseases. Although its accuracy and precision leave something to be desired, it still has provided a valuable and generally reliable technique for evaluating the safety of drinking water and warning of unsafe microbiological conditions when they occur. No other analytical method has been devised yet that does the job better.

Other Measures of Microbiological Quality

Fecal coliform bacteria are a subset of the total coliform group and are viewed as being more accurately related to organisms originating in human wastes. The analytical procedure for determining fecal coliform population is designed to reduce or eliminate the detection of coliforms that originate outside of the body by using selective media and higher incubation temperature.

The fecal coliform test has been used principally to assay microbiological quality of streams, lakes, and wastewater treatment plant effluents. The total coliform test remains the method of choice for evaluating drinking water quality, primarily because of its greater value for determining the effectiveness of treatment facilities. The rationale behind this was expressed by Kabler and Clark (1960) as follows:

> The presence of any type of coliform organism in treated drinking water suggests either inadequate treatment or access of undesirable materials to the water after treatment. . . . The presence of coliform organisms in treated water calls for definitive action for their elimination.

For many years, investigators have searched for better indicator organisms. One that has been suggested is the fecal streptococci group, but its role has been largely restricted to that of a supplementary tool in certain special microbiological investigations. Others have been considered but found even less acceptable, including *Clostridium perfringens, Klebsella pneumoniae,* several species of *Aeromonas, Pseudomonas* species, *Bifidobacterium,* and some pathogenic *Salmonella* species (McFeters, 1978; NAS, 1977).

A different approach to evaluate microbiological quality in waterworks practice has been based on bacteria plate counts conducted at 20 to 35°C, using nutrient agar and incubating for 1 to 2 days. This has been viewed favorably in some quarters because it includes many more types of bacteria than the coliform tests, making it the least restrictive microbiological measurement system. Its use is based on the judgment that *most* bacteria in drinking water, regardless of origin, should be at or very near zero after treatment in a properly operated system.

Current U.S. drinking water standards allow the measurement of chlorine residuals in water distribution systems as a substitute for some of the required coliform bacteria tests. The rationale behind that is based on observations that many disease outbreaks result from inadequate chlorination and that an inverse correlation usually is observed between chlorine residuals and coliform bacteria in distribution systems. This is not actually a method for assaying the biological quality of water, but it often does serve the same goal of protecting the public health by providing a measure that is useful in evaluating the probability of disease transmission.

WATER AS A VECTOR HABITAT IN DISEASE TRANSMISSION

Several important diseases are transmitted by organisms that require water as a habitat for at least part of their life cycles. Of those "vectors," the mosquito is best known and is responsible for transmitting some of the deadliest human diseases, including malaria, yellow fever, filariasis, and encephalitis.

Today malaria remains the greatest killer—over 1.5 million deaths per year, worldwide, with many times that number of cases (Gillett, 1973). The process for transmitting malaria begins with the withdrawal of blood, including the parasite that causes the disease, from a malaria victim by an anopheles mosquito. Thereafter, part of the life cycle of the parasite must be spent in the body of the mosquito, after which it may be transmitted to another person.

The role of water in this scenario has nothing to do with the causative organism itself or with the adult mosquito that transmits it. It is simply that the mosquito life cycle requires water to receive the eggs and to allow the growth of larvae until adult mosquitoes emerge. Without a proper water environment, development of the mosquito would be impossible, breaking the chain for transmitting the disease. Eradicating or preventing the development of mosquitoes is the basis for practical malaria control programs. That is equally true for other mosquito-borne diseases, including yellow fever, filariasis, and human encephalitis. A similar situation also exists for onchocerciasis, a disease transmitted by certain types of flies.

This role is radically different from the direct one involved in transmitting a disease organism or toxic chemical to one who drinks or otherwise contacts the water, but water is no less critical in the process. The mosquito and the water required in its life cycle provide viable points for attacking the public health problems associated with malaria or other diseases transmitted through those water-dependent vectors.

Schistosomiasis transmission requires the presence of specific types of snails in the watercourse receiving discharges from infected persons. The snail serves as "host" during development of the organism before it enters the body of another victim. As with mosquito-borne diseases, both the host and water required for its survival provide practical points of attack; when the snails are eliminated, effective control results. Other marine organisms, including fish and crabs, may be involved in the transmission of other diseases.

A somewhat different type of situation exists with respect to the role of shellfish in transmitting infectious agents, infectious hepatitis viruses, for example. The shellfish may not be infected by the organisms or otherwise involved in its life cycle. Their importance often lies simply in the fact that they pump large quantities of water through their systems to extract food. In doing so, they serve as filters and concentrate large numbers of the pathogens, increasing the risk of infection to a shellfish consumer.

WATER AS A VEHICLE FOR TRANSMITTING TOXIC CHEMICALS

A combination of circumstances during the past two or three decades has forced on us the realization that our activities are causing spectacular increases in numbers, types, and quantities of potentially toxic and suspicious chemicals added to the environment. Water is an important vehicle for transmitting those agents to potential victims. Although this aspect of the water-health relationship has been recognized for many years, it only recently has become the focus of greatly increased attention and concern. In fact, it now is the primary area of health concern in the United States and other countries where the waterborne diseases discussed earlier have been brought under control.

It must be recognized, however, that water is only one route by which persons may be exposed to toxic chemicals; other major sources include the food chain and air pollution. In evaluating health hazards, contributions must be considered in the context of the total exposure received by an individual from all sources. In setting maximum allowable concentrations of chemicals in drinking water, for example, dosages normally received through other routes must be considered, and the amounts that might otherwise be permitted in the water must be reduced.

Types of Toxicity Problems

There are at least three different types of problems to be considered here. The first is acute toxicity, in which adverse impact of the agent is observed quickly, usually within minutes, hours, or at most a few days. This type of effect can occur with high dosages of cyanide, arsenic, fluoride, or other so-called "poisons." It also is characteristic of the methemaglobenemia—the "blue babies" problem—that results from the consumption of water containing excessive concentration of nitrates by infants. Based on experience from many years of observing acute effects and facilitated by the short time between exposure and response, we have accumulated reasonable knowledge about them, their causative agents, and dosages at which harmful impacts may occur. These represent the simplest toxicity problems to evaluate and the easiest ones to control.

The second type of problem is chronic toxicity, in which effects may not be observed until exposure to the chemical has continued for an extended period. This often is true of metals and some organic chemicals that accumulate in the victim over months or years before reaching a body burden high enough to cause observable symptoms or illness. Thereafter, some of the problems are mostly irreversible, or only slowly so, because the chemical may not be readily released from areas of the body in which it has accumulated. Specific examples that have received considerable attention over the past several years include lead, mercury, cadmium, arsenic and several types of chlorinated hydrocarbons.

We know some things about these types of problems and the chemicals that can cause them but far less than we do about acute toxicity—partly because the responses may be imperceptibly gradual and separated from exposures by many years. Our knowledge is especially inadequate with respect to the thousands of new synthetic organic chemicals being introduced into our environment at a constantly increasing pace. Most have not yet exerted detectable harmful effects, but the fear is that sometimes that may be true only because they have not yet existed long enough to do so.

The third (and distinctly different) type of effect is genetic. Here, there may be no impact on someone exposed to the chemical, but it may surface as an effect on the offspring of that person. The terrible examples of those situations in our society, or worries about them because of grossly inadequate knowledge, are increasing daily. Examples that come to mind immediately include questions about the genetic effects of radiation, birth defects from using various specific drugs, and the current concerns and controversies that appear almost daily in the news media about agent orange, other defoliants, pesticides, PCBs, and other new chemicals.

Generally, we know little about these effects. Worse, we know virtually nothing about them within the context of the very low chemical concentrations that usually are found in drinking waters and the daily exposures over many years by persons who drink them.

Metals

Most inorganic chemicals specified in current drinking water standards are metals, as shown in Table 9.6 (U.S. EPA, 1976a). The potential chronic effects of lead on the human nervous system have been recognized for many years, and it is known that harmful amounts may accumulate in the body by drinking water containing lead. That is the basis for limiting the metal to less than 0.05 mg/liter in the U.S. drinking water standards.

Many other metals have attracted special interest during the past two decades because of a series of highly publicized events and concerns about their presence in water and effects on the public health. In the late 1960s, for example, the Minimata tragedy in Japan—resulting from the consumption of fish containing methylmercury—and several other less publicized outbreaks, focused public and regulatory attention on mercury. Canadian studies revealed that mercury levels in fish taken from the Great Lakes often were above those considered acceptable by food regulatory agencies, even though concentrations in the water were very low. Further nationwide studies discovered similar situations elsewhere, including mercury levels in tuna from the oceans.

The relatively high concentrations occurred because of biomagnification in aquatic food chains, similar to the effects noted with persistent chemicals, discussed briefly in Chapter 8. The process began with low concentrations of mercury in the water, from industrial and many other sources, which were concentrated to higher levels in aquatic organisms. Their subsequent consumption by other organisms in the food chain produced successively higher concentrations,

TABLE 9.6
Inorganic Chemicals Included in U.S. Drinking Water Standards

Constituent	Maximum Permissible Contaminant Level (mg/liter)	Maximum Advisable Contaminant Level (mg/liter)
Arsenic	0.05	
Barium	1.0	
Cadmium	0.010	
Chromium	0.05	
Lead	0.05	
Mercury	0.002	
Nitrate (as N)	10	
Selenium	0.01	
Silver	0.05	
Fluoride	1.4–2.4	
Chloride		250
Copper		1
Iron		0.3
Manganese		0.05
Sulfate		250
Total dissolved solids		500
Zinc		5

culminating in levels viewed as unacceptable in predators at the top of the trophic structure. Although often misunderstood by the public, the concern of health officials was not based on mercury concentrations found in drinking waters—most were far below the standard of 0.002 mg/liter.

Since then, investigators have moved through a series of studies of other metals, their occurrences, and their effects. Those of greatest concern included lead, mercury, cadmium, arsenic, selenium, chromium, barium, silver, copper, and zinc, most of which are listed in the current standards. All have been studied to evaluate permissible levels in drinking waters and their potential for accumulating in aquatic life and crops irrigated with waters, wastewaters, or sludges (Craun, 1975). Table 9.7 summarizes concentrations (in micrograms per liter) of several metals in surface waters of the United States, and Table 9.8 shows concentrations found in studies of 380 finished municipal waters (NAS, 1977).

Organic Chemicals

It is difficult to convey accurately the current situation with respect to organic chemicals in drinking waters because of the many types involved and the abys-

TABLE 9.7
Metals in 1577 Surface Waters in the United States (1962–67)

Element	Frequency of Detection (%)	Concentration (μg/liter) Minimum	Maximum	Mean
Zinc	76.5	2	1183	64
Cadmium	2.5	1	120	9.5
Iron	75.6	1	4600	52
Molybdenum	37.7	2	1500	68
Manganese	51.4	0.3	3230	58
Aluminum	31.2	1	2760	74
Beryllium	5.4	0.01	1.22	0.19
Copper	74.4	1	280	15
Silver	6.6	0.1	38	2.6
Nickel	16.2	1	130	19
Cobalt	2.8	1	48	17
Lead	19.3	2	140	23
Chromium	24.5	1	112	9.7
Vanadium	3.4	2	300	40
Barium	99.4	2	340	43

Source: National Academy of Sciences (1977).

TABLE 9.8
Metals in 380 Finished Waters in the United States (1962–67)

Element	Frequency of Detection (%)	Concentration (μg/liter) Minimum	Maximum	Mean
Zinc	77.0	3	2010	79.2
Cadmium	0.2	12	12	12
Iron	83.4	2	1920	68.9
Manganese	58.7	0.5	450	25.5
Copper	65.2	1	1060	43
Silver	6.1	0.3	5	2.2
Lead	18.1	3	139	33.9
Chromium	15.2	1	29	7.5
Barium	99.7	1	172	28.6
Molybdenum	29.9	3	1024	85.9
Aluminum	47.8	3	1600	179.1
Beryllium	1.1	0.02	0.17	0.1
Nickel	4.6	1	490	34.2
Cobalt	0.5	22	29	26
Vanadium	3.4	14	222	46.1

Source: National Academy of Science (1977).

mally poor state of knowledge about their long-term effects on humans. Table 9.9 lists a few organics that have caused special concern because they are known or suspected to be carcinogens* or mutagens† under some circumstances; more are being added to such lists as time passes. However, none has been demonstrated to be either a carcinogen or a mutagen at concentrations actually encountered in drinking waters.

Many of the organics that have caused concern enter watercourses in discharges from chemical manufacturing plants. On the other hand, many others are released by operations that use them for various purposes. For example, the pesticides that have received so much attention over the past 20 to 30 yr (Carson, 1962) enter streams and lakes principally in runoff from farmlands, not in wastewaters from manufacturing plants. Their concentrations in watercourses depend principally on the amounts used by farmers and others, patterns in which they are applied, rainfall occurrences after their application, and losses through spills. Pesticide control must be accomplished mostly by improving practices in their selection, handling, and application—not through industrial wastewater treatment.

A different source of potentially toxic organics that is receiving special attention today is in reactions between innocuous organics present in water naturally and chlorine added to it for disinfection (Singer, 1980, U.S. E.P.A., 1979b). That is the origin of most trihalomethanes (THMs) in drinking waters, including chloroform, for example, which has been demonstrated to be a carcinogen at concentrations and under conditions substantially different from those in municipal water supplies. Different exposure conditions notwithstanding, the association has generated enough concern to cause the EPA to establish a THM standard for drinking waters at 0.10 mg/liter (U.S. EPA, 1979b).

The current situation suggests that many potentially serious problems could be encountered with dozens or even hundreds of other organics in water, whether discharged in wastewaters or produced during treatment. The epidemiological problems involved in evaluating relationships between low concentrations of organics in waters and the health of persons drinking them are awesome because of the many variables that the investigator always must contend with. The few studies that have been carried to points where investigators were willing to make definitive conclusions invariably have been highly controversial.

Finally, in some instances chemicals discharged into the environment subsequently may undergo reactions that transform them into substances with dif-

---

*Carcinogen—an agent that causes cancer.

†Mutagen—an agent that increase the rate of mutation in a species.

Note: Another term with which the reader should be familiar is teratogen—an agent that causes birth defects.

TABLE 9.9
Organic Chemicals of Special Interest in Drinking Waters

Carcinogens and Suspected Carcinogens	Mutagens and Suspected Mutagens
Benzo(a)pyrene	1,1,1-Trichloroethane
Carbon tetrachloride	Bromomethane (methyl bromide)
Chloroform	Methyl chloride
Vinyl chloride	Bromochloromethane
1,4-Dioxane	Methylene chloride
Methly iodide	Bromoform
DDE	Bromodichloromethane
DDT	2-Chloropropane
Chlordane	1,1-Dichloropropane
Lindane	1-Chloropropene
Dieldrin	1,2-Dichloroethane
Benzene	Bis(2-chloroisopropyl) ether
Vinylidene chloride	Chlorodibromomethane
Heptachlor	1,3-Dichloropropene
1,1,2-Trichloroethane	2,6-Dinitrotoluene
1,1,2-Trichloroethylene	Dichloroacetonitrile
Bis(2-chloroethyl) ether	Methylene bromide
Simazine	Chlordane
Tetrachloroethylene	Vinylidene chloride
Heptachlor epoxide	n-Butylbromide
Acrylonitrile	Bis(2-chloroethyl)ether
Aldrin	Acrylonitrile
Butyl bromide	Benzo(a)pyrene
	Methyl iodide
	Vinyl chloride
	1,3-Butadiene
	1,2-Bis(chloroethoxy)ethane
	Pyrene
	1,1,2-trichloroethylene
	Tetrachloroethylene

Source: Appendix of Proposed Regulations for Control of Organic Chemical Contaminants in Drinking Water, *Federal Register,* Volume 43, pp. 29,149–29,150 (July 6, 1978).

ferent toxicological characteristics. For example, inorganic mercury released into a stream or lake may be transformed into methylmercury, a chemical of much greater toxicity than that originally discharged. The impacts of other toxic substances may be reduced rapidly and drastically through chemical or biochemical reactions in the environment. These possibilities may have great significance in setting standards for allowable discharges of some types of materials.

Other Toxic Constituents

There are several constituents of water other than metals and organics that can be toxic to humans. An important one encountered occasionally, and subject to a specific limit in the drinking water standards, is nitrate. High concentrations of that chemical may contribute to hypertension in adults and methemoglobenemia in babies. The latter, as mentioned earlier, commonly is referred to as the "blue baby" problem because of reactions with blood hemoglobin and interference with the transmission of oxygen through the body.

Fluoride now is added to the drinking water of over 50 percent of the U.S. population to reduce the incidence of dental caries, as discussed briefly in the section on beneficial effects of water. At the recommended dosages—about 1.0 mg/liter—it is beneficial, but at higher levels it can be harmful. At moderately high concentration—above 3 to 5 mg/liter—fluoride may discolor or cause structural damage to some teeth (AWWA, 1971; NAS, 1977). At very high dosages, it can be an active toxicant and cause severe illness or death. These occurrences have been rare, but a few have been reported.

Radioactivity has caused concern for many years and is specifically limited by the drinking water standards (U.S. EPA, 1976b). However, it seldom has caused practical problems in that context because of tight controls imposed by the Nuclear Regulatory Commission and its predecessor, the Atomic Energy Commisssion. A problem that has been encountered occasionally, however, is the biomagnification of radioactive chemicals in fish and other aquatic life, as described earlier for toxic organic chemicals and metals. This sometimes can cause accumulations of radioactivity through the food chain to reach objectionable levels.

This brief discussion would be incomplete without at least mentioning some of the current uncertainity about the effects of sodium in water on hypertension and heart disease. This has been studied by epidemiologists over several years, producing widely varying and often conflicting results (NAS, 1977). A major problem in trying to answer this seemingly simple question is that potentially harmful effects of the sodium may be camouflaged, or even offset, by potentially beneficial effects of calcium, magnesium, or other chemicals. Questions also have arisen periodically about potentially harmful or beneficial effects of other constituents of waters.

Although many of them are not mentioned specifically in standards and regulations, the waterworks profession must constantly guard against potential toxicants that may enter watercourses. Many originate in chemical manufacturing operations or because of unacceptable practices in using chemical products, for example, improper or excessive use of pesticides or herbicides. Although some potential toxicants are mentioned specifically in stream or drinking water standards, many others are not because they may have been developed only recently or their potential for causing harmful effects are not yet known.

Dangerous situations may occur with little or no notice through spills of chemicals during shipment or in industrial accidents. This is of great concern today and has led to the development of sophisticated procedures for combating problems that arise from the spills. Those sometimes include telephone hotline assistance by agency or industrial personnel and special teams of experts prepared to react quickly with advice and assistance in cleanup.

Dose–Response Relationships

A fundamental problem in trying to evaluate the health effects of very low concentrations of potentially toxic chemicals in drinking waters is the same one that has plagued those investigating effects of exposure to low levels of radioactivity—uncertainty about the shape of the dose-response curve. Without becoming too involved in technical aspects of the question, its general nature may be described with the help of Figure 9.2.

Data available for constructing a curve that shows the relationship between doses of a chemical (or radioactivity) and the degree of reaction to those doses by a population invariably are limited to dosages much higher than those needed to evaluate threshhold effects. Figure 9.2 shows that conclusions about responses, if any, at very low dosages may depend heavily on the shape of the dose-response curve. Unfortunately, that shape must be assumed because data do not exist that show accurately the actual curve shape at very low doses.

Assuming a shape similar to curve *A* could lead to a conclusion that a safe threshhold (no response) exists at dose *X;* in other words, any dose less than *X* would not be harmful. This curve shape is viewed as the most reasonable assumption by some authorities. On the other hand, different authorities contend that the curve should be shaped similar to *B,* which would produce a response at dose *X*. With this type of curve, no threshhold could be inferred and one must conclude that there is no concentration of the chemical (or radioactivity) that could be assumed to be entirely safe. Strong arguments have been advanced to justify each viewpoint (National Research Council, 1980). Currently, the subject remains highly controversial, making it impossible to establish a consensus on what concentrations, if any, of potential carcinogens or mutagens could be relied on as safe in drinking waters.

One approach to this matter has been to apply risk-analysis techniques; that is, to calculate the estimated probability of contracting cancer as a function of chemical dosage, extrapolating from the best data available for the purpose

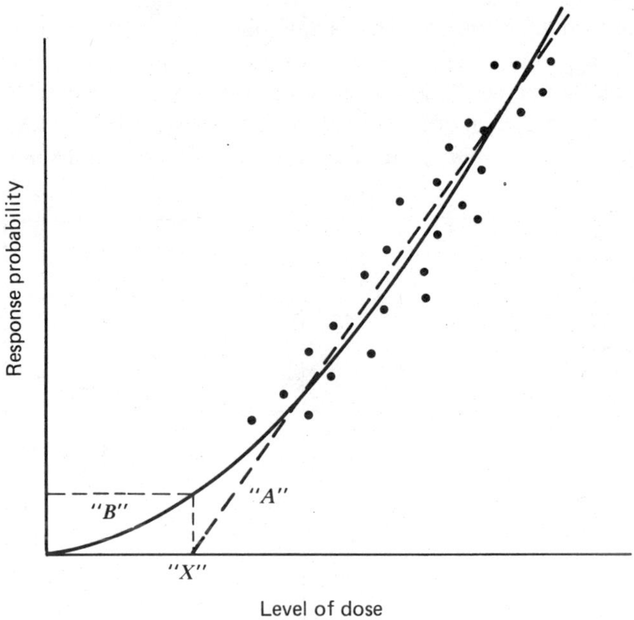

FIGURE 9.2
Two examples of dose-response relationships fitted to the same data.

(Christman, 1981; NAS, 1980b). Of course, this approach must be based on an inherent assumption about the shape of the dose-response curve—often that it is linear at low dosages.

BENEFICIAL HEALTH EFFECTS

After reading the preceding discussions of potential adverse health impacts, it is easy to overlook the many beneficial effects that can accrue from water and its constituents. The most obvious is in satisfying our metabolic needs for water, without which life can continue only for a short time. Little water is necessary for that purpose—perhaps about 3 percent of a person's body weight per day—but meeting that requirement is crucial for survival.

Among several chemicals in water known or suspected to be beneficial to humans, fluoride has been studied extensively under carefully controlled conditions—both in the laboratory and in the field. The benefit of fluoride in reducing dental caries, referred to earlier, occurs principally among young people who drink water containing a low concentration of the chemical when their teeth are being formed. Figure 9.3 shows data obtained in field studies of drinking water fluoridation and indicates that the number of decayed, missing, and filled (DMF) teeth among 12- to 14-year-olds decreased sharply with a rise in fluoride concentrations to about 1.0 mg/liter (AWWA, 1971). More fluoride produced little additional benefit but, as discussed earlier, could increase the

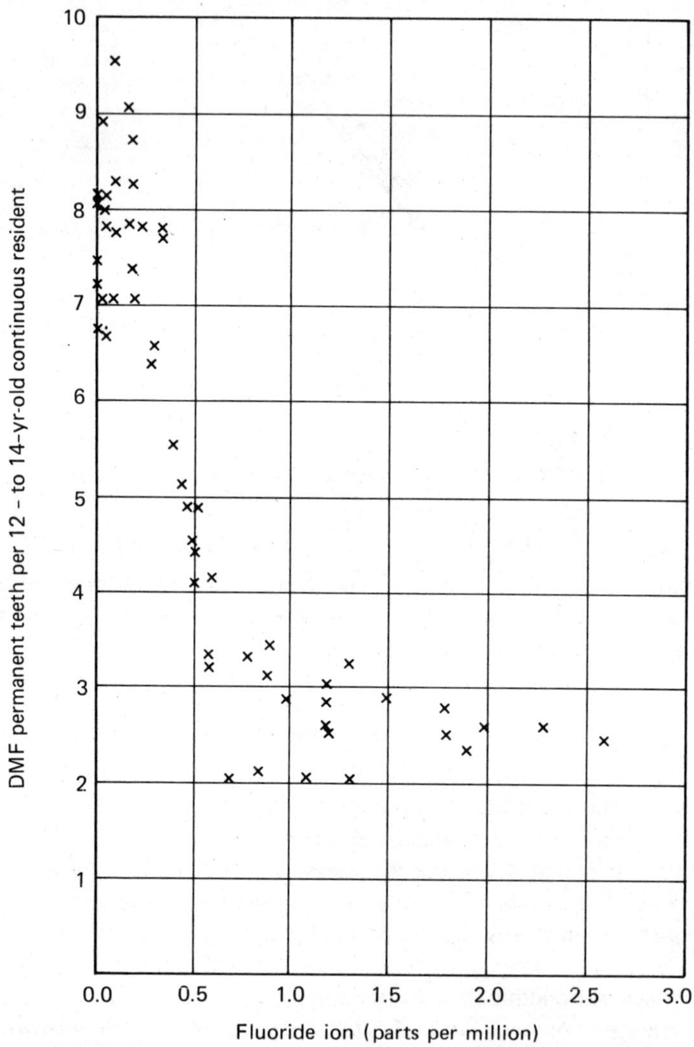

FIGURE 9.3
The effect of fluoride concentration on tooth decay. Reprinted From *Water Quality and Treatment,* Third Edition, by permission. Copyright © 1971, The American Water Works Association.

risk of discoloration or other damage to tooth enamel. Accordingly, the optimum concentration for fluoride is about 1.0 mg/liter.

The appendix to the U.S. drinking water standards recommends that fluoride concentrations should be within a range that varies slightly above and below 1.0 mg/liter, depending on average ambient temperatures (U.S. EPA, 1976a). The rationale is that the amount of water consumed by people varies with temperature, and the recommended dosage is changed with ambient temperature to produce the desired total daily addition of fluoride to the body. The standards also specify the maximum allowable levels for fluoride, which are double the midpoint values of the recommended ranges. This handling in the standards differs from that afforded other chemicals by specifying both a maximum level and a recommended fluoride concentration range.

Other chemicals also have been suggested as having possible beneficial effects in drinking waters. Iodine was considered as a possibility to be added to public water supplies many years ago to help prevent goiter. Calcium and magnesium may produce benefits by offsetting the negative effects of sodium in hypertension and heart disease. Also, there has been speculation that trace amounts of certain minerals in water may help insure that human nutritional requirements for them are met (NAS, 1977, 1980b).

Finally, and perhaps of greatest importance, one must not overlook benefits accompanying the use of water as an agent for improved cleanliness and general sanitation. About 1892, independent studies were conducted by Mills after a Lawrence, Massachusetts, epidemic of typhoid fever, and by Reincke after the Hamburg, Germany, cholera epidemic. Both reached the same conclusion about the benefits realized after installing sand filtration of drinking water supplies. In addition to reducing deaths from typhoid fever, cholera, and diarrhea, mortalities from nonwaterborne diseases also diminished. This observation became known as the Mills-Reincke phenomenon.

Subsequent studies by Hazen, in 1904, produced estimates of the reductions that could be expected in deaths from typhoid fever and other diseases after introducing pure water to a community. He concluded that for each typhoid death reduction, two to three deaths from other diseases also probably would be prevented.

The findings by Mills, Reincke, and Hazen subsequently were doubted and lost favor because of difficulties in separating the effects of several simultaneous improvements in sanitation. However, more recent studies continue to suggest substantial benefits in preventing various diseases through improvements in water supplies.

Water availability may be important as well as its quality. This is illustrated by data collected at migrant labor camps in California during 1952–53. Their occupants were found to exhibit 0.7 percent positive tests for presence of *Shigella* organisms in camps where all water faucets were installed inside the cabins. In camps where the faucets all were outside the cabins, the positivity rates averaged 7.2 percent. In camps where some were inside and others outside, the cabins with inside faucets produced a rate of 2.7 percent and those with outside faucets averaged 5.5 percent. These data imply that placing water supply fixtures in locations that facilitated their use resulted in better sanitation. Unfortunately, the question remains somewhat controversial because some other studies have failed to show similar beneficial effects.

PAST TRENDS AND FUTURE PROBLEMS

Because concerns about water and health initially were based on the transmission of waterborne diseases, especially cholera and typhoid fever, early drinking water standards were addressed principally to controlling microbiological content. Coliform bacteria and total bacteria counts were especially important as indicators of the extent of water pollution by human wastes. Only minimal attention was directed to chemical content—lead and arsenic, for example—because relatively little was known about the impacts of most chemicals on health. Also, the early standards included color and turbidity—primarily because of aesthetic considerations.

In the 1950s, professionals began to express concern about other aspects of water quality that had been largely ignored until then. In the microbiological area, attention focused on viruses and whether coliform tests provided adequate information about the safety of water in that respect. Also, there was increasing concern about whether conventional water treatment plants could provide adequate protection against viral waterborne diseases, especially infectious hepatitis and poliomyelitis.

During that period, there was little attention to specific organics in drinking water; actually, there were few analytical procedures capable of measuring specific organics quantitatively at the low concentrations usually present in water. Pesticides, for example, were not mentioned in the drinking water standards. Neither were metals, except for the few whose effects had been recognized for many years.

The 1960s brought the beginning of organized efforts to investigate more thoroughly the effects of metals and specific organics on human health. Much of the increased attention began with widespread publicity about the effects of mercury in the Minimata episode, elevated mercury concentrations in U.S. waters and aquatic life, and popular publications about pesticides and their potential effects (Carson, 1962). Many of the investigations carried out during and since that period were made possible by the development of new and highly sophisticated analytical techniques. These made it feasible, for the first time, to analyze many water samples quantitatively for very low concentrations of metals and specific organic chemicals.

The trend toward more specific identification of chemicals in water, their sources, and potential effects has continued and grown in momentum through the 1970s until today it represents a major area of U.S. work on water quality. Now it is well known that hundreds of organic chemicals and trace metals are present in drinking waters, even after treatment in conventional plants. Methods for removing those chemicals from waters have been studied and new treatment technology developed. The principal controversies and problems about the use of that technology now relate to questions about needs and justification of costs.

In my opinion, the most critical health effects problem facing the water profession today is the lack of adequate knowledge about effects and threshold concentrations of the chemicals that we know to be present. We now can identify and measure many organics at exceedingly low concentrations. What is missing is a sound basis for evaluating the significance of the data so obtained. We know much more, for example, about trihalomethane occurrence, formation, and removal from drinking waters than we do about the effects that those chemicals have on health at concentrations found in practice. Without such knowledge, we have little or no rational basis for deciding on what concentrations should be allowed, the types of treatment facilities that should be built, the degree of removal that must be required, or the health and economic justifications for any of those decisions.

Until that information can be developed—a long and expensive process—those decisions must be made for thousands of chemicals, based primarily on uninformed judgment. The situations inevitably become entangled in sensationalized news coverage, emotional responses, and public and political reactions to pressure groups. The decisions that are reached ultimately may bear little relationship to the true needs of society.

STUDY QUESTIONS

1. Outline and briefly describe four different types of relationships between water and human health.
2. Identify six diseases for which water is an important direct vehicle of transmission.
3. Describe three studies of waterborne diseases that are of significant historical importance, including where the studies were conducted, the circumstances surrounding them, and the lessons learned from them.
4. Identify several important variables involved in determining the probability that a person might become a victim of a waterborne disease.
5. Explain the rationale behind using coliform organisms to evaluate the microbiological quality of drinking waters.

6. Why is the enumeration of pathogens considered less desirable than the coliform test for evaluating the risk of contracting waterborne disease?
7. What are "indicator" organisms? Outline six desirable characteristics of indicator organisms.
8. Identify the two groups of indicator organisms most often used in evaluating the microbiological quality of water and explain the difference between them.
9. Do either or both of those groups meet the criteria for desirable indicator organisms referred to in question 7. If not, what are their shortcomings?
10. Compare and contrast fecal coliforms, total coliforms, and total bacteria counts as methods for evaluating the quality of drinking waters. What analytical technique(s) does the drinking water field use now? Explain why.
11. Which microbiological analysis is used to evaluate the quality of river waters for recreational purposes? Why is that the method of choice?
12. In establishing a quality standard for drinking water in a country with a high incidence of waterborne disease, the permissible count of coliform organisms could be set higher, lower, or the same as the U.S. standard. What would be the key factors that you would consider in making the decision among those choices? Explain the rationale behind your answers.
13. Explain the rationale behind the current policy of permitting measurements of chlorine residuals throughout distribution systems in lieu of some of the coliform tests otherwise required in drinking water systems.
14. Identify five diseases in which water plays an indirect role, as vector habitat, in their transmission.
15. Explain the role of water in the transmission of malaria.
16. Differentiate between the roles of water in transmitting (*a*) schistosomiasis, and (*b*) hepatitis through shellfish contamination.
17. Differentiate between the acute, chronic, and genetic effects of toxic materials.
18. Identify six metals that are considered important as potential toxicants in drinking waters and state the permissible concentration of each.
19. What is "biomagnification" of constituents in water? How does it occur? Why is it important?
20. Identify six substances that have been observed to bioaccumulate in food chains.
21. Explain the bases for the major public health concerns about the new organic chemicals that are being developed by the chemical industry today.
22. Explain the basis for the current controversy about evaluation of dose-response relationships at low dosages of radioactivity or toxic chemicals, including the public health signficance of the question.
23. Describe the effects of fluoride in drinking waters on public health of the populations served.

24. Describe the effects of providing adequate water supplies on the incidence of nonwaterborne diseases.
25. Identify three examples in which the chemical content of drinking water could be beneficial to health of humans drinking that water, and explain briefly the nature of those potential benefits.

REFERENCES

Allen, M. J., and E. E. Geldreich, "Evaluating the Microbial Quality of Potable Waters," in *Evaluation of the Microbiology Standards for Drinking Water,* pp. 3–11, Publication EPA-570/9-78 OOC U.S. Environmental Protection Agency, Washington, D.C. (1978).

American Water Works Association, *Water Quality and Treatment,* McGraw-Hill, New York (1971).

Batik, O., et al. "An Epidemiologic Study of the Relationship Between Hepatitis A and Water Supply Characteristics and Treatment," *American Journal of Public Health,* Vol. 70, p. 167 (1980).

Berg, G., "The Indicator System," *Indicators of Viruses in Water and Food,* Ann Arbor Publishers, Ann Arbor, Mich. (1978).

Budd, W. *Typhoid Fever: Its Nature, Mode of Spreading, and Prevention,* London (1874). Reprinted by the American Public Health Association (1931).

Carson, R. L., *Silent Spring,* Houghton Mifflin, Boston (1962).

Christman, R. F., "Toxic Chemicals in Water: A Public Health Perspective," *Proceedings, Southeast Conference on Toxic Chemicals,* Charleston, S.C. (April 14–15, 1981).

Clark, H. F., and P. W. Kabler, "Reevaluation of the Significance of the Coliform Bacteria," *Journal American Water Works Association,* Vol. 56, pp. 931–936 (1964).

Craun, G.F., and L. J. McCabe, "Review of the Causes of Waterborne Disease Outbreaks in the U.S.," *Journal American Water Works Association,* Vol. 65, pp. 74–84 (1973).

Craun, G. F., and L. J. McCabe, "Problems Associated with Metals in Drinking Water," *Journal American Water Works Association,* Vol. 67, p. 593 (1975).

Craun, G. F., L. J. McCabe, and J. M. Hughes, "Waterborne Disease Outbreaks in the U.S.: 1971–1974," *Journal American Water Works Association,* Vol. 68, pp. 420–424 (1976).

Craun, G. F., "Impact of the Coliform Standard on the Transmission of Disease," in *Evaluation of the Microbiology Standards for Drinking Water,* pp. 21–35, Publication EPA 570/9-78-00C, U.S. Environmental Protection Agency, Washington, D.C. (1978).

Craun, G. F., and R. A. Gunn, "Outbreaks of Waterborne Disease in the United States: 1975–1976," *Journal American Water Works Association,* Vol. 71, pp. 422–428 (1979).

Craun, G. F., "Outbreaks of Waterborne Diseases in the United States: 1971–1978," *Journal American Water Works Association,* Vol. 73, pp. 360–369 (1981a).

Craun, G. F., "Disease Outbreaks Caused by Drinking Water," *Journal Water Pollution Control Federation,* Literature Review Issue Vol. 73, pp. 1134–1138 (1981b).

Craun, G. F., "Waterborne Diseases in the United States and Canada," committee report in *Journal American Water Works Association,* Vol. 73, pp. 528–529 (1981c).

Dennis, J. M., "1955–56 Infectious Hepatitis Epidemic in Delhi, India," *Journal American Water Works Association,* Vol. 51, pp. 1288–1298 (1959).

Dufour, A. P., "Disease Outbreaks Caused by Drinking Water," *Journal Water Pollution Control Federation,* Literature Review Issue, Vol. 54, pp. 980–983 (1982).

Dufour, A. P., and W. Jakubowski, "Drinking Water and Legionnaire's Disease," *Journal American Water Works Association,* Vol. 74, pp. 631–637 (1982).

Fair, G. M., J. C. Geyer, and D. A. Okun, *Water and Wastewater Engineering,* Vol. 1, Wiley, New York (1966).

Fair, G. M., J. C. Geyer, and D. A. Okun, *Water and Wastewater Engineering,* Vol. 2, Wiley, New York (1968).

Frost, W. H., *Papers of Wade Hampton Frost, M.D.: A Contribution to Epidemiological Methods,* Kenneth F. Maxey, Editor, Commonwealth Fund, New York (1941).

Geldreich, E. E., "Bacterial Populations and Indicators Concepts in Feces, Sewage, Storm Water, and Solid Waste," in *Indicators of Viruses in Water and Food,* by G. Berg, Ann Arbor Publications, Ann Arbor, Mich. (1977).

Geldreich, E. E. *Sanitary Significance of Fecal Coliforms in the Environment,* Publication WP-20-3, Federal Water Pollution Control Administration, U.S. Department of the Interior, Washington, D.C. (November 1966).

Gillett, J. D. "The Mosquito: Still Man's Worst Enemy," *American Scientist,* Vol. 61, pp. 430–436 (1973).

Greeson, P. E., *Infectious Waterborne Diseases,* Circular 848-D, U.S. Geological Survey, Alexandria, Va. (1981).

Hendricks, C. W., Editor, *Evaluation of the Microbiology Standards for Drinking Water,* Publication EPA-570/9-78-00C, U.S. Environmental Protection Agency, Washington, D.C. (1978).

Interstate Commission on the Potomac River Basin, "Potomac's Tide of Coliforms Traced to Animal Population," *Newsletter, Interstate Commission on the Potomac River Basin,* Vol. 22, No. 8–9 (July–August 1966).

Kabler, P. W., and H. F. Clark, "Coliform Group and Fecal Coliform Organisms as Indicators of Pollution in Drinking Water," *Journal American Water Works Association,* Vol. 52, pp. 1577–1579 (1960).

Lippy, E. C., "Tracing a Giardiasis Outbreak in Berlin, New Hampshire," *Journal American Water Works Association,* Vol. 70, pp. 512–520 (1978).

McCabe, L. J., "Chlorine Residual Substitution Rationale," in *Evaluation of the Microbiology Standards for Drinking Water,* pp. 57–63 Publication EPA-570/9-78-00C, U.S. Environmental Protection Agency, Washington, D.C. (1978).

McFeters, G. A., J. E. Schillinger, and D. G. Stuart, "Alternative Indicators of Water Contamination and Some Physiological Characteristics of Heterotrophic Bacteria in Water," in *Evaluation of the Microbiology Standards for Drinking Water,* pp. 37–48 Publication EPA 570/9-78-00C, U.S. Environmental Protection Agency, Washington, D.C. (1978).

McJunkin, F. E., *Enginnering Measures for Control of Schistosomiasis,* A report for the Office of Health, Bureau of Technical Assistance, Agency for International Development, U.S. Department of State, Washington, D.C. (1970).

McJunkin, F. E., *Water and Human Health,* U.S. Agency for International Development, U.S. Department of State Washington, D.C. (1982).

Miller, A. P., *Water and Man's Health,* Publication No. 5, Technical Series, Community Water Supply Program, Agency for International Development, U.S. Department of State, Washington, D.C. (1962).

National Academy of Sciences, *Drinking Water and Health,* Vol. 1, National Academy Press, Washington, D.C. (1977).

National Academy of Sciences, *Drinking Water and Health,* Vol. 2, National Academy Press, Washington, D.C. (1980a).

National Academy of Sciences, *Drinking Water and Health,* Vol. 3, National Academy Press, Washington, D.C. (1980b).

National Academy of Sciences, *Drinking Water and Health,* Vol. 4, National Academy Press, Washington, D.C. (1982).

National Research Council, Committee on the Biological Effects of Ionizing Radiation, *The Effects on Populations of Exposures to Low Levels of Ionizing Radiation: 1980,* National Academy Press, Washington, D.C. (1980).

Pfeiffer, K. R., "The Homestead Typhoid Outbreak," *Journal American Water Works Association,* Vol. 65, pp. 803–805 (1973).

Reasoner, D. J., "Microbiology: Detection of Bacterial Pathogens and Their Occurrence," *Journal Water Pollution Control Federation,* Literature Review Issue,, Vol. 53, pp. 1112–1134 (1981).

Singer, P. C., J. J. Barry, G. M. Paleń, and A. E. Scrivner, "Trihalomethane Formation in North Carolina Drinking Waters," *Journal American Water Works Association,* Vol. 73, pp. 392–401 (1981).

Sobsey, M. D., *Source Document on Enteric Viruses,* Agency for International Development, U.S. Department of State, Washington, D.C. (1979).

United Nations, *Demographic Yearbook: 1979,* New York (1980).

U.S. Bureau of the Census, *Historical Statistics of the United States, Bicentennial Edition,* Part 2, U.S. Department of Commerce, Government Printing Office, Washington, D.C. (1975).

U.S. Centers for Disease Control, *Morbidity and Mortality Weekly Report: Annual Summary: 1978,* U.S. Public Health Service, Atlanta (September 1979).

U.S. Environmental Protection Agency, *National Interim Primary Drinking Water Regulations,* Publication EPA/570/9-76-003, Government Printing Office, Washington, D.C. (1976a).

U.S. Environmental Protection Agency, "Interim Primary Drinking Water Regulations: Promulgation of Regulations on Radionuclides," *Federal Register,* Vol. 41, No. 133, pp. 28,402–28,409 (July 9, 1976b).

U.S. Environmental Protection Agency, "National Secondary Drinking Water Regulations; Final Rule," *Federal Register,* Vol. 44, No. 140, pp. 42,195–42,202 (July 19, 1979a).

U.S. Environmental Protection Agency, "National Interim Primary Drinking Water Regulations; Control of Trihalomethanes in Drinking Water; Final Rule," *Federal Register,* Vol. 44, No. 231, pp. 68,624–68,707 (November 29, 1979b).

U.S. Environmental Protection Agency, "Interim Primary Drinking Water

Regulations; Amendments," *Federal Register,* Vol. 45, No. 168, pp. 57,332–57,357 (August 27, 1980).

World Health Organization, *Principles and Practice of Cholera Control,* Public Health Paper No. 40, World Health Organization, Geneva (1970).

World Health Organization, *Weekly Epidemiological Record,* No. 38, p. 295 (September 17, 1976a).

World Health Organization, *Weekly Epidemiological Record,* No. 50, p. 382 (December 10, 1976b).

CHAPTER 10

Oxygen Balance in Streams

It was indicated earlier that current U.S. water pollution control activities are directed principally to protecting aquatic life and preventing undesirable conditions in streams that could arise as a result of low dissolved oxygen (DO) concentrations. To that end, most regulatory programs are heavily oriented toward controlling oxygen-demanding materials (BOD) in wastewater discharges. Therefore, the rationale behind planning, designing, operating, and monitoring wastewater treatment facilities is based largely on understanding the DO balance in watercourses and maintaining desirable DO concentrations.

A brief introduction to the subject was presented in Chapter 8. This chapter presents more details on fundamental concepts and practical methods for evaluating DO assets and liabilities in streams. Also, a few techniques often employed to predict conditions downstream from wastewater discharges are discussed.

BIOLOGICAL DECOMPOSITION OF ORGANICS IN STREAMS

Many types of chemicals enter watercourses in wastes from municipalities and industries as well as in land runoff and from other sources outlined in Chapter 8. Some are biodegradable and supply the carbon, nutrients, trace elements, and energy needed for maintenance and growth of organisms there.

Usually, the biological community in an unpolluted stream is highly diverse, as discussed in Chapter 3; that is, there are many types of organisms present but relatively few of each. Each species is highly specialized in its metabolic capabilities, depending on its genetic makeup and environmental conditions in the system. It is limited in the types of chemicals that it can assimilate and the reactions that it can carry out with those chemicals. When an organic is added to water, the microorganisms capable of using it do so, receiving energy and chemicals needed for growth (multiplication) of the cells. The end results of the biochemical reactions are the production of new cell materials (synthesis), reduction in concentration of the chemical being used (biodegradation), and release into the water of by-products, some of which may be simpler organic compounds that serve, in turn, as food supplies for other organisms. Those organisms carry out similar reactions, producing more cells and further biodegradation of the organics.

Repetition of the process often proceeds through several successive steps, satisfying the metabolic needs of several species, degrading complex organics to simpler ones, and releasing by-product gases and chemicals into the water. In this biochemical "chain reaction," each species derives its food supply from preceding members of the chain and releases by-products for use by those that follow.

Invariably, there also are some organisms in the stream that cannot use the specific organics added to it because their metabolic needs require different chemicals. Those organisms are not supplied with food by that discharge and are not influenced by its addition to the ecosystem unless it produces inhibitory or stimulatory effects on them.

OXYGEN DEMAND IN STREAMS

The biochemical reactions also utilize some chemicals and trace elements dissolved in the water to satisfy metabolic requirements of the organisms. One of the most important of these is dissolved oxygen, which is the focus of attention in this chapter.

There are various mechanisms by which the limited oxygen resources in water can be depleted. In this discussion, it is convenient to divide the demands for DO into three categories: (1) "immediate" oxygen demand, (2) "carbonaceous" oxygen demand, and (3) "nitrogenous" oxygen demand. Because it is important to distinguish carefully among them, they are discussed separately in the following sections.

Immediate Oxygen Demand

This demand is caused by relatively rapid reactions between certain types of chemicals—sulfites, for example—and DO in the water. The amount of immediate oxygen demand imposed on a watercourse depends on the types and amounts of chemicals added and the amount of oxygen required to react with them.

Immediate oxygen demand is purely chemical in character. It should not be confused with that originating in the biochemical reactions just discussed and should be measured and evaluated separately. It is exerted through very different mechanisms and follows different kinetic relationships—usually being imposed in minutes instead of the hours or days required for BOD. Most of the chemicals that produce this immediate oxygen demand are inorganic compounds.

Carbonaceous Oxygen Demand

This demand is incurred in connection with the metabolism of organic chemicals by living organisms. When microorganisms use organics, the end products of the biochemical reactions often include carbon dioxide (CO_2) and water (H_2O); the oxygen in these chemical formulas demonstrates the need for that element. These gross reactions are similar to ones carried out in humans, who inhale oxygen to satisfy their metabolic needs and exhale carbon dioxide as a by-product.

The carbonaceous oxygen demand is referred to as biochemical oxygen demand, or BOD. The analytical procedure for measuring BOD should be carried out in ways that permit measuring this fraction of the total oxygen demand separately from the immediate and nitrogenous demands, which have different causes and follow different reaction kinetics. In fact, difficulties in conducting BOD tests in practice often can be traced to analyst failure to achieve clear separation among the three types of oxygen demand. The term BOD, as used here, refers *only* to the carbonaceous oxygen demand, as separate and distinct from the other two types.

The amount of BOD exerted in a watercourse and the rate at which it is imposed depend on many factors, including the types and amounts of chemicals in the water, types and numbers of organisms present, and many environmental characteristics of the system, including temperature, pH, availability of nitrogen and phosphorus, and availability of necessary trace elements. These will be discussed in more detail shortly.

Nitrogenous Oxygen Demand

Nitrogenous compounds released in discharges may include nitrogen in organic chemicals (organic nitrogen) and nitrogen present as ammonia, nitrites, and nitrates. Under appropriate environmental conditions, many nitrogenous compounds undergo a series of biochemical reactions that convert organic nitrogen to ammonia and oxidize ammonia to nitrites and nitrates:

$$\text{Organic nitrogen} \rightarrow \text{Ammonia} \rightarrow \text{Nitrite} \rightarrow \text{Nitrate}$$

The term "nitrification" is used to refer to the last two steps, which are accomplished in sequential biochemical reactions carried out by nitrosomonas and nitrobacter bacteria, respectively:

$$2\,NH_4^+ + 3\,O_2 \xrightarrow{\text{nitrosomonas}} 2\,NO_2^- + 4\,H^+ + 2\,H_2O$$

$$2\,NO_2^- + O_2 \xrightarrow{\text{nitrobacter}} 2\,NO_3^-$$

Oxygen demands from these two reactions are in addition to the immediate and carbonaceous demands outlined earlier. Theoretically, based on stoichiometry of the equations, they require a total of 4.6 g of oxygen for each gram of nitrogen that is oxidized from ammonia to nitrate. However, the actual demand is somewhat less, ranging from an upper limit of perhaps 4.3 g to substantially lower values that often are experienced under field conditions. The lower oxygen requirements encountered in practice may be attributed to several factors, including the incomplete conversion of ammonia nitrogen to nitrate, the recovery of some oxygen through subsequent decomposition of the nitrates (denitrification), and the utilization of some nitrogen in producing new cells.

The oxygen required in nitrification is referred to as "nitrogenous oxygen demand," or "NOD." This also is biochemical in nature but should be considered separately from the BOD because it is imposed by different types of organisms, metabolizing different types of chemicals, with different responses to environmental factors and different growth and kinetic relationships.

The potential importance of NOD in practice can be illustrated by pointing out that a municipal wastewater usually has a nitrogen content (organic and ammonia) of approximately 30 to 40 mg/liter. Based on a concentration of 35 mg/liter nitrogen and a ratio of 4.3 g oxygen per g nitrogen, referred to earlier, its NOD approximates 150 mg/liter. That same wastewater may contain organics producing about 200 mg/liter BOD, indicating that the NOD can be a very substantial portion of the total oxygen demand.

The differences between BOD and NOD cause major problems in developing and using mathematical models for oxygen demand. The same types of relationships may not represent accurately the substantially different kinetics of carbonaceous and nitrogenous oxygen demand systems, and major errors often are encountered when similar models are used for both. Furthermore, the state of knowledge about nitrification in streams does not yet permit the development of models that are accurate for that purpose.

MEASUREMENT OF OXYGEN DEMAND

Measurement of oxygen demand should be done under conditions approximating as closely as possible those in the practical systems that are of principal concern. Test procedures often followed in practice are described in *Standard Methods* (Water Pollution Control Federation, 1980) and by Sawyer and McCarty (1978).

The first step in measuring immediate oxygen demand is to determine the DO content of dilution water before mixing it with the sample being analyzed. After combining the two in proper proportions and allowing appropriate reaction time—perhaps 30 min—the DO content of the mixture is measured and its oxygen demand calculated.

Measurement of BOD is somewhat more complicated because the amount of oxygen required by microorganisms during the stabilization of organics can be determined accurately only by using living organisms in the test. The rationale behind the usual analytical procedure is simple and is outlined here through an example.

1. Two 300-ml BOD bottles are filled completely with the water sample of which the oxygen demand is to be measured.
2. The bottles are stoppered carefully to exclude all air bubbles.
3. One bottle is analyzed immediately to determine its DO content.
4. The other bottle is incubated at 20°C (68°F) in total darkness (usually for 5 days), after which its DO is measured.
5. The difference between the two DO values is the amount of oxygen that was consumed by organisms during the 5 days and is reported as the BOD_5 (5-day BOD) of the sample.
6. Data from this type of test might appear as follows.

$$\begin{aligned} \text{Initial DO (first bottle)} &= 7.8 \text{ mg/liter} \\ \text{Final DO (second bottle)} &= 3.6 \text{ mg/liter} \\ BOD_5 \text{ (5-day BOD)} &= 7.8 - 3.6 \\ &= 4.2 \text{ mg/liter} \end{aligned}$$

In practice, BOD measurements are more complicated than this example, for several reasons. First, the sample may have a BOD high enough to deplete all DO in the bottle before the end of 5 days, making the test result indeterminate. That makes it necessary, often, to dilute the sample to prevent the total depletion of DO. Upon completion of the test, the BOD value obtained for the bottle contents must be reduced by the dilution water BOD and the result then multiplied by the dilution factor to calculate the BOD of the original sample before dilution, as shown in the following example.

Assume that a wastewater sample is diluted 150:1 in the BOD bottle and produces DO values before and after incubation of 8.1 and 4.3 mg/liter, respectively. Dilution water samples, incubated separately, produce values of 8.3 and 7.7 mg/liter DO before and after incubation.

BOD of the dilution water	= 8.3 − 7.7
	= 0.6 mg/liter
BOD of the waste and dilution water	= 8.1 − 4.3
	= 3.8 mg/liter
BOD of diluted waste only	= 3.8 − 0.6
	= 3.2 mg/liter
BOD of undiluted waste	= 3.2 × 150
	= 480 mg/liter

Enough nutrients and trace elements must be in the bottle during incubation to allow unrestricted microbial growth and full development of the sample BOD. BOD dilution water is prepared with high-purity water to which certain reagent grade chemicals are added to insure presence of the necessary constituents, as specified in *Standard Methods* (Water Pollution Control Federation, 1980). The specific chemicals added in preparing dilution water are selected, also, to provide a chemical buffer system to insure that the pH in the BOD bottle does not change enough during incubation to inhibit the action of the organisms.

BOD and NOD should be measured separately to prevent confusing the two different types of demand and to avoid errors that could arise from nitrification of ammonia added to the BOD dilution water as a nutrient and buffer. Therefore, steps should be taken in measuring BOD to suppress the bacteria that could exert NOD simultaneously during incubation. That usually is accomplished by adding to the dilution water a chemical that inhibits the nitrosomonas and nitrobacter organisms.

Measurement of the NOD presents additional problems because no reliable system has been developed to produce nitrification in laboratory tests in amounts and with kinetics comparable to those in receiving streams. Laboratories often try to approximate the NOD by runnning BOD tests with and without the nitrification inhibitor and without adding ammonia to the dilution water. The difference between oxygen demands in the uninhibited and inhibited BOD tests then is assumed to be caused by nitrification. This approach probably is as good as any now available but, nevertheless, is of dubious accuracy because it does not produce nitrification kinetics similar to those in watercourses. Data obtained in that fashion therefore should be viewed with skepticism and used with caution.

Even under the best of conditions, results obtained in tests that use living organisms usually are poor in reproducibility and questionable in accuracy.

Variations of 10 to 25 percent in repeated BOD analyses of the same sample are not uncommon in field laboratories. Correlation between the result of laboratory measurements and oxygen demands that actually occur in streams often are even poorer. The complexity of these types of tests and the numerous opportunities for obtaining erratic results require great care in conducting them and caution in interpreting their results.

PURPOSES OF OXYGEN BALANCE MODELS

The oxygen balance in a watercourse is the result of combined deoxygenation and reoxygenation reactions. As used in this discussion, deoxygenation refers to the utilization of oxygen by various chemical and biochemical reactions, excluding the consideration of any reoxygenation that might occur simultaneously. The reactions may include immediate oxygen demand, carbonaceous oxygen demand (BOD), nitrogenous oxygen demand (NOD) and conceivably others as well. The deoxygenation produced by all these reactions determines the overall impact on stream DO, assuming no replacement of DO by any mechanism.

On the other hand, reaeration (reoxygenation) of the stream represents the asset side of the DO ledger—that is, the additions of oxygen to the stream through various mechanisms that include gas transfer from the atmosphere (atmospheric reaeration) and photosynthesis by green plants.

An oxygen balance "model" permits one to describe those phenomena and the effects of variables in concise mathematical equations. Without a model, system complexity and interrelationships among the many variables would virtually prohibit accurate summary and prediction of oxygen balance in a stream. The existence of a good model makes it possible to summarize in convenient form the net effects of many variables on the system. A model for oxygen balance must include at least the expressions for deoxygenation and reoxygenation of the stream.

Once developed and verified, a good mathematical model for oxygen balance in a watercourse can be used to compare and contrast different streams or various reaches of a single one in meaningful terms. It also facilitates the application of computer technology to evaluate, interpret, and extend field data for planning and design purposes.

Reliable models are of special value in regulatory activities because they can provide methods for predicting DO concentrations under various conditions of loading, flow, temperature, or combinations of variables that may never have been experienced in the recorded history of that stream. In turn, that can provide a rational basis that the agency can use in defining the degree of treatment that would be necessary to achieve some desired level of quality in the stream and to develop cost-benefit relationships for different cleanup tactics. Accordingly, they can be used as a rational foundation for sound regulatory decisions in water pollution control programs.

DEOXYGENATION OF WATER

The accurate evaluation and use of deoxygenation models requires understanding of their origins, assumptions, characteristics, and limitations. The most important point to grasp immediately is that no simple mathematical relationship actually can describe the complex biological phenomena occurring in streams.

The overall oxygen demand imposed by deoxygenation reactions results from the activities of many types of organisms metabolizing a wide variety of organic compounds. Any one biochemical reaction may, in fact, conform reasonably closely to some simple mathematical relationship (or perhaps not!). However, there is no scientific basis for supposing that all of them must follow the same type of relationship, or that the same kinetics must apply to all of them, or that the same relationship must hold true at other times under different environmental conditions. The complex, dynamic, ever-changing characteristics of natural biological systems explain the often-demonstrated fact that there are not now, and never will be, relatively simple models capable of describing those systems accurately.

That is the basis for the most important single characteristic of all deoxygenation models: they may be useful for *approximating* the gross impact of oxygen balance reactions in a watercourse, but they actually have no basis in scientific fact for being viewed as rigorous descriptions of deoxygenation processes. A corollary is that often the same models that are so valuable in some practical situations may be grossly inaccurate when applied under different circumstances.

Several types of relationships have been proposed for describing BOD exertion in streams. One of the earliest, and widely favored even now, was proposed by Streeter and Phelps (1925). They observed that the BOD concentrations measured at successive points along the Ohio River below major discharges decreased as a function of distance traveled (or time of flow). They discovered that the rate of BOD exertion could be described with reasonable accuracy by applying a relatively simple, but flexible, mathematical relationship.

$$-\frac{dL}{dt} = K_1 L \tag{10-1}$$

Where: L = BOD remaining at time t, mg/liter (total BOD, not that measured in 5 days)

t = time, days

K_1 = deoxygenation constant, days^{-1}

Integration of that equation between the limits of time equal to zero and t produces

$$L_t = L_0 e^{-K_1 t} \tag{10-2}$$

or the alternate

$$L_t = L_0 10^{-k_1 t} \tag{10-3}$$

Where: K_1 = deoxygenation constant (log base e)
e = natural logarithm base (2.7183)
k_1 = deoxygenation constant (log base 10)
= $0.434\ K_1$
L_t = BOD remaining in the water at time t
L_0 = BOD in the water at time zero

Figure 10.1 shows the type of curve produced when Equation 10-2 or 10-3 is plotted on arithmetic graph paper. Taking logarithms of Equation 10-3 gives:

$$\log L_t = -k_1 t + \log L_0 \tag{10-4}$$

A straight line is obtained by plotting the logarithmic form—Equation 10-4—on arithmetic paper (Figure 10.2). An alternative approach uses semilogarithmic paper, as shown by Figure 10.3, to produce a straight line for Equation 10-3.

This type of mathematical relationship—referred to by chemists as a "first-

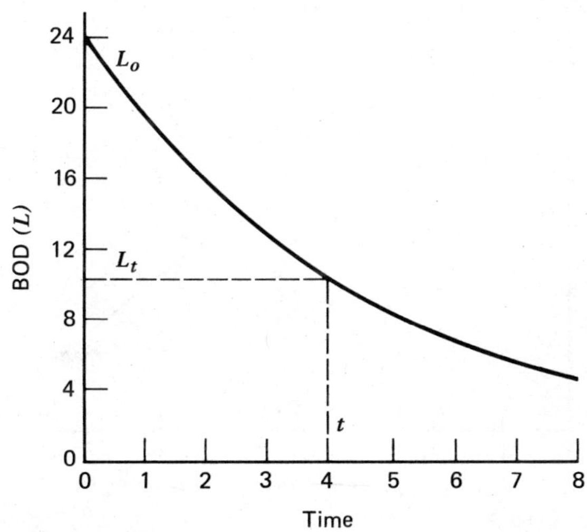

FIGURE 10.1
First-order reaction plotted on arithmetic graph paper.

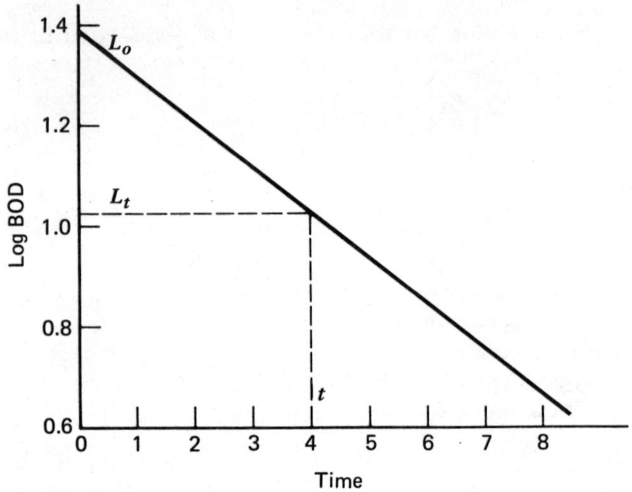

FIGURE 10.2
Logarithmic form of first-order reaction plotted on arithmetic graph paper.

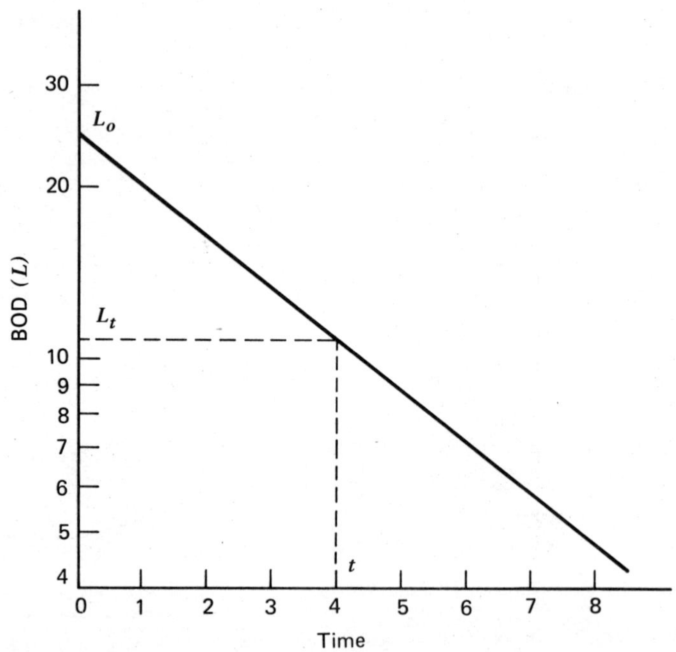

FIGURE 10.3
First-order reaction plotted on semilogarithmic graph paper.

order reaction"—has been used to describe many phenomena. Some applications are based on scientific evidence that the equation actually is correct for that system. For example, many chemical reactions, radioactive decay, and heat transfer rates follow first-order kinetics and may be described by applying the same basic equation. The first-order equation also has been applied to many phenomena that do not follow the relationship rigorously but can be conveniently rationalized and summarized by using it. Examples include the multiplication of bacteria, fruit flies, or humans, the death of bacteria during disinfection, gas transfer processes, and many others, including BOD.

Its application to BOD by Streeter and Phelps (1925) and others was purely empirical; they observed that this relatively simple mathematical equation sometimes could be used with reasonable accuracy to summarize the exertion of BOD. There is no scientific basis for assuming that the first-order equation must fit heterogenous biochemical processes rigorously, nor should its widespread use in practical deoxygenation models be misconstrued as evidence that exertion of BOD really is a first-order reaction. In fact, BOD reactions usually do not follow that relationship closely, except maybe for short periods of time. Sulzer (1960) has presented an interesting explanation of how this type of equation may provide a reasonable approximation for biochemical actions, even though none of those in his example were first order in character. Other types of mathematical expressions also have been employed to describe BOD exertion and, in some instances, have proven superior to the first-order model.

Equation 10-1 states that the rate at which BOD is exerted at any given time is proportional to its concentration in the water. Figure 10.1 shows that immediately after a discharge of oxygen-demanding chemicals the rate of BOD exertion—represented by the slope of the curve—will be maximum. As time passes thereafter, the rate of BOD exertion decreases in response to the reduction in BOD concentration in the stream. Accordingly, the milligrams per liter of oxygen demand imposed on the stream in any time interval will be less than that during the preceding equal time interval.

The nature of the first-order model can be illustrated with the simple example in Table 10.1, which is calculated for a value of $k_1 = 0.1 \text{ day}^{-1}$. The amount of BOD in the water at the beginning of each day is expressed as the percent of that initially present at the beginning of day 1 (100 in the example). The solution of Equation 10-3, using 100 for L_0, 1.0 for t, and 0.1 for k_1, produces a BOD remaining at the end of the first day of 79.4 percent of that present at the beginning of the day. This indicates that 20.6 percent of the initial BOD was exerted during the first day. Furthermore, the calculation of $10^{-k_1 t}$ indicates that this same percentage of the BOD present at any time will be exerted during the succeeding one-day interval.

Solutions for successive days produced the results in Table 10.1. The BOD remaining at the beginning of the second day is 79.4. During the second day,

TABLE 10.1
An Example Illustrating the Exertion of BOD, Assuming a First-Order Reaction

Day	BOD at Beginning of Day	BOD Destroyed	Percent Left at End of Day
1	100.0	20.6	79.4
2	79.4	16.3	63.1
3	63.1	13.0	50.1
4	50.1	10.3	39.7
5	39.7	8.2	31.5
6	31.5	6.5	25.0

Note: Assumed values: $k_1 = 0.10 \text{ day}^{-1}$ (base 10). Initial BOD = 100.

20.6 percent of the 79.4, or 16.3, is exerted. This leaves 63.1 percent of the original remaining at the end of day 2. The rest of Table 10-1 is similarly constructed.

A note of interest is that, with $k_1 = 0.1$, 50 percent of the BOD is exerted by the end of the third day (the "half-life" in this example). In another 3 days 50 percent of the BOD present at the beginning of that period (50) is exerted, leaving 25 percent of the original remaining in the water after a second half-life.

Table 10.2 shows the effect of k_1 on the proportion of BOD exerted each day. Also shown is the proportion of the total BOD that is exerted within the five day incubation period commonly used in conducting the BOD test. With a value of 0.05 for k_1, only 44 percent of the total oxygen demand is exerted during the standard five-day BOD test, but 94 percent is exerted in that length of time when the sample has a k_1 equal to 0.25.

Figure 10.4 shows BOD curves representing a group of samples having the same total BOD, but with different values of k_1 (Sawyer, 1978). Oxygen

TABLE 10.2
The Effect of the Deoxygenation Constant on BOD Exerted Daily and in Five Days

Value of k_1	Percent Exerted per Day	Percent Exerted in Five Days
0.05	10.9	44
0.10	20.6	68
0.15	29.2	82
0.20	36.9	90
0.25	43.8	94

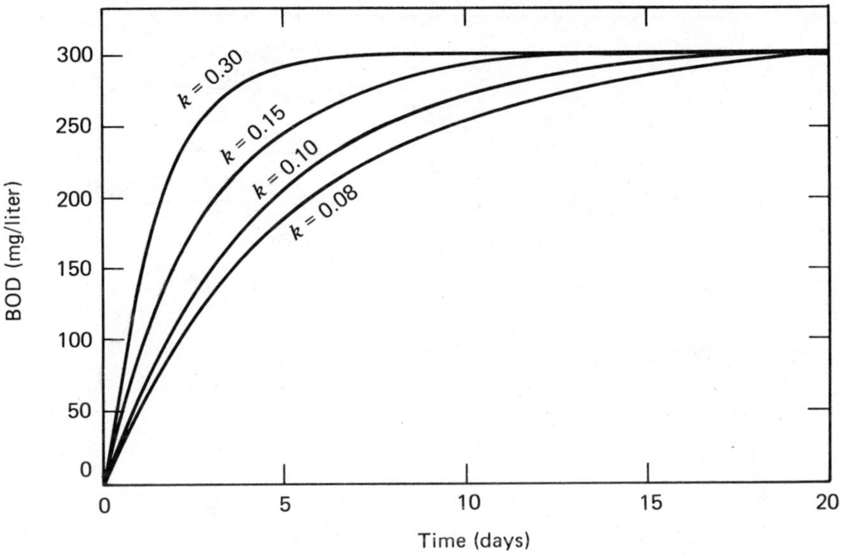

FIGURE 10.4
Effect of reaction constant, k_1, on the progression of the BOD curve to the same 20-day BOD. From *Chemistry for Environmental Engineering,* C.N. Sawyer and P.L. McCarty, Copyright © 1978 by McGraw-Hill Book Company, New York, Page 427. Reprinted by permission.

demands of the samples that would be recorded at five days in laboratory analyses differ widely, in spite of having the same total BOD. Figure 10.5 shows BOD curves for samples with the same five-day BOD but different k_1 values. Wastewaters represented by these samples would have very different impacts on receiving streams, in spite of having the same five-day BOD, because of the different total oxygen demands and rates at which they would be exerted. These figures illustrate the importance of knowing the rate constant (k_1), as well as amount of BOD, if one wishes to assess the impact of a wastewater on a receiving stream.

The value of k_1 may be influenced greatly by the temperature of the water because biological reactions are sensitive to that variable. The exact nature of the impact is impossible to generalize because of the varied organisms involved and the different temperature effects on each. However, several investigators have used empirical expressions similar to Equation 10-5 to adjust values of k_1 to produce reasonably valid estimates in practice.

$$k_{1(T^\circ C)} = k_{1(20^\circ C)} \times 1.047^{(T-20)} \qquad (10\text{-}5)$$

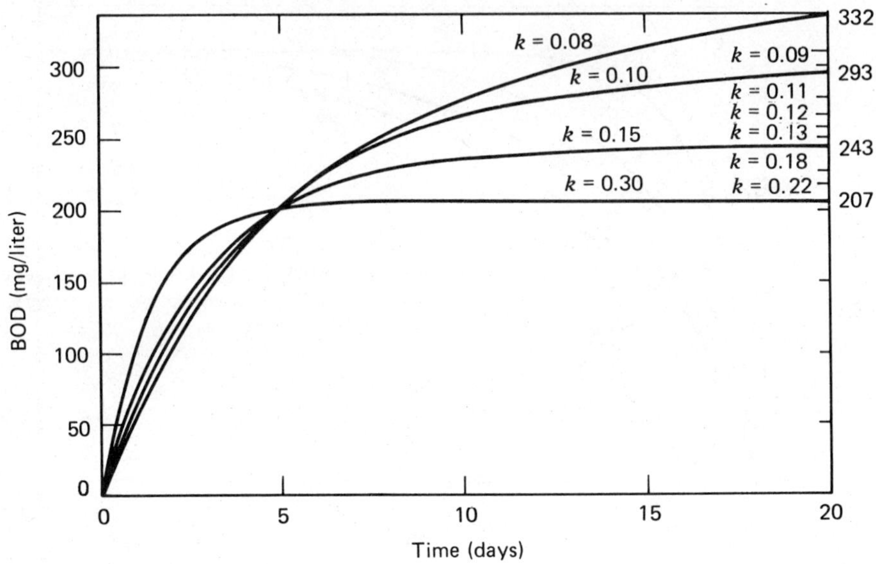

FIGURE 10.5
Effect of reaction constant, k_1, on the progression of BOD in samples with the same BOD_5 value. From *Chemistry for Environmental Engineering,* C.N. Sawyer and P.L. McCarty, Copyright © 1978 by McGraw-Hill Book Company, New York, Page 428. Reprinted by permission.

Where: $k_{1(T^\circ C)}$ = k_1 at temperature $T^\circ C$
$k_{1(20^\circ C)}$ = k_1 at temperature $20^\circ C$
1.047 = an empirical constant

A different type of BOD model that has been used from time to time is Equation 10-6, in which BOD exertion is treated as being independent of its concentration of solution.

$$-\frac{dL}{dt} = K \tag{10-6}$$

this integrates to yield

$$L_t = -Kt + L_0 \tag{10-7}$$

This relationship produces a straight line on arithmetic paper when L is plotted as a function of t. In chemistry, this is referred to as a "zero-order reaction."

Equation 10-8 is a different proposed model in which the rate of BOD exertion is proportional to the square of its concentration (a "second-order reaction").

$$-\frac{dL}{dt} = K_1L^2 \tag{10-8}$$

These relationships have been used much less than the first-order expression.

OXYGENATION OF STREAMS

On the asset side of the DO balance sheet are the mechanisms by which oxygen is added to water to offset losses incurred through deoxygenation. Three methods by which oxygenation may occur are (1) atmospheric reaeration, (2) photosynthesis, and (3) artificial oxygenation. These methods are discussed in this section; the next section describes a model commonly used to summarize oxygenation of streams.

Atmospheric Reaeration

Although the term commonly used is reaeration, it really would be more accurate to refer to it as oxygenation because, of course, oxygen is the gas of primary concern. This process for adding oxygen to watercourses is based on transferring the gas from air into water through its surface. Gas transfer is a physical-chemical process that occurs continuously at liquid-gas interfaces. High-velocity movement allows oxygen molecules to penetrate the interface, producing a transfer of that gas from air into water. Simultaneously, some molecules of oxygen dissolved in the water escape through the interface into the atmosphere. Movements in both directions proceed at their own rates, which are determined by temperature and other variables.

When there is no oxygen demand, these processes ultimately reach a dynamic equilibrium in which the rate of transfer from air into water is equal to the rate of transfer in the opposite direction. This produces a steady oxygen concentration in the liquid at its "saturation" value. That DO concentration depends on several variables, the most important ones being pressure of the oxygen gas and temperature of the water. Saturation values for oxygen in water in equilibrium with air at 1 atm pressure and different temperatures are given in Table 10.3 (Water Pollution Control Federation, 1980) and range from 8 to 14 mg/liter at temperatures commonly encountered in natural freshwater streams.

When oxygen concentration in water is depressed below the saturation level

TABLE 10.3
Solubility of Oxygen in Fresh Water at Equilibrium with Air at 1.0 Atmosphere

Temperature (°C)	Solubility (mg/liter)	Temperature (°C)	Solubility (mg/liter)
0	14.6		
1	14.2	26	8.2
2	13.8	27	8.1
3	13.5	28	7.9
4	13.1	29	7.8
5	12.8	30	7.6
6	12.5	31	7.5
7	12.2	32	7.4
8	11.9	33	7.3
9	11.6	34	7.2
10	11.3	35	7.1
11	11.1	36	7.0
12	10.8	37	6.9
13	10.6	38	6.8
14	10.4	39	6.7
15	10.2	40	6.6
16	10.0	41	6.5
17	9.7	42	6.4
18	9.5	43	6.3
19	9.4	44	6.2
20	9.2	45	6.1
21	9.0	46	6.0
22	8.8	47	5.9
23	8.7	48	5.8
24	8.5	49	5.7
25	8.4	50	5.6

Source: Standard Methods for the Examination of Water and Wastewater, 15th Edition. Reprinted by permission. Copyright © 1981, American Public Health Association, American Water Works Association and Water Pollution Control Federation.

by oxygen demand, that causes a reduction in the rate of escape of oxygen from water into the air. However, the rate of entry of oxygen from the atmosphere remains essentially the same because the pressure of that gas in the atmosphere has not changed; only its concentration in the water has changed. Accordingly, the transfer from air into water then exceeds that from water into the air, producing a net addition of oxygen to the water and replacing some of that lost through deoxygenation.

Several models have been proposed to describe this process, the simplest and most widely used one being Equation 10-9.

$$\frac{dc}{dt} = K_L \frac{A}{V}(S - C) \tag{10-9}$$

Where: C = DO concentration in the water
S = DO concentration at saturation
t = time
K_L = transfer coefficient
A = interfacial (stream surface) area
V = liquid (stream) volume

Equation 10-9 states that the net change in oxygen concentration in water is directly proportional to the area of air-water interface (A) through which transfer occurs, inversely proportional to the water volume (V), and directly proportional to the *difference* between DO in the water and that which would be there at saturation ($S - C$). The effect of area to volume (A/V) ratio indicates that the rate of change of oxygen concentration in a stream is inversely proportional to water depth. The transfer coefficient (K_L) is influenced by many factors, including temperature, films on the water surface, turbulence, and constituents dissolved in the water. The application of this gas transfer model to reaeration of streams, as used in practice, will be discussed in more detail shortly.

Photosynthesis

This mechanism for adding oxygen to watercourses depends on biological action of green plants, especially algae. They contain chlorophyll, making it possible to derive energy from sunlight and produce cell material using carbon dioxide as a carbon source and releasing oxygen into the water.

$$CO_2 + 2H_2O \xrightarrow{\text{light}} (CH_2O)_x + O_2 + H_2O$$

The term $(CH_2O)_x$ represents new algal cells. Algae also consume oxygen through respiration.

$$(CH_2O)_x + O_2 \rightarrow CO_2 + H_2O$$

The overall production of oxygen by algal cells generally is about 1.5 times the rate at which they use it in respiration. However, when algae die, they are decomposed biochemically and exert an oxygen demand on the stream that sometimes may have a very substantial impact in addition to that imposed by other BOD in the water.

Because of the crucial role of light, oxygen is produced only during daylight hours, but the cells use oxygen in respiration around the clock. This causes a diurnal variation in DO in natural waters, especially in lakes, similar to that shown in Figure 10.6. Minimum dissolved oxygen is attained in the early morning. Algal photosynthesis causes it to rise to a peak concentration during the afternoon, after which it drops as reduced sunlight decreases oxygen production. This phenomenon is especially important in planning and executing stream and lake sampling programs because widely different data could be collected on DO conditions, depending on when in the diurnal cycle measurements are made.

This mode of oxygen production is biochemical and does not follow the same type of model as that used for atmospheric reaeration, a physical-chemical process. Although models for the photosynthetic reoxygenation process have been proposed, none has yet attained general, or even broad, acceptance.

Artificial Reoxygenation

These methods for adding oxygen to streams often are not mentioned in discussions of reoxygenation phenomena, presumably because they are not "nat-

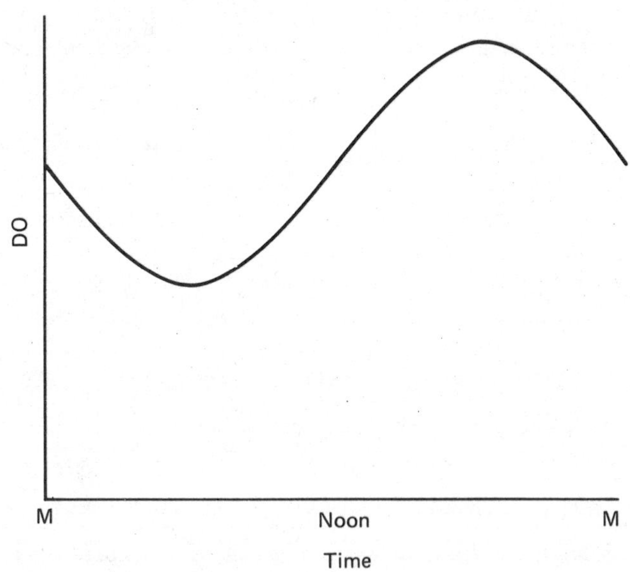

FIGURE 10.6
Illustration of diurnal changes in DO in streams and lakes.

ural" processes. Nevertheless, they are used sometimes in practice and can be important in maintaining desirable oxygen balance in some streams.

There are several methods available for artificial reoxygenation of streams. They are not discussed in detail here because of space limitations, but they are identified and described briefly.

1. Mechanical Aeration. There have been many studies on feasibility of using various types of aeration devices to add oxygen directly to streams when needed, but they have been seldom used in practice.
2. Turbine Venting. At some hydroelectric power installations oxygenation of water passing through turbines has been accomplished by allowing air to enter at low-pressure points in the system.
3. Reaeration on Spillways. Sometimes spillways with roughened surfaces have been constructed to enhance the reaeration of water passing over them.
4. Addition of Oxygen. Occasionally, the intermittent addition of oxygen on an as-needed basis could be economical for insuring maintenance of desired DO in some streams.
5. Addition of Hydrogen Peroxide. This chemical decomposes to yield DO and is another method for adding oxygen to a stream, especially when needed intermittently.

PRACTICAL REAERATION MODELS

Models used widely in practice to describe the reoxygenation of streams have been based on the gas transfer equation (Equation 10-9). Because of inadequate understanding of its complex processes, photosynthesis usually has been omitted in practical models. Artificial reaeration techniques have been used only infrequently and have been evaluated on an ad hoc basis in those instances.

In Equation 10-9, the terms $(S - C)$ is the difference between the saturation concentration of DO (S) and the concentration actually present in the water (C). This often is referred to as the oxygen "deficit" (D) because it is the amount by which the water misses being at oxygen saturation.

$$D = (S - C) \qquad (10\text{-}12)$$

For a specific section of stream and temperature, the area (A), volume (V), and saturation concentration (S) are constant. The change in concentration (C) is equal to the change in deficit, allowing Equation 10-9 to be rewritten

$$-\frac{dD}{dt} = K_2 D \qquad (10\text{-}13)$$

which integrates to the form

$$D = D_0 \times e^{-K_2 t} \quad (10\text{-}14)$$

or

$$D = D_0 \times 10^{-k_2 t} \quad (10\text{-}15)$$

Where: D = oxygen deficit ($S - C$) at time t
D_0 = initial oxygen deficit at time zero
K_2 = reaeration coefficient (log base e), including constants K_L, A, and V
k_2 = reaeration coefficient (base 10).

Of course, this relationship is mathematically similar to that for deoxygenation (another first-order equation), as shown by comparison with Equations 10-1, 10-2, and 10-3. Figure 10.7 shows the effects of various values of k_2 on the rates at which oxygen concentrations approach the saturation value, analogous to the presentation in Figure 10.4 for BOD exertion.

Reaeration is affected by the temperature of the water because that variable changes the value of S (and D) and also the value of k_2. Temperature effect on S is shown by the data in Table 10.3, and its effect on k_2 may be described as follows.

$$k_{2(T°C)} = k_{2(20°C)} \times 1.0238^{(T-20)} \quad (10\text{-}16)$$

Where: $k_{2(T°C)}$ = reaeration constant at T°C
$k_{2(20°C)}$ = reaeration constant at 20°C
1.0238 = empirical constant

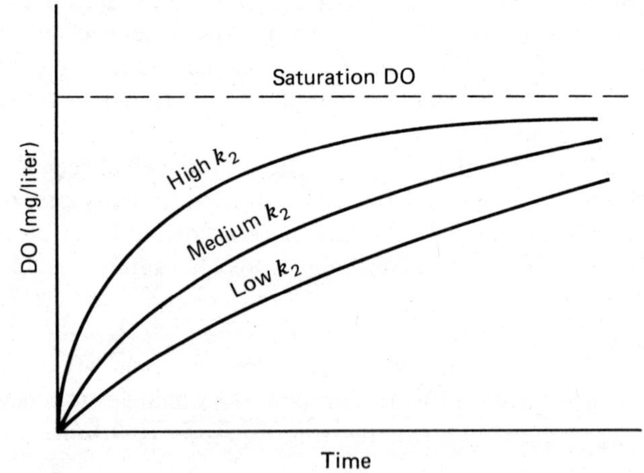

FIGURE 10.7
Effect of reaeration constant, k_2, on changes in DO in streams.

Practical values for the reaeration coefficient have been reported by many investigators for various types of watercourses, and some taken from a recent text (Metcalf and Eddy, 1980) are shown in Table 10.4. Langbein and Durum (1967) summarized the results of several investigators and produced the following empirical equation for determining K_2 (base e).

$$K_2 = 3.3 \frac{V}{H^{1.33}} \tag{10-17}$$

Where: K_2 = reaeration coefficient, days^{-1}
V = velocity, ft/sec
H = depth, ft

Other equations and approaches for determining reaeration characteristics of streams have been described by several authors (Churchill, 1962; Velz, 1970; Hydroscience, 1971; U.S. EPA, 1972; Tsivoglou, 1972).

THE STREETER-PHELPS FORMULA

Having discussed the deoxygenation and reoxygenation of streams separately, it now remains to combine the models for those processes into a single expression that describes the oxygen balance. That is done very often in practice through using the Streeter-Phelps formula, named after two investigators who developed the rationale in the 1920s (Streeter and Phelps, 1925; Phelps, 1944).

The method is simple in concept. Beginning with oxygen concentration at a point and moving downstream, the approach merely subtracts the cumulative oxygen demand on the stream by BOD exertion and adds the oxygen replenished through reaeration. As developed earlier, the deoxygenation may be summarized by the expression

$$-\frac{dL}{dt} = K_1 L \tag{10-1}$$

TABLE 10.4
Some Practical Values of K_2

Type of Watercourse	k_2, to base 10	K_2, to base e
Ponds	0.05–0.10	0.12–0.23
Lakes	0.10–0.15	0.23–0.34
Streams (low velocity)	0.15–0.20	0.34–0.46
Streams (medium velocity)	0.20–0.30	0.46–0.69
Streams (swift velocity)	0.30–0.50	0.69–1.15
Rapids	>0.50	>1.15

Source: Adapted from Metcalf and Eddy (1980).

Ignoring reaeration for the moment and expressing the BOD impact in terms of how it would change the DO deficit if acting alone, we may rewrite Equation 10-1.

$$\frac{dD}{dt} = K_1L \qquad (10\text{-}18)$$

The reaeration, of course, may be expressed as

$$-\frac{dD}{dt} = K_2D \qquad (10\text{-}13)$$

Adding these two algebraically produces an expression that gives the net effect of the two phenomena.

$$\frac{dD}{dt} = K_1L - K_2D \qquad (10\text{-}19)$$

Substituting $L_0e^{-K_1t}$ for L and $D_0e^{-K_2t}$ for D and integrating, Equation 10-20 results.

$$D = \frac{K_1L_0}{K_2 - K_1}(e^{-K_1t} - e^{-K_2t}) + D_0e^{-K_2t} \qquad (10\text{-}20)$$

This expression describes the oxygen deficit (D) at any downstream point based on (1) the initial deficit at the upstream point (D_0), (2) the intitial total BOD (L_0), (3) the deoxygenation coefficient (K_1), (4) the reaeration coefficient (K_2), and (5) the time of travel (t) in the stream between the two points.

Solution of the Streeter-Phelps equation for different values of t produces an "oxygen sag curve" similar in appearance to Figure 10.8. Beginning with the addition of BOD at the upstream point, at the left, and moving downstream to the right, the DO decreases initially because deoxygenation occurs at a rate

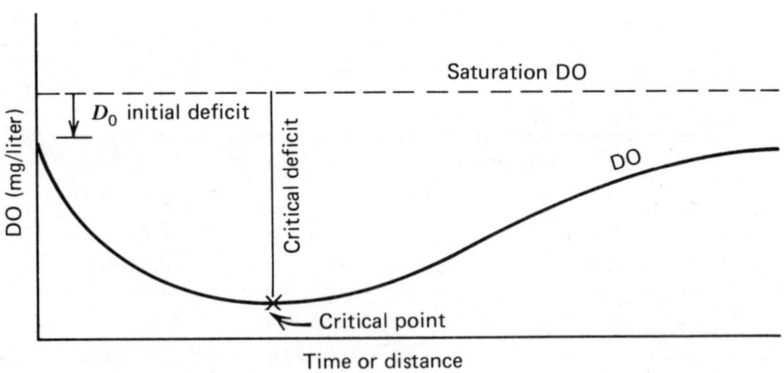

FIGURE 10.8
Example of an oxygen sag curve below a point-source discharge.

exceeding reaeration. The rate of deoxygenation decreases as the BOD is reduced with the passage of time, and the reaeration rate increases in response to the increased oxygen deficit. As the flow proceeds downstream, the two finally become equal at the "critical point," and the DO reaches its minimum concentration. Thereafter, the reaeration rate exceeds deoxygenation and the stream DO gradually recovers to higher levels. The critical deficit is considered to be the most important point on the curve because it represents the minimum level of DO, making it the zone where difficulties caused by low dissolved oxygen should be most acute.

Figure 10.9 illustrates the types of sag curves that could be observed in a stream with three different levels of BOD discharge from a point-source. A small amount of BOD produces a minor oxygen sag at the critical point. An increase in the amount of BOD discharged reduces the DO at the critical point to a lower level, and a high-discharge BOD may result in an extended period of zero DO in the flowing water.

Successive solutions of the Streeter-Phelps equation for a series of different upstream additions of BOD could allow one to determine how much oxygen demand could be assimilated by the stream without having DO at the critical point drop below some specified concentration. That is the general approach usually used by regulatory agencies to determine allowable discharges from point-sources of BOD and represents a key tool in most regulatory programs today.

The equation can be solved manually, but several successive solutions are time consuming, so digital or analog computers usually are employed to good advantage for that purpose (Cohen, 1963; Hydroscience, 1971; U.S. EPA, 1972; Sinkoff, 1958). Other convenient methods that can be used for relatively rapid solutions include specially constructed tables, graphs, and nomographs (McBride, 1982).

The Streeter-Phelps formula is widely used for regulatory purposes and rep-

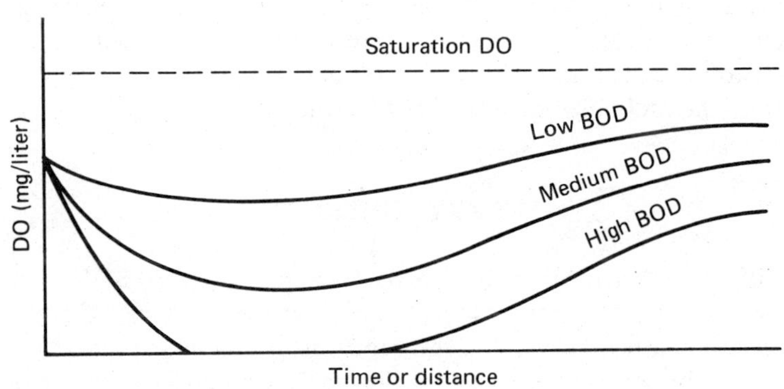

FIGURE 10.9
The effect of changing the BOD discharged into a stream on the oxygen sag curve.

resents the principal method now available for predicting the effects of various loadings on streams. However, it has several limitations, not the least of which is the underlying assumption that deoxygenation in streams can be described accurately by using a first-order equation (Equation 10-1). When that is not true, the equation is invalid and serious errors may result, which happens too often.

Other limitations inherent in using the Streeter-Phelps formula include its lack of consideration of the effects of nitrification and photosynthesis on DO in streams, although additions to the basic formula have been proposed to help compensate for those omissions. It is applicable principally to freshwater streams and is of dubious validity for estuaries or lakes. Perhaps the most serious problem connected with its use is the fact that a shortage of good alternative techniques often leads to the temptation to use the relationship too generally, even when it is inappropriate for the section of stream in question.

ALTERNATIVE APPROACHES

Other approaches that have been used successfully to model dissolved oxygen relationships in streams are based on the statistical evaluation of data collected in the field over extended periods of time (Churchill, 1954, 1956; Okun, Lamb, and Wells, 1963). These have the advantage that they do not rely on preconceived ideas about the types of mathematical relationships involved; they depend only on actual relationships observed in the specific stream under study.

Frequently, regression techniques can produce much more accurate relationships among BOD, DO, temperature, flow, and other variables in the stream than can the Streeter-Phelps approach. However, the mathematical relationships developed by these methods usually are highly specific to the stream under study and should not be transferred to others. Also, it is risky to extrapolate them beyond the limits of the observed data in attempts to predict situations that might exist under different conditions. Equations developed in this way are excellent for some purposes, but they are of dubious validity for evaluating the probable impacts of new discharges.

STUDY QUESTIONS

1. Differentiate clearly among imediate oxygen demand, BOD, and NOD.
2. Describe the occurrences in a stream after the discharge of biodgradable organic matter that lead to the depletion of DO in the water.
3. Write the mathematical expression most commonly used for depicting the rate of BOD exertion. Briefly interpret the equation and explain the assumptions inherent in its use.

4. Sketch two curves showing BOD exertion as a function of time, with the same total BOD but different values of k_1. Explain how the differences in wastewaters represented by those curves could be significant in determining their impacts on a stream and in laboratory measurements of their BOD.
5. Describe the effect of temperature on BOD exertion in streams.
6. Describe the effects of temperature on the reaeration of streams.
7. Outline the effects of four variables on reaeration of streams.
8. Explain the diurnal variation of DO often observed in lakes and streams.
9. Distinguish clearly between the roles of bacteria and algae in determining oxygen relationships in streams.
10. Given the Streeter-Phelps formula, explain the identity and significance of each term in it.
11. Sketch a typical oxygen sag curve, label it, and explain why it assumes that shape.
12. Sketch and explain the differences between oxygen sag curves for a stream under low and high rates of BOD loadings.
13. Calculate the proportion of total BOD of a wastewater that would be exerted in five days with k_1 values of 0.07, 0.13, and 0.22, using the first-order equation model.
14. Calculate and plot the percentage of total BOD that is exerted during each day for the first week after discharge of a wastewater with a k_1 value of 0.18.
15. A wastewater is diluted 50 times in conducting a BOD test and the DO drops from 7.9 to 2.8 mg/liter during the five days of incubation. The dilution water blank changes during incubation from 8.1 to 7.7 mg/liter. What is the BOD of the wastewater?
16. If the wastewater in question 15 has a k_1 of 0.10 and contains 25 mg/liter ammonia (as N), what would be a reasonable estimate for the maximum total oxygen demand of the wastewater?
17. The temperature of a medium-velocity stream containing 5.1 mg/liter DO is 24°C. After exactly 1 day's flow downstream, the water DO and temperature are measured again and found to be the same. Explain how this situation could occur. What would be a reasonable estimate of the oxygen demand that was exerted in the stream during that 1 day interval?

REFERENCES

Churchill, M. A., "Analysis of a Stream's Capacity for Assimilating Pollution," *Sewage and Industrial Wastes,* Volume 26, pp. 887–904 (1954).

Churchill, M. A., and R. A. Buckingham, "Statistical Method for Analysis of

Stream Purification Capacity," *Sewage and Industrial Wastes,* Vol. 28, pp. 517–537 (1956).

Churchill, M. A., R. A. Buckingham, and H. L. Elmore, *The Prediction of Stream Reaeration Rates,* Tennessee Valley Authority, Chattanooga (July 1962).

Cohen, J. B., and R. L. O'Connell, "The Analog Computer as an Aid to Stream Self-Purification Computations," *Journal Water Pollution Control Federation,* Vol 35, pp. 951–961 (1963).

Gehm, H. W., and J. I. Bregman, *Handbook of Water Resources and Pollution Control,* Van Nostrand Reinhold, New York (1976).

Hydroscience, Inc., *Simplified Mathematical Modeling of Water Quality,* Management Report, U.S. Environmental Protection Agency, Washington, D.C. (March 1971). (See also U.S. EPA, 1972.)

Langbein, W. B., and W. H. Durum, *The Aeration Capacity of Streams,* Circular 542, U.S. Geological Survey, Washington, D.C. (1967).

McBride, G. B., "Nomographs for Rapid Solutions for the Streeter-Phelps Equations," *Journal Water Pollution Control Federation,* Vol. 54, pp. 378–384 (1982).

McGauhey, P. H., *Engineering Management of Water Quality,* McGraw-Hill, New York (1968).

Metcalf and Eddy, Inc. *Wastewater Engineering,* McGraw-Hill, New York (1980).

Nemerow, N. E., *Scientific Stream Pollution Analysis,* McGraw-Hill, New York (1974).

Okun, D. A., J. C. Lamb, and C. C. Wells, "A Waste Control Program for a River with Highly Variable Flow," Journal Water Pollution Control Federation, Vol. 35, pp. 1025–1043 (1963).

Phelps, E. B., *Stream Sanitation,* Wiley, New York (1944).

Sawyer, C. N., and P. L. McCarty, *Chemistry for Environmental Engineering,* 3rd Edition, McGraw-Hill, New York (1978).

Sinkoff, M. D., C. D. Gailker, and J. G. Rennerfelt, "An Analog Computer for the Oxygen Sag Curve," *Journal of the Sanitary Engineering Division,* Vol. 84, No. SA6, American Society of Civil Engineers (November 1958).

Streeter, H. W., and E. B. Phelps, *A Study of the Pollution and Natural Purification of the Ohio River. III. Factors Concerned in the Phenomena of Oxidation and Reaeration,* Public Health Bulletin No. 146, U.S. Public Health Service, Washington, D.C. (1925).

Sulzer, F., "Discussion; Fundamental Considerations in High Rate Digestion,"

Journal of the Sanitary Engineering Division, Vol. 86, No. SA5, pp. 81–83, American Society of Civil Engineers, (1960).

Thomas, H. A., "Pollution Load Capacity of Streams," *Water and Sewage Works,* pp. 409–413 (November 1948).

Tsivoglou, E. C., and J. R. Wallace, *Characterization of Stream Reaeration Capacity,* Report EPA-R3-72-012, U.S. Environmental Protection Agency, Washington, D.C., (1972).

U.S. Environmental Protection Agency, *Addendum to Simplified Mathematical Modeling of Water Quality,* water quality management planning report, EPA Washington, D.C. (May 1972). (See also Hydroscience, 1971.)

Velz, C. J., *Applied Stream Sanitation,* Wiley, New York (1970).

Water Pollution Control Federation, *Standard Methods for the Examination of Water and Wastewater,* 15th Edition, p. 448, Washington, D.C. (1980).

CHAPTER 11

Municipal Water Systems

Water is used for many purposes to produce a wide variety of benefits, as discussed in Chapter 5. Although only about 7 percent of the water withdrawn from the U.S. resource is for municipal water supply, that use is vitally important, readily identified by the public and subject to exacting quality control. For those reasons, and because municipal water systems tend to have many similarities, they have been selected for detailed discussion.

It is not the intent in this chapter to dwell on details of technology for planning and designing water supply components such as reservoirs, treatment facilities, and distribution systems. The goal is to identify criteria that are important in those activities, with special attention to the rationale underlying decision-making processes, and to describe the types of elements frequently found in municipal water systems.

SOME PRELIMINARY CONSIDERATIONS

The fraction of municipal water used to satisfy human metabolic requirements is somewhat less than 1 percent. That notwithstanding, drinking water needs are vitally important, and its quality is of utmost concern in protecting the public health. Therefore, the requirements for drinking largely dictate the quality standards for municipal water and the characteristics of treatment and other facilities which insure that they are met reliably.

Reliability is viewed as nonnegotiable in municipal water supply, with respect to both the system's ability to supply water in required quantities and to insure that its quality is satisfactory when used by the consumer. It is a guiding principle behind the conservative approaches, thought processes, and actions of planners, designers, and operators of municipal systems. All these activities, and regulatory control of them, include the conscious realization that mechanical failures or negligence, even for short periods, can cause the illness and death of many people. It is not a responsibility to be taken lightly, and it isn't. If one is to understand the motivations and methods of municipal water supply practitioners, that psychological mind set must be recognized.

Quality requirements for municipal water are specified in regulations of the U.S. Environmental Protection Agency (U.S. EPA, 1976a, 1976b, 1979a, 1979b, 1980), sometimes with modifications and additions by state health agencies to meet problems of special concern to them. Some of the required and recommended limitations on concentrations of constituents specified in EPA drinking water standards are summarized in Table 11.1 "Primary" regulations relate to the constituents of water that may affect the health of consumers and are enforceable by the EPA or the states as mandatory regulations. "Secondary" regulations establish quality guidelines that deal principally with the esthetic qualities of drinking water and may not affect the health of consumers.

More exacting water quality goals have been specified by the American Water Works Association (AWWA, 1971) to serve as guidelines for practitioners in the field. They describe characteristics that are judged to be attainable in well-designed and properly operated municipal systems. Additional standards also have been developed by the AWWA and other organizations to regulate the design, construction, and use of specific elements in water treatment and distribution systems, with the goal of enhancing quality and reliability of service.

ROLES OF WATER IN MUNICIPALITIES

Household uses of water have been discussed in Chapter 5 (see Table 5.2). Although they represent a major and vital fraction of the total municipal supply, several other uses also are important to community welfare and usually account for most of the water demand.

Table 11.2 summarizes the broad categories of water uses in U.S. municipalities and typical ranges of amounts for each, but it should be noted that they vary widely among different communities. The residential (household) use var-

TABLE 11.1
Limitations on Specific Constituents from EPA Water Quality Regulations

Constituent or Characteristic	Maximum Level (mg/liter, unless otherwise noted)
Primary Regulations (related to health)	
Arsenic	0.05
Barium	1
Cadmium	0.010
Chromium	0.05
Lead	0.05
Mercury	0.002
Nitrate, as N	10
Selenium	0.01
Silver	0.05
Fluoride (<12°C)	2.4
(12.1–14.6°C)	2.2
(14.7–17.6°C)	2.0
(17.7–21.4°C)	1.8
(21.5–26.2°C)	1.6
(26.3–32.5°C)	1.4
Trihalomethanes	0.1
Endrin	0.0002
Lindane	0.004
Methoxychlor	0.1
Toxaphene	0.005
Chlorophenoxys, 2,4-D	0.1
Chlorophenoxys, 2, 4, 5-TP Silvex	0.01
Turbidity (monthly average units)	1
Coliform bacteria (average number per 100 ml)	1
Radium 226 and radium 228 (pCi/liter)[a]	5
Gross alpha particle activity (pCi/liter)	15
Beta particle and photon activity (mR/yr)[b]	4
Secondary Regulations (primarily esthetic effects)	
Chloride	250
Color (units)	15
Copper	1
Corrosivity	Noncorrosive
Foaming agents	0.5
Iron	0.3
Manganese	0.05
Odor (threshold odor number)	3
pH (units)	6.5–8.5
Sulfate	250
Total dissolved solids (TDS)	500
Zinc	5

[a]pCi = picocurie.
[b]mR = millirem.

TABLE 11.2
Municipal Water Uses

Use	Gallons per Person per Day		Percent of Total
	Range	Average	
Residential	25–100	65	15–60
Commercial	15–50	25	10–30
Industrial	25–100	65	15–60
Public	5–30	10	5–20
Unaccounted	5–40	15	5–25
Total	50–400	180	100

Note. Liters/person/day = gal/person/day × 3.785.

ies from a small fraction of the municipal total, in communities with heavy industrial demands, for example, to over one-half in small towns that are predominantly residential. On the average, residential use is about one-third of the municipal demand.

Commercial uses refer to the water needs of business establishments that provide services in most communities, including restaurants, hotels, laundries, and stores, and average about 15 percent of the municipal total. Industrial demands within municipalities average about one-third of the water used but vary widely—from zero in predominantly residential areas to very large proportions in communities with extensive "wet" manufacturing operations. Public uses include municipal buildings, street washing, fire fighting, and flushing waterlines.

The "unaccounted" water is especially important. The total of water uses recorded by individual meters in a community never equals the amount actually pumped into the system. Part of the discrepancy can be attributed to inaccuracies in meters, which fail to record all flow passing through them after they have been in service long enough to suffer wear. Another major cause for the difference is leakage from the distribution system, which can vary from less than 5 percent to over 40 percent, depending on the construction and maintenance of the piping network. The amount of leakage can be very important in determining system economy and whether it is adequate in capacity. In fact, the most economical solution for municipal water shortage or inadequate capacity often is merely to locate and correct leaks in the distribution system.

MUNICIPAL WATER USES AND VARIABILITY

The average U.S. municipal water use was 183 gal/person/day (693 liters/person/day) in 1980 (Solley et al., 1983) and varied among the states from about 100 to over 300 gal/person/day (380 to 1130 liters/person/day). Average uses in individual municipalities varied from less than 50 to over 400 gal/person/day (190 to 1500 liters/person/day). In planning and designing a water system, however, community average is not the most important flow characteristic because capacities of its elements must be adequate to cope with variations, expecially maximum demands.

Most municipalities exhibit seasonal variations. For example, many use more water in the summer than in the winter because of demands for air conditioning, lawn watering, swimming pool operation, and other summertime activities. On the other hand, an unmetered water system in a northern climate could have its maximum demand in the winter because customers may leave water taps flowing slightly to prevent freezing on cold nights. Many resort communities have radical seasonal changes in population. Some municipal systems supply canneries or other seasonal industries that can cause major impacts on water use patterns.

In some communities, significant changes in water use may occur throughout the week, depending on scheduling of business and industrial operations, public uses, or even laundry and other household activities. Weekly variations in flow, however, usually are minor when compared to seasonal and daily ones.

In a typical community, the average demand for the maximum day often is 1½ to 2½ times the average daily flow during the year.

Daily patterns vary widely, but Figure 11.1 illustrates the type of demand curve (hydrograph) sometimes encountered in municipal systems without major industrial effects or other unusual requirements. The flow in this exam-

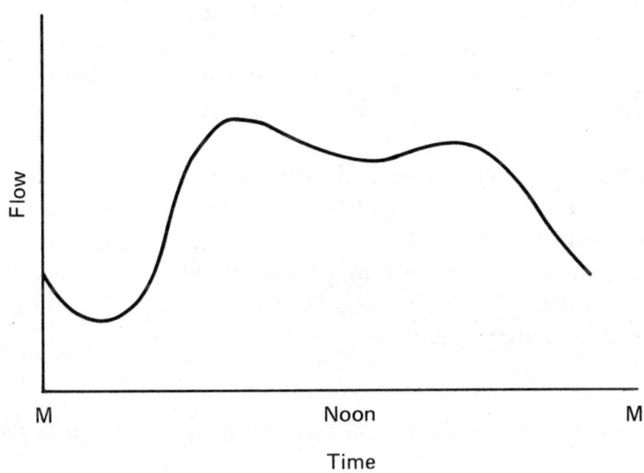

FIGURE 11.1
Illustration of a municipal water demand pattern.

ple declines to a minimum during the early morning hours; the minimum flow, incidentally, may be influenced greatly by system leakage. The demand increases sharply when the population begins the day's activities, and a peak often is reached during the morning. That may be followed by a slight decline in demand in the middle of the day and a rise to another peak in the late afternoon, caused by supper and other home activities. Thereafter, the flow declines gradually to the early morning minimum.

The ratio of peak demand during a day to the average for that day varies over a wide range, depending on many factors such as industrial use patterns, size of the community, and standard of living (Gehm and Bregman, 1976). Values for the peak may range from two to eight times the average flow.

Trends in average water uses have been upward for many years in most communities—because of both increasing urban population and higher use of water per person. Also, the ratio of peak to average demand has been increasing. This is of special importance because of its impact on the costs of distribution systems, which must be able to deliver peak flows without excessive pressure drops. Much of the increases in demand and flow variability can be attributed to a rising standard of living and to the growth in water uses for "luxury" purposes, such as lawn watering, air conditioning, swimming pools, and many types of appliances.

The cost structures for fire insurance in communities depend heavily on water system reliability and adequacy of flows and pressures at fire hydrants. The flows necessary for a favorable rating vary from 500 to over 10,000 gal/min (32 to 630 liters/sec), depending on the size and density of population, types and sizes of industries and businesses, and several other factors. Although the rates of fire flows are high when needed, the total amount of water used for that purpose is only a small fraction of municipal use and has little influence on the selection of water sources or design of treatment plants. The principal impacts on system design are in sizing the distribution network and storage for finished water.

The fire flow requirement actually causes it to be the factor of primary importance in designing many small distribution systems because the hydraulic capacity necessary to satisfy that need often is much higher than for all other municipal demands combined. In larger communities, flows needed to satisfy other municipal needs are high enough to make the fire fighting requirements less important in system design.

OUTLINE OF ELEMENTS IN MUNICIPAL WATER SYSTEMS

Water supplies differ widely in components used to produce the required flow and quality, making it impossible to define one system with characteristics that would apply under the varied circumstances found in practice. Nevertheless, it is helpful here to outline a reasonable group of elements often found in municipal water supplies to serve as a basis for discussing factors involved in decisions about the types of units used to meet planning and design goals and methods employed in selecting among available alternatives.

Clearly, there must be a source of water that is satisfactory and reliable both in quantity and quality. Where raw water quality is not satisfactory for direct delivery to consumer, corrections are required, and this is accomplished by appropriate treatment. Transmission lines and pumps must be provided to move the flow from the source to the treatment plant and, subsequently, from there to the distribution system in the community.

The function of the distribution system is to deliver treated water to each consumer when desired, in needed quantities, at acceptable pressures, and with unimpaired quality. Because of the extensive network of pipes and appurtenances necessary to meet those goals, the distribution system usually represents a major part of the total capital investment in the water system. A vital component is elevated or ground-level storage for finished water. That must be carefully designed and located to enhance water pressures throughout the network, to provide a reserve for use in emergencies, and to realize savings in capital and operating costs for the system.

Some water supplies include all of these elements, sometimes in rather elaborate forms, but others may be fairly simple. Each system must be tailored to fit the needs of the community, using resources available in that locality. The elements just outlined are discussed in more detail in the following sections.

SELECTING THE WATER SOURCE

Selection of the source is a crucially important decision by the engineers who plan and design a water system. That is where it all begins; the source determines the basic ability of the system to meet community needs, the types of treatment that may be required and risks to health, or other problems if something goes wrong. This cannot be taken lightly because the welfare of the entire community—often many thousands of people and organizations—is at stake.

Three factors provide the bases for decisions in source selection: quantity, quality, and cost. The source must be capable of supplying the amount of water needed for an extended period, perhaps 25 to 50 years, even during droughts and other unfavorable situations. Its quality must meet requirements established for municipal drinking water supplies, usually after appropriate treatment. Finally, where two or more sources are satisfactory both in quantity and quality, the enginer turns to cost for help in reaching a decision on selection.

Processes at practical treatment plants cannot remove all impurities from water before delivery to the public. That, of course, means that drinking water derived from sources that include complex and ill-defined mixtures of chemicals will continue to contain some of them even after treatment. Often those constituents are uncertain in identity and unknown with respect to potential effects on human health. Although the risk that treated water could contain constituents harmful to health may be low, it is greater when the supply is taken from a source that receives a wide variety of unknown pollutants in land runoff or wastewaters.

This explains and justifies the rationale behind a cardinal rule in water supply engineering—drinking water must always be taken from the best-quality source that can be developed feasibly. That is more than just tradition or a reflection of conservative waterworks practice. It is a requirement stated clearly in the regulations of federal and state water agencies. Anything less is poor professional practice and open to severe criticism because it could result in unnecessary risks to the public health.

EXAMPLES OF WATER SOURCES

Precipitation

Although unusual, there are instances in which communities have resorted to direct collection of precipitation as a water supply source. For example, a municipality in the Virgin Islands, where freshwater sources are exceedingly limited, has paved a large area for direct catchment of rainfall, which is diverted into storage basins until needed. This approach also has been used on other islands and in some areas of South America.

In coastal areas of the eastern United States, adequate freshwater supplies sometimes are very difficult to obtain because of saltwater intrusion into wells. Some homes are equipped with storage tanks to collect rainwater from the roof. That may be used for washing, for example, while bottled water satisfies drinking needs and brackish well water could serve to flush toilets. Roof catchment also is employed elsewhere for individual homes where surface water and groundwater supplies are not readily available.

In the Arctic, it is not uncommon to obtain fresh water by melting snow or ice—another form of precipitation. Periodically, studies have been undertaken to evaluate the feasibility of towing icebergs from Arctic or Antarctic regions to water-short areas and subsequently melting them to produce fresh water. One with widely publicized interest in this approach was Prince Faisal, of Saudi Arabia, who sponsored studies and meetings on the subject from 1977 to 1980 (Weeks and Mellor, 1979).

Surface Freshwater Sources

This type of supply includes flows withdrawn from lakes and rivers and is the source of about two-thirds of the U.S. freshwater withdrawal (see Table 5.7).

River supplies may be considered in two broad categories: those not requiring storage ("run-of-the-river"), and those that do require storage to meet system demands. The first category refers to situations in which flow that is available in the river always is greater than the system demand for water. This allows the installation of a water intake and pumping directly from the flowing stream to the treatment and distribution systems. This situation prevails for most municipalities along major streams, such as the Ohio and Mississippi rivers, but also applies to small communities near many lesser streams.

When available stream flow sometimes is less than that required by the community but is adequate when considered over extended periods, storage may be provided to insure adequate and reliable water "yield" for the system. Under this approach, some of the excess water during times of high river flow is stored in a reservoir. Later, when system demand exceeds flow that can be withdrawn from the stream, the needed water is taken from the stored inventory.

One way to provide storage is to construct a dam that forms an impounding reservoir through which the stream flows (an "on-stream" impoundment). A different approach, sometimes employed when good on-stream sites are not available, is to build a separate off-stream reservoir. During wet seasons, some of the excess river flow is pumped into the storage facility and later, when river flow is inadequate to meet demand, the stored water is withdrawn from the reservoir or released back into the river to meet needs of downstream users. Both approaches have been widely used by communities to meet their water needs.

Some of the better-known examples of lake supplies include those for Chicago, Cleveland, and Duluth on the Great Lakes. However, smaller natural lakes also are used to supply water for many communities. When available, they offer the advantage of storing large quantities of water to meet demands in drought years at little cost.

Groundwater

Although surface sources supply most of the water used in the United States, groundwater is the source for a larger number of supplies. That can be attributed to the many individual homes and small communities that use groundwater from wells and springs.

In many instances, wells simply penetrate into the aquifer (see Chapter 4) and water below the groundwater table flows to them through permeable soil to replace that removed by pumping. Some wells are located near surface streams or lakes to facilitate the replenishment ("recharge") of the water withdrawn. Sometimes special underground structures, called "infiltration galleries," may be installed near streams to allow the use of water from the surface source after natural filtration through the soil.

Some wastewater disposal facilities employ "spreading" by constructing surface ponds that allow wastewater to percolate down through the soil for treatment and reclamation later as groundwater. This approach has been used to produce water for various purposes in Long Island, California, The Netherlands, Israel, and many other locations. When feasible, it offers the dual benefit of wastewater treatment and disposal combined with enhanced water supply. Usually water from this type of reclamation system is used for irrigation and industrial or other nonpotable purposes, thereby easing the demands on high-quality supplies that are more suitable for satisfying drinking water needs.

Saline Waters

The production of fresh water from saline sources now is well know and has been proven practical in many instances. A wide array of technology has been developed over the past generation for this purpose, including processes such as distillation (several versions), reverse osmosis, electrodialysis, and freezing. Saline waters that are available as supplies include oceans, brackish wells near the sea, saline wells in the interior, and some streams with a high content of dissolved solids.

With technology now readily available, the main deterrent to saline water conversion is economic. Unfortunately, basic thermodynamics and the types of processes that have proven feasible result in energy-intensive systems with high capital and operating costs. Decisions about using saline sources must carefully consider the costs of competitive methods for supplying the needed water. In practice, saline waters have been adopted principally where adequate freshwater sources either are not available at all (for example, on islands) or where they are located at distances requiring expensive transportation through excessively long pipelines or by other means.

Wastewater Reuse

Wastewater has been reused for many years. This is evident upon considering that any community taking water from a stream below wastewater discharges from another community or industry is using a mixture composed partly of wastewater. Accordingly, in the sense of the successive use in lakes and along streams, wastewater reuse is common.

During the past generation, the possibilities for "direct reuse" of wastewater in municipalities have received much attention and publicity. Technology now is available that makes it possible to use the effluent from a sewage treatment plant as a source of supply for municipal water; as one example, groundwater recovered in spreading operations, mentioned earlier, could be used for that purpose. The controversial aspect of direct reuse involves the judgment as to whether that actually is the wisest way to marshal our water resources. It has been debated heatedly for several years and probably will continue to be controversial for many more.

The heart of the debate lies in the conservative posture of water supply personnel and regulations, outlined earlier, which stipulate that drinking water should be taken from the best-quality source that can be developed feasibly. Viewed in the context of minimizing risk of transmitting waterborne diseases, the best-quality source clearly cannot be the sewage from a municipality. That is why many rank sewage at or near the bottom of priority lists for desirable municipal supplies.

A different approach has received much wider acceptance—the reuse of treated wastewater to supply uses in which there is less concern about transmitting waterborne diseases. For example, the past few years have seen substantial growth in the use of treated municipal sewage for irrigating nonedible crops. That accomplishes three desirable goals: (1) treatment and disposal of the wastewater, (2) satisfying a need for irrigation water, and (3) reducing the use of higher-quality water for irrigation, making it available to satisfy potable water needs.

This approach also extends into the reuse of municipal wastewaters by industries for cooling or other purposes. For example, much of the effluent from a municipal treatment plant in Baltimore is given further treatment and used for cooling water in a large steel mill near that city. Many similar applications could be cited for other locations, especially in the arid West. One even can visualize increased multiple use of water, for example, use of the best-quality source as a municipal supply, followed by industrial cooling with treated sewage from that municipality and subsequent irrigation with the same water.

WATER QUALITY AS RELATED TO SOURCE

Several aspects of resource characteristics and variability in the quality of water from some of the preceding sources have been discussed in Chapters 4,7, and 8. Because most water supplies are derived from surface water and groundwater sources, selected differences in quality often noted between the two are worth summarizing here.

Stream flows may be predominantly surface runoff during periods of rainfall but, at other times, they are composed of groundwater to which constituents have been added, perhaps in wastewater discharges. As might be expected, the quality characteristics of streams vary widely from time to time. For example, turbidity usually ranges from very high levels during storms to much lower values when the flow is low and relatively quiescent, composed mostly of groundwater inputs. Dissolved solids content varies with stream flow, as outlined in Chapter 7. Also, temperatures may vary widely from season to season and even day to day because stream flow is exposed to the atmosphere and sunlight. Furthermore, streams are subject to the addition of potential pollutants, including many types of chemicals and microorganisms, in land runoff and municipal and industrial wastewaters.

Because of those influences, a surface water seldom is found that is acceptable for use as a municipal supply without treatment. The quality defects usually include excessive turbidity, color, and microbiological content. Perhaps of greatest importance, the quality reliability of a surface supply usually is dubious because of rapid changes that may follow storms, accidental discharges, and contacts with humans and animals on the contributing watershed. Almost all surface supplies now are assumed to require treatment at least for the removal of turbidity, color, and microorganisms. A type of treatment plant often used is described later in this chapter.

In contrast, groundwater is relatively consistent in quality because of underground storage and long travel time before reaching the point of use. Its temperature is much more uniform, turbidity and color are low because of filtration through the soil, and microorganisms are often absent because of long storage and filtration. The extended contact with various types of soils often causes higher dissolved solids in groundwaters than in surface flows. On the other hand, contact with solids also sometimes can result in exchange reactions that remove some types of potential pollutants during flow between sources of pollution and points of use.

It is common to be able to use groundwater supplies with little or no treatment to correct quality defects. However, sometimes the character of the aquifer may produce short times of travel of water through it, in limestone regions, for example. That, or improper construction of wells, may negate many otherwise positive aspects of groundwater supplies.

Perhaps the most negative aspect of groundwater supply is the sparse information usually available on quantity and quality of the water in a specific locality.

WATER TREATMENT

The purpose of water treatment is to correct deficiencies in quality. Those include characteristics that could impact unfavorably on human health, are objectionable esthetically, could cause corrosion or scale formation in pipelines, or result in other problems in household, commercial, or industrial uses of the water. The goal in municipal water treatment is both to make the water safe for human consumption and to produce a finished product that is attractive and pleasing to the consumer.

The first step in determining the type and extent of necessary water treatment usually is to assess the quality of the raw water. The specific analyses conducted include those for quality characteristics and constituents that are specified in the drinking water standards and define the minimum acceptable finished water quality for that purpose. Additional analyses may be conducted for other constituents known or suspected to be important in various uses to which the water will be put. Comparison of raw water characteristics with those required for the finished water indicates the areas in which deficiencies exist that must be corrected. When no deficiencies exist or are likely to occur, no treatment is necessary. When deficiencies are found, appropriate types of treatment must be undertaken to correct them.

For each problem in water quality, there are treatment processes capable of correcting it. Some examples are listed in Table 11.3, and others are described in a manual published by the U.S. Environmental Protection Agency (1977) and in various books on water treatment (AWWA, 1971; Fair et al., 1968; Gehm and Bregman, 1976).

Selection of the specific treatment system to be used involves organizing a combination of processes appropriate for removing the undesirable constituents to produce a coherent whole that accomplishes the required overall performance reliably and economically. The selection can be highly complex because of interactions among the many constituents of water and the effects of those interactions on various combinations of the processes.

To save time and effort and minimize misunderstandings, regulatory agencies often establish "design standards" to shortcut some of the decision processes. They may specify, for example, certain minimum types of treatment systems that are recommended or required in various practical situations. Sometimes, they also may identify design parameters for sizing specific treatment elements. For example, one design standard that is commonly used requires that treatment of surface water must at least include provisions for removing turbidity, color, and microorganisms; the specific processes often are coagulation, settling, filtration, and chlorination. Other design standards could specify permissible detention times and hydraulic loadings for each process ele-

TABLE 11.3
Examples of Processes for Removing Contaminants from Water

Contaminant	Treatment Methods
Arsenic	
As^{+3}	Oxidation to AS^{+5} is required before treatment
As^{+5}	Ferric sulfate coagulation, pH 6–8
	Alum coagulation , pH 6–7
	Excess lime softening
Barium	Lime softening, pH 10–11
	Ion exchange
Cadmium	Ferric sulfate coagulation, pH >8
	Lime or excess lime softening
Chromium	
Cr^{+3}	Ferric sulfate coagulation, pH 6–9
	Alum coagulation, pH 7–9
	Excess lime softening
Cr^{6}	Reduction to Cr^{+3} is required
	Ferrous sulfate coagulation, pH 7–9.5
Fluoride	Ion exchange with activated alumina or bone char media.
Lead	Ferric sulfate coagulation, pH 6–9
	Alum coagulation, pH 6–9
	Lime or excess Lime softening
Mercury	
Inorganic	Ferric sulfate coagulation, pH 7–8
Organic	Granular activated carbon
Nitrate	Ion exchange
Selenium	
Se^{+4}	Ferric sulfate coagulation, pH 6–7
	Ion exchange
	Reverse osmosis
Se^{+6}	Ion exchange
	Reverse osmosis
Silver	Ferric sulfate coagulation, pH 7–9
	Alum coagulation, pH 6–8
	Lime or excess lime softening

ment. These and other standards in waterworks practice, including some of the advantages and disadvantages of each type, will be discussed in Chapter 14.

EXAMPLES OF TYPICAL WATER TREATMENT SYSTEMS

Figure 11.2 is a flow sheet for a "typical" plant, similar to many used for treating surface waters without special quality problems, from rivers and lakes. The principal raw water characteristics corrected in this type of facility are excessive turbidity and color, unacceptable numbers of microorganisms and, perhaps, problems with tastes and odors, iron, and manganese.

Chemical Additions

The first step in treatment could be chlorination for any of several purposes. Addition of the disinfectant at this point provides a maximum time of contact between the chemical and organisms, assuring efficient bactericidal action. Chlorination at this point also sometimes enhances the subsequent removal of tastes and odors, iron, manganese, and color by oxidizing some organics in the raw water. Furthermore, the addition of chlorine inhibits growth of algae and other objectionable organisms during the several hours required for water to pass through the treatment plant.

On the other hand, recent investigations have disclosed that the early addition of chlorine can maximize the production of trihalomethanes (THMs), discussed in Chapter 9. For that reason, there is movement now to postpone this step to a later point in the treatment, after most of the humics, which are precursors for the formation of THMs, have been removed (Singer, 1981). An alternative could be to use a different oxidant—say ozone—at the beginning of the process because THMs are not formed by that chemical. However, there are some reservations about that, as well, because of uncertainty about the possibility of forming other objectionable chemicals through reactions between ozone and water constituents.

The plant shown in Figure 11.2 is equipped to remove organics by adding activated carbon at the beginning of the treatment process. This approach often is quite effective in removing, through adsorption, many chemicals that can cause tastes and odors. Subsequently, the carbon and its adsorbed organics are removed along with other suspended matter during coagulation, sedimentation, and filtration.

Alum (Aluminum sulfate) also may be added here as the first step in the "coagulation" process. Alum reacts with alkalinity in the water to form a gelatinous precipatate that is effective in promoting coagulation (agglomeration) of small suspended particles into larger ones that separate more rapidly in the

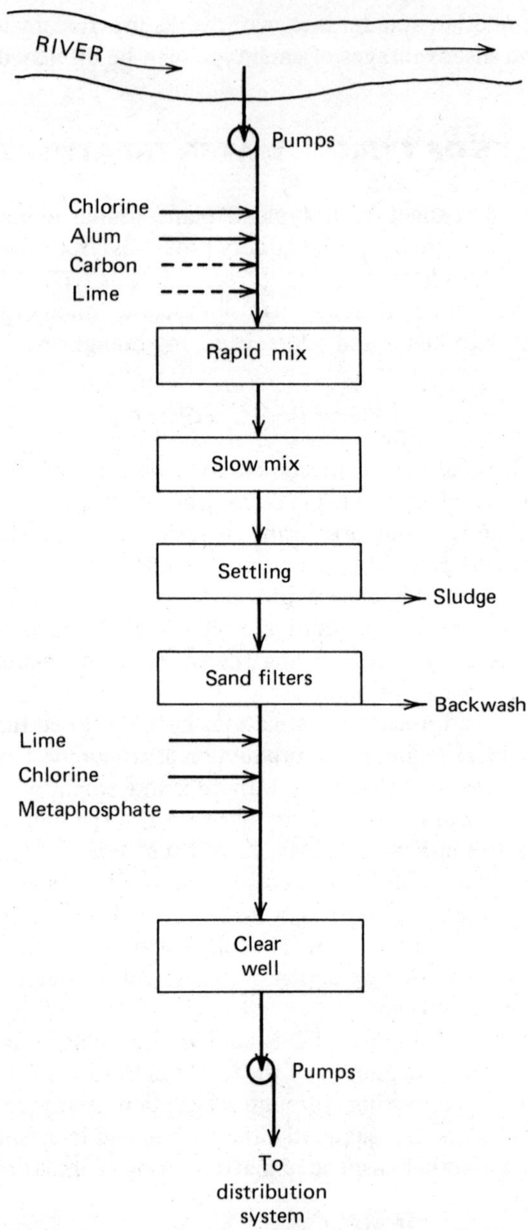

FIGURE 11.2
Flow sheet showing typical municipal treatment facilities for a surface water supply.

settling basin. Also, alum changes other characteristics of particles and enhances their removal during sand filtration. Other chemicals often used to accomplish the same purpose include iron compounds, especially ferric sulfate or ferric chloride. Because alum or iron compounds react with alkalinity, the water may be rendered acid in character, making it necessary to add lime at this point for pH adjustment.

Coagulation

The rapid mix step is a short, violent agitation that distributes added chemicals throughout the water. This may be acomplished in 1 to 5 minutes in a small tank equipped with mechanical mixers or by using mixers installed in the pipeline delivering raw water to the plant.

After the chemicals have been thoroughly mixed into the flow, a longer (15 to 30 minutes, usually) period of gentle agitation through slow mixing is provided. This allows the chemical reactions to proceed to completion and insures that precipitate formed by the alum contacts suspended matter so that the two adhere to produce larger particle sizes. The slow mix also prevents premature settling of the floc before entering the settling basin. This portion of the process often is referred to as "flocculation," and the agglomerated particles that are formed are called "floc."

The overall process of chemical addition, rapid mix, and flocculation is referred to as "coagulation." Other chemicals, called "coagulant aids," also may be added during the rapid mix or flocculation to help form larger particles or to increase floc strength and prevent it from breaking up before settling can occur.

Settling

The settling tanks are designed to provide minimum turbulence conditions for, perhaps, 2 to 6 hours during which particles formed in coagulation can separate from the water by settling. Sludge accumulations at the bottom of the tank may be removed periodically by emptying and flushing the basin, or they may be removed continuously by mechanical scrapers if the rate of accumulation is rapid enough to justify the additional cost.

In a properly operated plant, effluent from the settling tank is of high quality. Most turbidity and microorganisms should have been removed, leaving only a few particles suspended in the liquid and providing quality that is greatly improved but still unacceptable as a drinking water. The next step is to "polish" the water by removing that relatively small amount of residual suspended matter.

Filtration

Filtration involves passing the water through a bed of carefully selected and graded sand, usually in a downward direction. The remaining suspended particles are removed and accumulate in interstices between the sand grains until they have been largely filled. Then the filter must be cleaned to renew its capacity for storing floc particles. This is done by reversing the flow of water through the bed, expanding the sand, and releasing the accumulations of solids into the "washwater," which is discharged as a process waste. Subsequently, filtration may be resumed through the clean sand.

Posttreatments

The addition of lime after filtration in Figure 11.2 reduces the acidity that results from adding alum and may be done here, or at the influent to the plant, or both, depending on process needs. This reduces the corrosiveness of finished water to the distribution system and to plumbing in homes and businesses. The metaphosphate added in this example is a corrosion inhibitor that is used for further protection of the system and to reduce calcium carbonate scale formation in the pipes.

Chlorine is added as needed for disinfection. When prechlorination is practiced, as discussed earlier, further additions might not be necessary. In fact, sometimes high dosages at the plant influent actually may make it necessary to add sulfur dioxide or other chemicals after filtration to reduce the chlorine residual to a level acceptable in drinking water. Because of present concern about formation of THMs, postchlorination now is becoming the favored point of addition. The "clear well" is for storage of finished water and provides the contact time needed to insure the killing of microorganisms remaining after coagulation, settling, and filtration.

Regulatory agencies commonly require maintenance of a chlorine residual throughout the distribution system at all times as a public health preventive measure. The intent is to reduce the potential for health impacts in the event of microbiological contamination of the treated water through cross connections with polluted supplies or faulty plumbing. This requirement has been controversial from time to time, but it remains conventional wisdom among conservative professionals in the water supply field. In fact, current drinking water standards even allow the substitution of chlorine residual measurements in the distribution system for some of the coliform tests that otherwise would be required, as outlined in Chapter 9.

The clear well is a storage tank from which pumps can withdraw finished water for delivery into the distribution system. It also provides time for the completion of disinfection reactions and furnishes an emergency reserve of treated water in case of failure of pumps or other treatment plant elements.

A Groundwater Supply

Figure 11.3 illustrates a very different type of water system, one frequently encountered in small communities using good wells. In this example, water removed from the well is low in turbidity and color, virtually free of microorganisms, and has no other significant defects or potential quality problems, making processes for removing those constituents unnecessary. No corrective treatment is indicated, and the only one provided is the preventive measure of chlorination for disinfection and to maintain a chlorine residual in the distribution system.

Additional Observations about Treatment

The water treatment systems just outlined are only for illustrative purposes. Elements actually included in a practical system depend on specific quality defects that must be corrected and may result in a flow sheet very different from either of those discussed. Some of the other processes that may be used

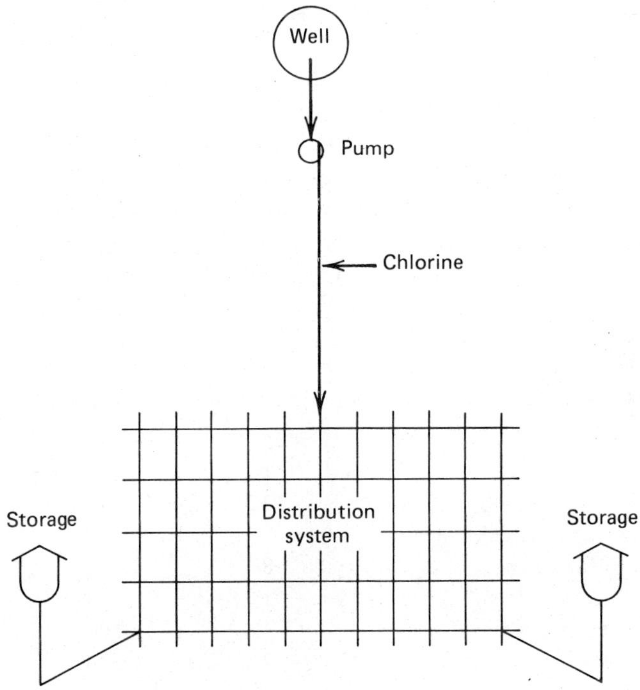

FIGURE 11.3
Flow sheet for a municipal system using a good quality groundwater source.

are described in texts and other publications (AWWA, 1971; U.S. EPA, 1977; Fair et al., 1968; Clark et al., 1977; Gehm and Bregman, 1976).

Table 11.4 summarizes the types of water sources used by the 100 largest cities in the United States (Durfor and Becker, 1964) and types of treatment used in their supplies. They are divided into systems having only surface sources (66), those using only groundwaters (20), and those employing both surface water and groundwater sources (14). Ninety-eight of the 100 cities

TABLE 11.4
Public Water Supplies of the 100 Largest Cities in the United States

Source and Treatment	Number of Cities	Population Served: Thousands	Population Served: % of Total
Surface Water	66		
Chlorination	66	39,939	65.9
Sedimentation and coagulation	54	27,772	45.8
Rapid sand filtration	51	26,511	43.8
Slow sand filtration	7	2,536	4.2
Pressure filtration	2	356	0.6
Iron removal	4	1,475	2.4
Softening, lime	10	4,649	7.7
Softening, lime-soda ash	9	3,359	6.0
Groundwater	20		
No treatment	1	150	0.2
Chlorination	19	5,565	9.2
Sedimentation and Coagulation	7	2,147	3.5
Rapid sand filtration	7	2,970	4.9
Iron and manganese removal	5	1,474	2.4
Softening, lime	3	1,055	1.7
Softening, lime-soda ash	1	320	0.5
Mixed Surface Water and Groundwater	14		
No treatment	1	72	0.1
Chlorination	13	14,015	23.1
Sedimentation and coagulation	7	8,770	14.5
Rapid sand filtration	8	1,921	3.2
Slow sand filtration	1	33	0.1
Iron and manganese removal	1	500	0.8
Softening, lime	2	340	0.6
Softening, lime-soda ash	1	259	0.4
Softening, cation exchange	2	698	1.2

Source: Durfor and Becker (1964).

used chlorination. Other methods of treatment varied substantially, depending on the types of water sources and needs for quality corrections. Table 11.5 summarizes the finished water quality characteristics for those systems, including minimum, median, and maximum concentrations for each constituent.

WATER DISTRIBUTION SYSTEMS

The water distribution system is a network of pipes and appurtenances, as illustrated in Figure 11.3, that delivers finished water to the consumers. Constraints under which the system must be planned, designed, and operated require delivery of water when and where it is wanted by the consumer, with adequate residual pressure, and without impairment of quality.

Water quality must be satisfactory upon delivery to the consumer, not merely when it leaves the treatment plant. One of the most important distribution system characteristics for preventing biological or other contamination of water is maintenance of internal pressure constantly at all points. All dis-

TABLE 11.5
Finished Quality in Public Water Supplies of 100 Largest U.S. Cities

	(mg/liter, unless otherwise noted)		
Constituent or Property	**Maximum**	**Median**	**Minimum**
Silica (Sio_2)	72	7.1	0.0
Iron (Fe)	1.30	0.02	0.00
Manganese (Mn)	2.50	0.00	0.00
Calcium (Ca)	145	26	0.0
Magnesium (Mg)	120	6.25	0.0
Sodium (Na)	198	12	1.1
Potassium (K)	30	1.6	0.0
Bicarbonate (HCO_3)	380	46	0
Carbonate (CO_3)	26	0	0
Sulfate (SO_4)	572	26	0.0
Chloride (Cl)	540	13	0.0
Fluoride (F)	7.0	0.4	0.0
Nitrate (NO_3)	23	0.7	0.0
Dissolved solids	1580	186	22
Hardness as $CaCo_3$	738	90	0
pH (units)	10.5	7.5	5.0
Color (units)	24	2	0
Turbidity (units)	13	0	0

Source: Durfor and Becker (1964).

tribution systems have leaks, and adequate pressure insures that flow always is outward from the pipes. Low or zero pressures at any points could permit polluted water outside the pipes to enter them at leaks, contaminating the drinking water before its delivery to consumers. That is why drinking water standards require constant maintenance of pressure in distribution systems.

Reliability of the system, including constant maintenance of pressure, requires careful attention to the interconnection of pipes to provide closed loops. That allows the flow to approach any point in the network from more than one direction and helps insure a continued water supply and the maintenance of pressure everywhere in the event of pipeline failures. Proper placement of valves permits sections of pipes suffering failures to be isolated for repairs without losing service in the rest of the system.

Also, the piping layout must include fire hydrants at reasonable intervals and provide adequate flows and pressures to serve them in accordance with the fire-fighting requirements referred to earlier. Otherwise, fire insurance rates for the community may be substantially higher.

TREATED WATER STORAGE

Elevated or ground-level storage for finished water fulfills several roles in the distribution system, the first of which is to provide a method for reducing the capital cost of the treatment plant. Without any finished water storage, the plant would have to be large enough to satisfy peak demand in the com-

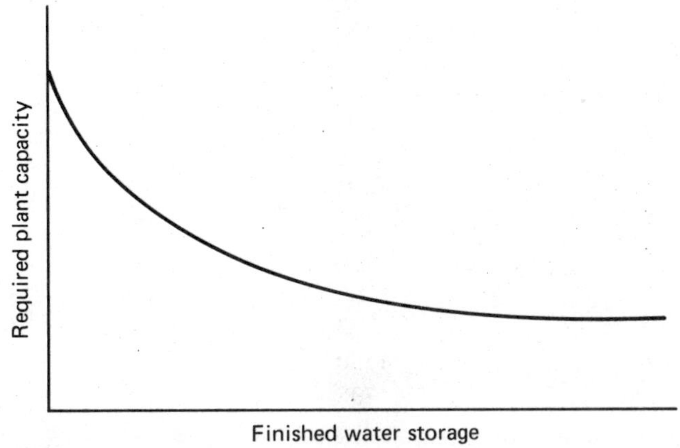

FIGURE 11.4
Effect of treated water storage on required size of the treatment plant.

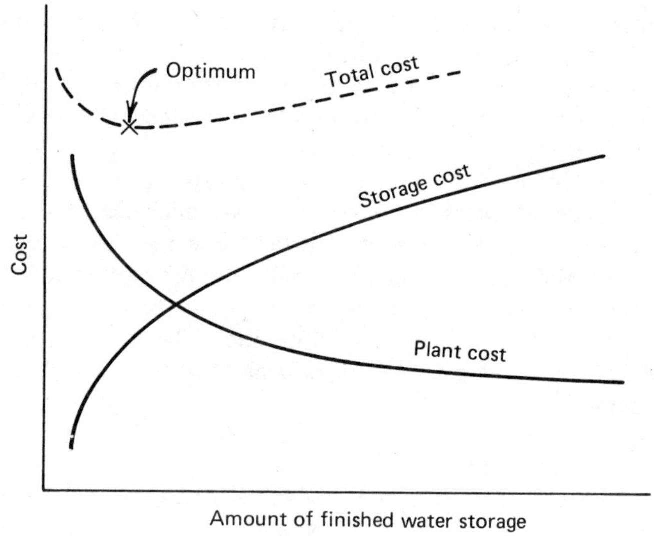

FIGURE 11.5
Optimization of the system cost, considering both storage and treatment plant.

munity, which is several times the average flow. The provision of storage, on the other hand, makes it possible for water stored in the tanks when municipal uses are lower than plant capacity to be returned to the distribution system later to meet peak demands. This allows the construction and operation of a smaller treatment plant without compromising system integrity or reliability. As the amount of finished water storage is increased, the size of the treatment facility can be reduced, as illustrated by Figure 11.4.

Savings in the capital cost of the treatment plant can offset much or all of the expense for storage. Actual storage needs are determined through an economic optimization study in which the volume of storage is adjusted to produce a total cost for both storage and treatment plant that is minimum, as illustrated by Figure 11.5.

Another purpose of finished water storage is to insure an adequate supply of water to fight fires. The Insurance Services Office has standards for this feature of the system that are in addition to the specifications for flow rates mentioned earlier. The amount of storage required to obtain favorable insurance rates depends on the size and characteristics of the community and usually must provide for sustained fire flow over a period of 2 to 10 hr.

Some of the finished water storage also meets community needs in emergencies. For example, the failure of a transmission main or pumping equipment between the treatment plant and community could could cause a major water supply failure unless sufficient storage is available and located where it does not require the use of the same pipelines and pumps to deliver water to the distribution system. For that reason, preferred locations for some of the emergency storage are in areas well removed from the incoming supply mains, as shown in Figure 11.3.

Another factor that is considered in locating storage is the maintenance of uniform and adequate pressures through the system during peak demands. For example, the strategic location of elevated tanks near the areas of highest demands can reduce pressure losses through the piping system or permit the installation of smaller pipes without producing unacceptable pressure drops. Also the location of storage on the opposite side of the community from incoming water mains often is helpful in equalizing pressures throughout the distribution system.

SOME HEALTH IMPLICATIONS IN DISTRIBUTION AND STORAGE

The importance of insuring distribution system integrity to prevent contamination of water before it reaches the consumer has been discussed earlier. This is the principal reason for requiring residual pressure throughout the system at all times: to prevent the entry of contaminated water into leaking pipelines.

It also is the justification for establishing codes that regulate consumer plumbing to prevent cross connections between the municipal system and different, potentially unsafe, water supplies and to minimize the opportunities for chance contamination by defective plumbing fixtures. The importance of careful controls in distribution and storage of treated water has been demonstrated repeatedly through the detection of unsatisfactory water quality and the disease outbreaks that often have resulted where poor practices have been followed.

Unsatisfactory quality also may result from excessively long detention times in pipelines or storage tanks, which may allow the growth of unacceptable populations of coliform or other bacteria, sometimes producing tastes and odors. Furthermore, in many instances the corrosion of pipelines may produce objectionable concentrations of iron in the water, causing precipitates, discoloration, and perhaps growths of "iron bacteria" in the pipes.

Objectionable quality produced during long storage is part of the basis for concern about the use of commercial bottled waters for drinking, which is popular in some communities. However, the greatest worry among health professionals about bottled waters has been caused by the often poor sanitary practices in the bottled water industry and the lack of adequate regulatory control over it.

STUDY QUESTIONS

1. Explain the difference between "primary" and "secondary" regulations in the EPA drinking water standards. Identify five specific constituents that fall into each category and the maximum permissible concentration for each.
2. Indicate the proportion of total municipal water demand that usually can be accounted for by each of the five major use categories described in the text.
3. Explain the significance of "unaccounted" water in municipal systems.
4. What is the average per capita use of water in U.S. municipalities and over what range does it vary among different municipalities?
5. Describe variations commonly observed in municipal water demand, including the shape of a typical daily flow hydrograph and the magnitude of the maximum day and peak flow rates, expressed in terms of the average flow.
6. The amount of water used to fight fires in a municipality is only a small fraction of the total demand. Why, then is the fire flow requirement so important in determining the hydraulic capacity of some distribution systems? In what types of systems is it most important? Why?
7. Describe and explain current trends in municipal water use patterns.
8. Identify the main elements in a typical municipal water supply system and briefly explain the function of each.
9. Briefly explain the key factors involved in selecting a municipal water supply source and the rationale used in that process.
10. Identify and briefly describe five types of sources commonly used for municipal water supplies.
11. Explain the circumstances that could make it necessary to install a raw water impoundment in developing a stream to serve as water source for a municipality. Describe how the impoundment would accomplish its role in the system.
12. Outline various ways in which municipal wastewaters can be reused and identify the advantages and disadvantages of each approach.
13. Differentiate between water quality characteristics of surface water and groundwater sources.
14. Draw a flow sheet for a typical plant for treating a surface water and indicate (*a*) major elements in the plant, (*b*) the function of each element, (*c*) points of chemical additions, and (*d*) the purposes for adding each chemical.
15. What are the principal water quality problems that can be solved by the typical plant outlined in question 14? Identify several water quality problems that would not be solved by the plant as shown.

16. Identify and explain four reasons for including storage for finished water in the distribution system.
17. Briefly explain the importance of maintaining adequate pressure in all parts of the distribution system at all times.
18. Describe a reasonable rationale for selecting the types of treatment that should be applied to a water from a given source before its delivery to the consumers.
19. A homeowner in an area with a poor water supply potential plans to install a cistern to store rainwater from the roof to meet some of the family needs. With a roof area of 1800 ft^2 (167 m^2), an annual rainfall of 35 in. (890 mm), and 75 percent recovery of the water, what average per capita flow could be provided by this approach for a family of five?
20. Assuming an average alum dosage of 30 mg/liter and a cost of $140/ton ($127/metric ton), what would be the annual cost for that chemical in a 5.0 million gal/day (219 liters/sec) water treatment plant? What would be the average chemical cost per 1000 gal of water produced?

REFERENCES

American Water Works Association, *Water Quality and Treatment,* McGraw-Hill, New York (1971).

Clark, J. W., W. Viessman, and M. J. Hammer, *Water Supply and Pollution Control,* 3rd Edition, Harper & Row, New York (1977).

Durfor, C. N., and E. Becker, "Public Water Supplies of the 100 Largest Cities in the United States, 1962," *U.S. Geological Survey Water Supply Paper 1812,* Government Printing Office, Washington, D.C. (1964).

Fair, G. M., J. C. Geyer, and D. A. Okun, *Elements of Water Supply and Wastewater Disposal,* Wiley, New York (1971).

Fair, G. M., J. C. Geyer, and D. A. Okun, *Water and Wastewater Engineering,* Vol. 1, Wiley, New York (1966).

Fair, G. M., J. C. Geyer, and D. A. Okun, *Water and Wastewater Engineering,* Vol. 2, Wiley, New York (1968).

Gehm, H. W., J. I. Bregman, *Handbook of Water Resources and Pollution Control,* Van Nostrand Reinhold, New York (1976).

Okun, D. A., "Alternatives in Water Supply," *Journal American Water Works Association,* Vol. 61, pp. 215–224 (1969).

Singer, P. C., J. J. Barry, G. M. Palen, and A. E. Scrivner, "Trihalomethane Formation in North Carolina Drinking Waters," *Journal American Water Works Association,* Vol. 73, pp. 392–401 (1981).

Solley, W. B., E. B. Chase, and W. B. Mann, *Estimated Use of Water in the United States in 1980,* Circular No. 1001, U.S. Geological Survey, Washington, D.C. (1983).

U.S. Environmental Protection Agency, "Interim Primary Drinking Water Regulations: Notice of Proposed Maximum Contaminant Levels for Radioactivity," *Federal Register,* Vol. 40, No. 158, pp. 34,324–34,328 (August 14, 1975).

U.S. Environmental Protection Agency, *National Interim Primary Drinking Water Regulations,* Publication EPA/570/9-76-003, Government Printing Office, Washington, D.C. (1976a).

U.S. Environmental Protection Agency, "Interim Primary Drinking Water Regulations: Promulgation of Regulations on Radionuclides," *Federal Register,* Vol. 41, No. 133, pp. 28,402–28,409 (July 9, 1976b).

U.S. Environmental Protection Agency, *Manual of Treatment Techniques for Meeting the Interim Primary Drinking Water Regulations,* Report No. EPA-600/8-77-005, Cincinnati (May 1977).

U.S. Environmental Protection Agency, "National Secondary Drinking Water Regulations, Final Rule," *Federal Register,* Vol. 44, No. 140, pp. 42,195–42,202 (July 19, 1979a).

U.S. Environmental Protection Agency, "National Interim Primary Drinking Water Regulations; Control of Trihalomethanes in Drinking Water; Final Rule," *Federal Register,* Vol. 44, No. 231, pp. 68,624–68,707 (November 29, 1979b).

U.S. Environmental Protection Agency, "Interim Primary Drinking Water Regulations; Amendments," *Federal Register,* Vol. 45, No. 168, pp. 57,332–57,357 (August 27, 1980).

Weeks, W. F., and M. Mellor, "The Iceberg Cometh," *Technology Review,* Vol. 81, No. 8, pp. 67–75 (1979).

CHAPTER 12

Municipal Wastewater Systems

Facilities for handling and disposing of wastewater comprise the "other half" of a municipal water system. Most municipal water is used for carriage of wastes; about 75 to 80 percent of it ultimately is discharged as wastewater, as outlined in Chapter 5. Therefore, provision of a municipal water supply requires arrangements and facilities for disposing of the used water, which must be done properly because of potential adverse impacts on human health and aquatic resources.

Municipal wastewater systems have basic components and a scope that are comparable in many respects to the water supply systems discussed in Chapter 11. This chapter describes the amounts and characteristics of wastewater flows and the types of facilities commonly used to collect, transport, treat, and dispose of them.

COMPONENTS OF MUNICIPAL WASTEWATER SYSTEMS

The sewerage system is a network of pipes that parallels the water distribution system to collect the used water and transport it away from homes and businesses. Usually it must be accessible to virtually every user of municipal water.

Design details of wastewater collection systems differ in several important respects from those of water distribution networks. Water supply piping is full and under pressure at all times, but typical sewerage systems must never be pressurized because flows into them are by gravity to avoid the need for each discharger to use pumps, which would be expensive and require regular maintenance. If the flow in a sewer ever is high enough to cause it to be under pressure, sewage can flow backward through connecting pipes to flood homes and other buildings. That obviously would be unacceptable. Therefore, sewers *must* be designed with hydraulic capacities adequate to carry their peak flows without ever becoming surcharged (under pressure).

A sewer then functions as an open channel—flowing only partially full with a "free" and unconfined water surface. Because it cannot be full and under pressure, it must be laid at the proper slope to insure downhill flow at all times. Also, the slope must be adequate to provide velocities high enough to prevent the settling of suspended matter that might obstruct the pipe. Those constraints make sewerage systems more complicated to design and construct than water distribution systems, which can flow full and under pressure, uphill and down, following the contours of the land.

Because quality characteristics of municipal wastewaters usually are not acceptable for direct discharge, they must be treated to correct those deficiencies before release into watercourses. Wastewater treatment systems differ greatly from those used in water supplies because the constituents to be removed and the quality goals for the finished product are radically different. The types of processes employed in municipal wastewater plants will be discussed in subsequent sections of this chapter.

Provisions also must be made for the ultimate disposal of the treated wastewater. Often, that is by discharge into flowing streams or lakes—sometimes the same one from which the water supply is taken. Other possibilities include discharge onto land for irrigation, storage in holding ponds from which water may leach into the ground, percolation into soil from underground tile fields, and use of treatment plant effluent for industrial cooling water or other purposes.

MUNICIPAL WASTEWATER FLOWS

For the most part, municipal wastewater flows originate from the community water supply, as discussed earlier. Typically, the wastewater approximates 75 to 80 percent of the water use and is subject to the same types of variations among communities, and trends, as those outlined earlier for municipal water systems. Based on a national average municipal water use of 180 gal/person/day, the average wastewater wasewater flow is about 135 gal/person/day. In some communities, however, those figures may be influenced significantly by the use of water from private wells that subsequently is discharged into sewers, or the use of municipal water with separate treatment and discharge of wastes into streams or individual soil disposal systems.

Most communities are not average, and flows vary widely among various municipalities and at different times within any given one. The annual, seasonal, and weekly variations in wastewater flows usually are similar in character to those observed in the municipal water supply.

Daily Flow Hydrographs

Hourly wastewater flow variations differ somewhat from those in the community water system because of their different hydraulic characteristics. Distribution systems flow full and under pressure, making the overall system demand responsive immediately and fully to changes in rates of use throughout the community. On the other hand, sewers flow only partially full, leaving storage available in upper areas of the pipes. During flow increases, for example, the water level rises to fill that storage, making flow changes at the end of the system slower than the changes in inputs. This reduces the short-term impact of an input change on rate of flow through the pipe, spreading it out over a longer period. Thus, peaks in sewage flow are lower and later in time of arrival at the end of the pipeline and the flow hydrograph is flatter and delayed when compared with the municipal water use hydrograph.

Figure 12.1 shows a flow hydrograph for a community of about 50,000 population (Chapel Hill, N.C.) on a typical day. In that example, the flow varies

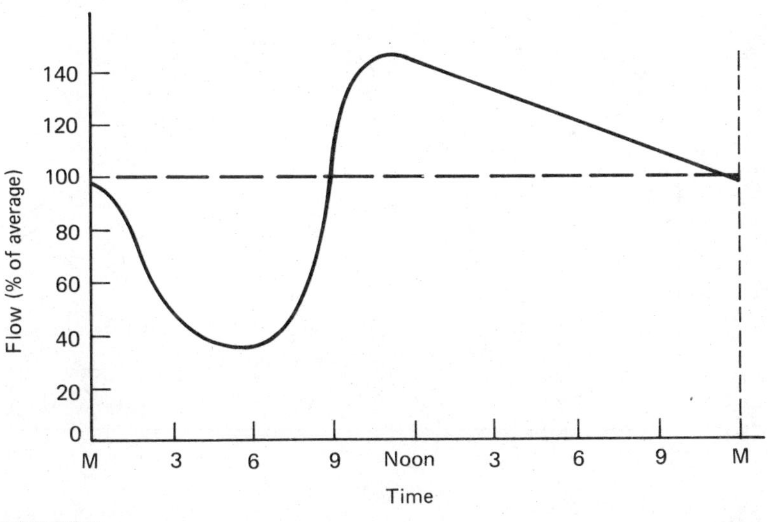

FIGURE 12.1
A 24-hour wastewater flow hydrograph for the town of Chapel Hill, North Carolina.

from about 40 percent of the average in early morning to 150 percent of average in midday.

Effect of Community Size

The shape of the flow hydrograph is influenced by many variables relating to the character of the community. Very large cities often have only minimal variations during the day because the time of travel through the sewers may be so long that peak flows from distant areas arrive at the same time as minimum flows from areas near the end of the system. Thus, hydrographs for large systems show less flow variations than those for small ones.

This effect is illustrated by Figure 12.2 (Keefer, 1940), showing wastewater flow patterns for small, medium, and large residential areas. The hydrograph for a small area has considerably wider range between maximum and minimum flows than those for larger areas. This reflects differences in amounts of storage in the sewerage systems as well as proximity of contributing areas to the point of measurement.

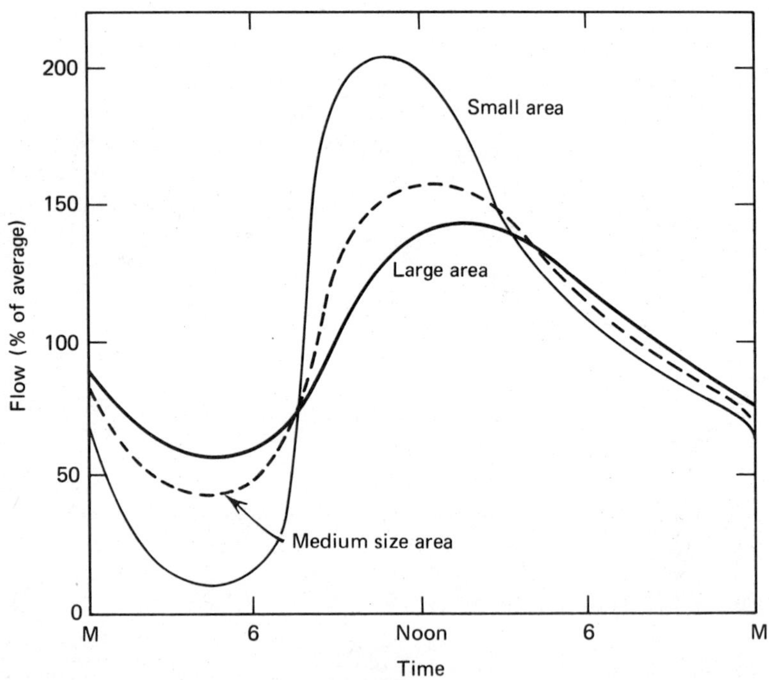

FIGURE 12.2
Hourly variation in sewage flow from different-size municipal areas.

Effects of Rainfall

Sewage hydrographs may be influenced radically by inflows of rainwater to the system. Connections of downspouts from roofs to sanitary sewers usually are forbidden by regulations; nevertheless, they are encountered often. Large numbers of those connections can produce sharp flow peaks during each rainfall and may contribute to serious hydraulic overloading of sewers and treatment facilities.

Recent practice has been based mostly on installing two "separate" sewerage systems in communities—one to carry sanitary wastes to the treatment plant, and the other to transport storm flow directly to discharge points, usually without treatment. However, many older communities still have "combined sewers," designed to carry both sanitary wastewaters and storm runoff in the same pipeline. Flow in a combined sewerage system during a storm may be many times that during dry weather, often exceeding the capacities of sewers and treatment plants and requiring the direct discharge of some mixed sanitary sewage and storm runoff to streams. The overflows are highly undesirable but continue to exist because of the great expense of converting "combined" systems into "separate" ones or building sewers and treatment plants large enough to handle the peak flows during storms.

Industrial Flows

Industrial operations in a community also may have a major influence on the shape of its wastewater hydrograph. In some instances, large water users operate around the clock, perhaps even 7 days a week. Hydrographs in those systems usually are flatter because the large and continuous industrial flow makes variations in household and other uses have less relative effect on the total flow from the community. Conversely, discharges from large water users that operate for perhaps only 8 hr a day may have the opposite impact on the flow pattern, accentuating daytime peaks.

Infiltration

Another important variable affecting the flow in a municipal sewerage system is "infiltration." That refers to leakage of underground water into the sewers, which is possible because they are not under pressure. Infiltration usually is relatively clean water and contributes mainly to hydraulic overloading of sewers and treatment facilities—not loadings of wastewater constituents.

The rate of infiltration depends on the location of the sewer relative to groundwater, the quality of sewer construction, and the effectiveness of their maintenance. Periodic surveys of infiltration rates and control of sewer leakage are vitally important in determining the size of the treatment plant required—similar in importance to leak control in water distribution systems.

WASTEWATER CHARACTERISTICS

Municipal wastewaters vary widely in character in response to many variables. The fractions derived from household wastes and typical commercial activities commonly are viewed as relatively uniform in contributions per person per day. Actually, that is not strictly correct because of variations in diets, household activities, use of garbage disposal units, and many other aspects of community life, including its social customs. Even wider variations are observed in the character of municipal wastewaters in other countries.

Perhaps the most important variable encountered among different U.S. municipal wastewaters is industrial input. That varies from none in predominantly residential communities to contributions far exceeding those from all other sources in towns with heavy industrial activities. The industrial discharges may include additions of "conventional" pollutants (that is, BOD and suspended solids) or contributions of radically different types of substances, such as metals or nonbiodegradable organic chemicals.

It is important to distinguish carefully between concentrations of pollutants in wastewaters and the total amounts present. The amounts of constituents discharged into a municipal system generally are dependent principally on contributing population and industrial operations, but they are relatively unaffected by water use in the community. On the other hand, their concentrations are inverse functions of flow; for example, for a given amount of suspended matter discharged by a household or industry, its concentration in the wastewater decreases with an increase in the amount of water into which it is discharged.

The physical, chemical, and biological characteristics of municipal wastewaters discussed here refer to constituents and concentrations from typical communities, without radical impacts by industrial discharges, atypical municipal water uses, or other unusual combinations of variables.

Physical Characteristics

The total solids content of municipal sewage usually is in the range of 500 to 1000 mg/liter, perhaps 25 percent of which was in the water supply and 75 percent added during its use in the community. Placing that in a different perspective, the total impurities are about 0.05 to 0.1 percent, by weight, making typical untreated municipal sewage more than 99.9 percent pure water.

The result is that it behaves hydraulically as water. Also, treatment systems used to correct quality deficiencies must begin with a liquid that already is high in quality (by common standards) and remove most of the traces of selected impurities before discharge to the environment. That must be accomplished by processing perhaps 135 gallons (about one-half of a ton) of wastewater per person per day at a cost of a few pennies. Viewed in this context, it should not be surprising that some unusual processes are employed in wastewater treatment systems.

Beginning with a wastewater containing 700 mg/liter total solids, a typical distribution of those impurities into different broad categories may be illustrated by Figure 12.3. The "suspended" solids account for about one-third of the total solids and are determined by measuring the amount of material removed from the wastewater during filtration through a specific type of laboratory filter and drying the residue at 103°C. Perhaps one-half to two-thirds of the suspended solids have particle size and specific gravity allowing separation from the liquid in about one hour of settling in a laboratory graduated cylinder. Those are reported as milligrams per liter of "settleable" solids, as shown in Figure 12.3, or as milliliters of sludge accumulation per liter of sewage settled, measured in a standard laboratory container called an Imhoff cone. The remaining "nonsettleable" solids are fine suspended matter or colloidal particles.

The 450 mg/liter of constituents passing through the filter are termed "filterable" solids. Often, in the jargon of the field, these are referred to as "dissolved solids," although strictly speaking that is not correct. Most are in solution as dissolved constituents, but a small fraction is composed of fine particles and colloidal aggregates.

Other subclassifications not shown in Figure 12.3 include "fixed" and "volatile" solids. The volatile fraction of total solids, for example, is determined by

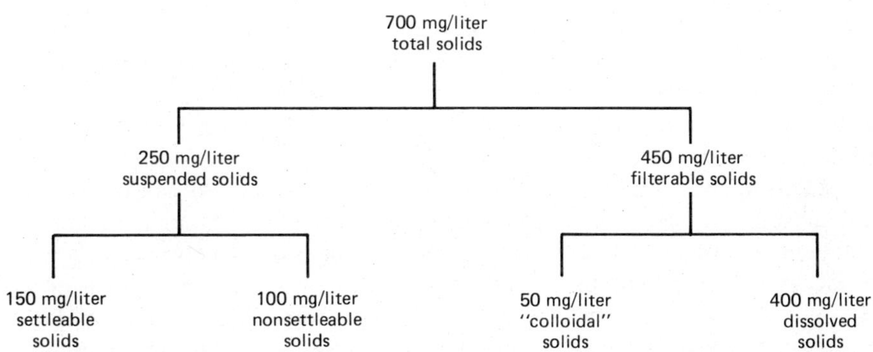

FIGURE 12.3
Distribution of solids in a typical municipal wastewater.

measuring the loss from the dried sample after further heating to 600°C; the fixed total solids is the residue that does not volatilize. Volatile fraction can be determined similarly for any of the various categories of solids just referred to, for example, volatile suspended solids and volatile dissolved solids. Often, the volatile fraction is assumed to represent organic content of the sample, although that usually is inaccurate because some chemicals and water remaining after drying at 103°C may be volatile at the higher temperature.

The suspended solids content of municipal wastewater averages about 200 to 250 mg/liter but often may range from less than 100 mg/liter to over 500 mg/liter. This produces an average contribution of about 0.20 to 0.30 lb (90 to 150 g) of suspended matter per person per day.

Fresh wastewater usually is gray and has a distinctive, slightly objectionable odor. After the depletion of all dissolved oxygen through biological action, the sewage becomes "septic," with a dark or black color and the disagreeable "rotten egg" odor of hydrogen sulfide. Other colors and odors may result from the presence of industrial wastes.

Chemical Characteristics

Organic chemicals in municipal wastewaters include nitrogenous compounds, such as proteins and their decomposition products, urea, and amino acids; carbohydrates, including sugars, starches, and cellulose; and various types of fatty materials in wastes from kitchens, laundries, and industries. Other types of organic constituents, usually present at lower concentrations, are detergents, phenols, and many chemicals that may be discharged by industries. Except for chemicals that offer special interest or concern, the organics usually are not measured separately. Instead, they are determined as a group by analyzing for total organic carbon content (TOC) of the sample or by evaluating gross polutional effects, such as BOD.

Inorganic constituents include grit, acids, alkalies, salts, and metals. Some are dissolved and others are present as colloidal particles or suspended matter. They may originate as constituents in the water supply of the community, in industrial discharges, and in wastewaters from washing, other household operations, and human wastes.

Usually there is little or no DO in municipal wastewater because of chemical and biochemical demands for it imposed during flow through the sewers. Nitrogen is present as a dissolved gas because of contact between the water surface and atmosphere, but usually it is of little or no interest because of its inert character. Gaseous decomposition products from biochemical breakdown of organics also often may be present, including carbon dioxide, methane, hydrogen sulfide, and ammonia.

Frequently, certain generic chemicals are of greater interest than the specific compounds in which they may be present. For example, the phosphorus content of a wastewater may be of concern because of its role as a limiting nutrient in watercourses. Accordingly, the wastewater may be analyzed for total phosphorus content, or dissolved phosphorus, without regard to specific identities or concentrations of compounds in which it occurs. Other generic constituents that frequently are of interest include various forms of nitrogen, types of metals, and chemicals that cause alkalinity.

The middle column of Table 12.1 summarizes typical concentrations of common constituents of municipal wastewaters. As mentioned earlier, they include substances present in the water supply before its use in the municipality and can vary widely depending on characteristics of that water. The right column of Table 12.1 shows typical increases in concentrations that might result from a single municipal use of water. It shows that the use adds virtually all of the suspended solids, BOD, total organic carbon (TOC), chemical oxygen demand (COD), nitrogen, and phosphorus, as well as substantial amounts of

TABLE 12.1
Constituents in Municipal Wastewater

Constituent	Typical Concentration (mg/liter)[a]	Concentration Increase During Use (mg/liter)
Solids		
Total	700	550
Suspended	250	250
"Dissolved"	450	300
BOD_5	200	200
TOC	200	200
COD	400	400
Nitrogen		
Total—N	40	40
Organic—N	15	15
Ammonia—N	25	25
$NO_2 + NO_3$—N	1	1
Phosphorus	7	7
Chlorides	90	75
Sulfates	50	25
Calcium	40	15
Magnesium	20	10
Sodium	75	60
Alkalinity as $CaCO_3$	200	100

[a]Data in this column represent total concentrations, including contributions by carriage water and additions during use.

several other constituents, including chlorides, sulfates, calcium, magnesium, and sodium.

Table 12.2 shows expected increases in concentrations of various constituents during domestic water use, without commercial and industrial contributions (Metcalf and Eddy, 1979).

Biological Characteristics

Municipal wastewaters contain wide varieties of microorganisms, many of which originate in household wastes. Several groups have been discussed in earlier chapters dealing with health effects and evaluation of water quality. Of

TABLE 12.2
Typical Mineral Pickup During Domestic Water Use[a]

Constituent	Concentration Increase (mg/liter)
Anions	
Bicarbonate	50–100
Carbonate	0–10
Chloride	20–50
Nitrate	20–40
Phosphate	20–40
Sulfate	15–30
Cations	
Calcium (as $CaCO_3$)	15–40
Magnesium (as $CaCO_3$)	15–40
Potassium	7–15
Sodium	40–70
Others	
Aluminum	0.1–0.2
Boron	0.1–0.4
Iron	0.2–0.4
Manganese	0.2–0.4
Silica	2–10
Alkalinity (as $CaCO_3$)	100–150
Total dissolved solids	150–400

Source: Wastewater Engineering: Treatment Disposal Reuse, Metcalf and Eddy, Copyright © 1979 by McGraw-Hill, New York, p. 66. Reprinted by permission.

[a]U.S. average range, without commercial and industrial.

primary importance, of course, are the pathogens that are capable of causing waterborne diseases, including bacteria, protozoa, viruses, and nematodes (see Table 9.1).

Although the pathogens are of special importance, they usually are few in number as compared with other organisms in wastewaters. One of the highest populations is the coliform group, which was discussed in Chapter 9 in connection with evaluating the health risks in water supplies. It has been estimated that the discharge of coliform bacteria in human wastes approximates 100 to 400 billion organisms per person each day. Concentrations of coliforms in municipal wastewaters vary widely—often in the range of 1 to 10 million/ml.

Another major group in sewage is fecal streptococci, which have been used occasionally as indicator organisms in water pollution investigations. The Salmonella group of bacteria includes several pathogens and also has been proposed for use as indicators.

Many other organisms are present in sewage in substantial numbers and play important roles in wastewater treatment processes. They originate in human discharges, kitchen wastes, and soils entering the sewerage system in washwaters, rainfall, and land runoff. The populations depend on several factors, including food supplies available in the flow, time of travel in the sewers and treatment plant, and many environmental variables. They include organisms that degrade organics in biological treatment plants and impose oxygen demand in those facilities and subsequently in streams.

Algae seldom are present in untreated sewage but may occupy very important roles in certain types of biological treatment facilities to be discussed later. Other forms that usually are found in smaller numbers include rotifers, crustacea, worms, and flies.

Wastewater Variability

Municipal wastewaters vary in character from community to community, as discussed earlier, because of differences in water use, diet, standard of living, social customs and, especially, industrial discharges. Beyond those differences among communities, however, there are substantial variations in strength in any given community during each day.

Figure 12.4 illustrates the diurnal variability in BOD and suspended solids in raw sewage from the town of Chapel Hill, North Carolina, a community of about 25,000 population at that time, with no industrial input (Hanson et al., 1970). It shows that pollutional strength of the wastewater, as represented by both parameters, declines to a minimum value during the early morning and rises to a peak in the afternoon. Diurnal changes in flow have been discussed earlier (see Figure 12.1). Variations in BOD and suspended solids can be explained based on minimal community activity during early morning hours

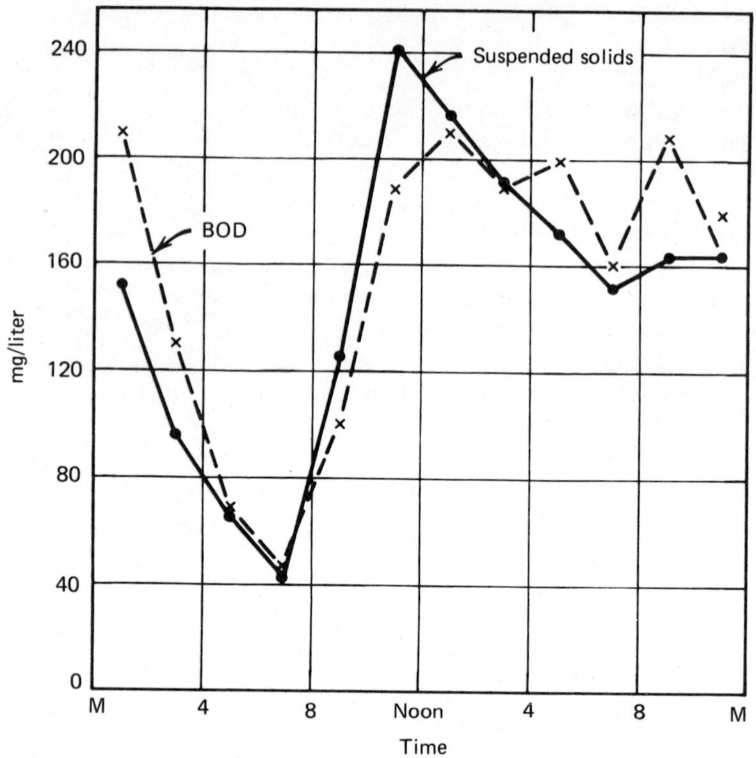

FIGURE 12.4
Variations in flow, suspended solids concentrations, and loadings, chapel Hill, North Carolina.

and increases in washing, cooking, and use of bathroom facilities during the day.

The simultaneous changes in flow and strength compound the impacts on treatment facilities designed to remove BOD and suspended solids, as illustrated in Figure 12.5. Combining the Chapel Hill flows and suspended solids concentrations to calculate pounds of suspended solids arriving at the treatment plant at different times of the day shows variations in loadings from about 12 lb/hr at the early morning minimum to 218 lb/hr at the noon peak—a range of over 18 to 1. Data for BOD loading would show similar ranges.

Unless plant design includes facilities that can cope with those radical swings in loading, one could expect serious diurnal impacts on performance. That provides an important constraint on the types of treatment processes that are acceptable for treating the wastewater and the parameters used in plant design.

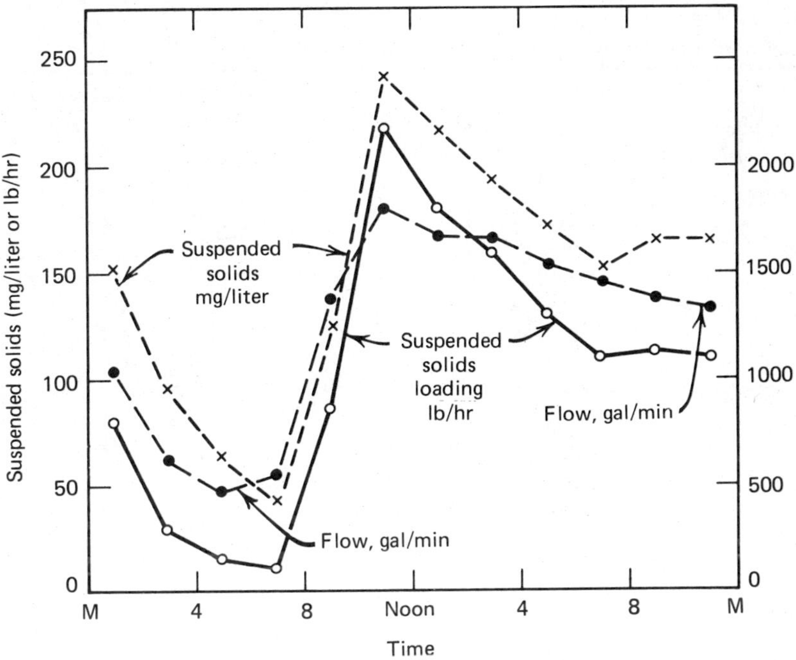

FIGURE 12.5
Variations in flow, suspended solids concentrations and loadings, Chapel Hill, North Carolina.

WASTEWATER TREATMENT RATIONALE AND SYSTEMS

The fundamental rationale in wastewater treatment, as in municipal water treatment, is based on correcting deficiencies in quality. The rational approach in selecting treatment systems begins with a critical comparison of raw wastewater quality and required effluent characteristics. The effluent quality that must be attained is defined by standards issued by the responsible regulatory agency; types of standards and methods for setting them will be discussed in more detail in Chapters 14 and 15. When a comparison of standards and wastewater characteristics indicates that improvements in effluent quality are necessary before discharge, treatment is indicated.

Many treatment processes are available, each of which can accomplish certain specific types of quality changes. Some examples of processes used in practice to remove potential pollutants from municipal wastewaters are listed in Table 12.3, along with information about the extent to which they are employed currently. The selection of the specific sequence of processes to be used in a given plant depends on the matrix of pollutants that must be removed,

TABLE 12.3
Treatment Process Available for Municipal Wastewaters

Constituent to Be Removed	Process	Status of Use in Practice		
		Frequently	Sometimes	Few or None
Large solids	Bar screens	X		
	Comminutors	X		
Suspended solids	Fine screens		X	
	Settling	X		
	Flotation		X	
	Coagulation		X	
	Sand filtration		X	
Biodegradable organics (BOD)	Storage ponds			X
	Aerated lagoons		X	
	Oxidation ponds	X		
	Activated sludge	X		
	Trickling filters	X		
	Activated carbon			X
Other organics	Activated carbon		X	
Nitrogenous oxygen demand (NOD)	Nitrification		X	
	Stripping			X
	Ion exchange			X
Nitrogen removal	Nitrification/denitrification			X
	Stripping			X
	Ion exchange			X
Oils	Gravity flotation		X	
	Air flotation		X	
Inorganics	Reverse osmosis			X
	Ion exchange			X
	Precipitation		X	
	Electrodialysis			X
	Distillation			X
	Freezing			X
Sludge digestion	Anaerobic	X		
	Aerobic		X	
Sludge dewatering	Flotation		X	
	Filtration	X		
	Centrifugation	X		
	Gravity thickening	X		
	Incineration		X	

efficiencies that must be attained, and capital and operating costs for the entire treatment system.

Several treatment processes and systems commonly used in municipal practice are described in subsequent sections of this chapter. In those discussions, wastewater treatment is divided into a few broad categories, each of which is directed toward attaining certain goals: (1) preliminary treatment, (2) primary treatment, (3) secondary or biological treatment; (4) advanced treatment, and (5) adjunct processes that are necessary to complete the plant functionally.

PRELIMINARY TREATMENT

Municipal wastewaters contain several types of materials that can interfere with operation of the treatment plant and must be removed at an early stage to assure reliable and economical plant functioning.

Figure 12.6 shows a simple device for removing large objects that can obstruct pipelines and pumps, jam equipment, interfere with proper functioning of flow metering units, or create other mechanical problems. The manually cleaned "bar screen" (at the right of Figure 12.6) consists of parallel bars between which the flow passes as it enters the plant. They are spaced perhaps one or two inches apart to prevent the passage of large objects and are cleaned

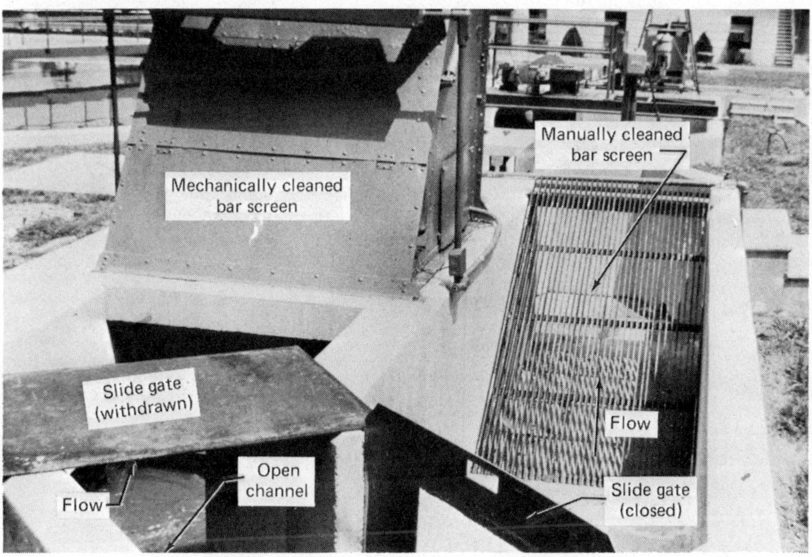

FIGURE 12.6
A typical manually cleaned bar screen installed in an emergency bypass channel.

FIGURE 12.7
Cutaway view of a mechanically cleaned bar screen. Used by permission of Envirex, Inc., a Rexnord Company.

periodically by raking trapped accumulations up the inclined bars. The screenings are removed for disposal with municipal refuse or are ground into smaller particles and released back into the flow for removal in subsequent processes. In many plants, the bar screen is cleaned automatically by mechanical devices that remove the screenings (Figure 12.7) or by cutters that chop the accumulations into particles small enough to pass through the openings.

An alternate method for solving this problem is to install a device called a "comminutor," which consists of a rotating, slotted drum with cutters mounted on it. The slots retain large particles upstream of the drum until they have been chopped into sizes small enough to pass through the plant without causing trouble.

Another constituent in raw wastewater that usually is removed in preliminary treatment is grit, which can cause excessive equipment wear and accumulate in deposits that interfere with the operation of treatment units. It is removed by installing a small settling unit designed to separate the grit without removing much of the organic suspended matter (Figure 12.8). That provides a relatively clean grit that is easy to handle and discard without creating odors by decomposition of putrescible organics.

Bar screens and grit removal are the two most common preliminary treatment devices in municipal plants, but sometimes others may be installed to cope with different types of problems. For example, communities with industries that discharge large amounts of grease may install flotation devices to remove it from the treatment plant influent. Municipalities and industries often find it desirable to install fine screens for the removal of smaller objects such as fibers, hair, or feathers.

FIGURE 12.8
A grit chamber, with screw conveyor to transport and clean grit.

Preliminary treatment devices usually have no significant effect on the pollutional characteristics of the wastewater. They remove little or none of the organics that are responsible for causing BOD, have no practical effects on suspended solids content, and remove no dissolved constituents. The objective in using them is simply to minimize maintenance problems and interruptions in plant operations.

PRIMARY TREATMENT

So-called primary treatment plants are designed principally to remove suspended matter from wastewater, often using the sequence of units shown in Figure 12.9. The first two processes in that plant, of course, constitute preliminary treatment to remove large objects and grit from the influent flow. The sole additional major treatment element is a settling basin, similar in principle to the ones used in water treatment plants, discussed in Chapter 11.

Suspended matter with a specific gravity greater than water settles to the bottom of the settling basin and is removed as sludge, while particles lighter than water rise to the surface and are skimmed off. The average time required for flow to pass through the basin is about two hours—shorter than in water treatment to minimize odors from the decomposition of putrescible solids in the wastewater. Also, for the same reason, sludge is not allowed to remain in the tank very long but is removed frequently or even continuously by mechan-

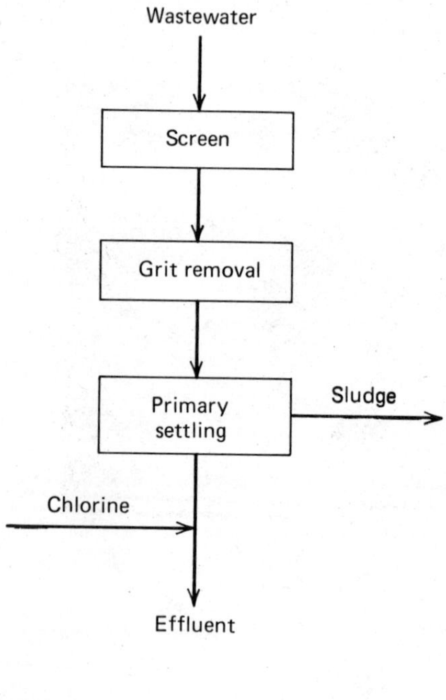

FIGURE 12.9
Flow sheet for a primary treatment plant.

ical scrapers or hydraulically. The unit usually is about 10 feet deep, and its influent and outlet structures are designed to minimize turbulence in the flow passing through it.

Figure 12.10 shows a circular primary settling basin. The influent enters a circular baffled compartment at the center of the tank and flows radially out-

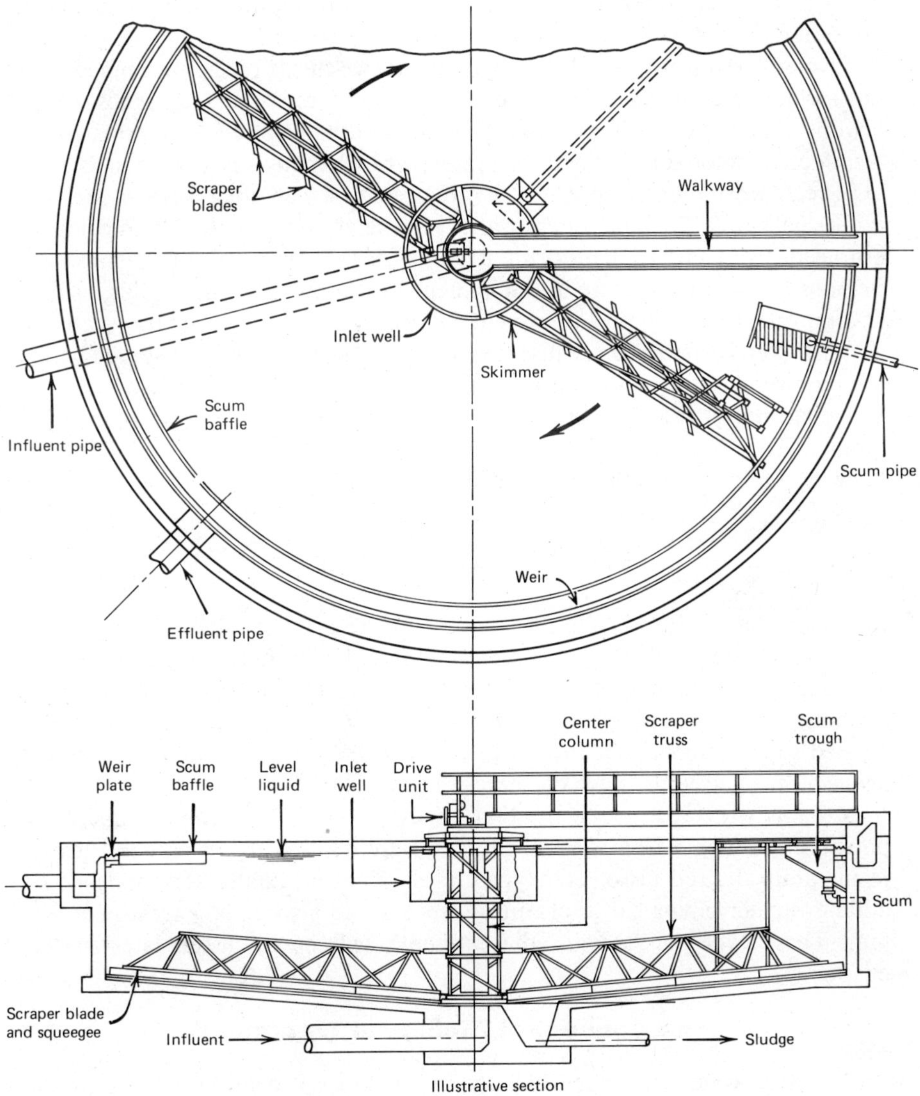

FIGURE 12.10
Circular primary settling tank. Drawing courtesy of Infilco Degremont, Inc.

ward to overflow over a weir extending around its periphery. A baffle near the weir retains floating scum in the tank, and a rotating skimmer moves it to a point (the scum trough) at which it is withdrawn for treatment and disposal. Sludge settling to the bottom is moved to the center of the tank by plows (scraper blades) suspended from a rotating structure and is pumped to other units for treatment and disposal. Other designs of settling basins differ in flow configuration and may be rectangular or square, with different types of sludge removal equipment.

Primary treatment usually removes about two-thirds of the suspended matter, including about one-third of municipal wastewater BOD, and most of the floating matter. The effluent from primary treatment often is chlorinated for disinfection before discharge. This system was for many years the backbone of U.S. regulatory requirements for the treatment of municipal wastewaters, but it does not provide adequate removal of suspended solids or BOD to meet modern needs. Standards for most municipal systems now require the removal of perhaps 85 percent or more of the influent BOD and suspended solids, usually accomplished by biological treatment. Many plants employ primary treatment as a first step, sometimes because that system already had been installed before the requirement for biological treatment was imposed. In other instances, primary settling is included in new plants simply because it is an economical method for removing a substantial fraction of the suspended matter and BOD, allowing the use of smaller biological treatment facilities.

SECONDARY TREATMENT

In the jargon of water pollution control, the term "secondary" treatment is synonomous with biological treatment and is the minimum requirement today for virtually all municipal wastewater discharges. Most secondary treatment systems that are properly designed and operated can produce BOD and suspended solids removals in the range of 75 to 95 percent, which satisfies the needs of most streams in which there is reasonable dilution.

Streams with little or no dilution, or that receive substantial amounts of potential pollutants from several sources, may require higher effluent quality to maintain desired dissolved oxygen levels. That is usually accomplished by adding further biological or chemical processes to a secondary treatment system. Those types of "advanced treatment" will be discussed later in this chapter.

The Concept of Biological Treatment

Earlier discussions of oxygen balance in Chapter 10 pointed out that biodegradable organics discharged into watercourses are used by organisms as sources of energy and chemicals necessary to maintain life processes and synthesize new cell material (growth). In completing those reactions, the organisms use DO and release decomposition products that include simpler organics, carbon dioxide, and other chemicals. Reducing the amount of oxygen demand (BOD) is the principal goal in wastewater treatment under current pollution control regulations.

Provided biochemical action in the stream does not deplete DO sufficiently to create or threaten objectionable conditions, it represents a viable method for destroying some wastewater constituents. Accordingly, it could be viewed as an elementary form of biological wastewater treatment and illustrates the basic mechanisms involved in all of those systems. The fact is that the biochemical reactions in secondary treatment plants would occur eventually in the stream anyway if the waste were discharged without treatment. The principal difference is that they are made to occur in the plant before discharge instead of in the stream afterward.

Biological treatment processes, then, are "natural" systems in which growth occurs commensurate with the food supply available to the organisms and environmental conditions under which they operate. The successful design and operation of those systems depend principally on maintaining the proper environment to encourage the growth and metabolism of organics that are to be removed from the wastewater.

Storage Lagoons

Provisions of simple storage to retain wastewater long enough for completion of the biochemical reactions can accomplish the destruction or removal of organics that cause BOD. However, this approach also can result in objectionable odors because oxygen transfer through the basin surface often is inadequate to meet the needs of the organisms, resulting in zero DO ("anaerobic" conditions) in the liquid and the generation of hydrogen sulfide or other odorous gases. For that reason, and to eliminate the need for large land areas, simple holding basins seldom are employed for biological wastewater treatment.

The odor problem can be avoided by taking steps to insure that oxygen is supplied at a rate high enough to maintain DO in the water ("aerobic" conditions) at all times. Biological treatment plants used in municipal and industrial wastes practice do that by any of several different approaches, some of which are outlined in the following sections.

Aerated Lagoons

This type of aerobic biological treatment system uses mechanical aeration to supply oxygen to the wastewater during storage. Two types of aeration devices commonly are used: bubble diffusers or mechanical mixers located at the surface of the liquid (surface aerators). Mixing provided by either type of device helps to keep the living organisms in suspension and facilitate their contact with dissolved organics. Oxygen supply is enhanced by increased turbulence of the pond surface, transfer of oxygen from air bubbles into the water, and transfer of oxygen from air into droplets formed by mixers.

The mixing and improved oxygen transfer permit aerated lagoons to attain good BOD removal efficiency in ponds that are substantially smaller than those that would be necessary using simple storage lagoons, as well as minimizing or eliminating odor production. For municipal wastewaters, aerated lagoons often are designed with average detention times in the range of 5 to 10 days and attain 75 to 85 percent BOD removal. System efficiency can be improved by increasing the aeration time and by installing settling tanks after the aerated lagoon to remove some of the biological growth that contributes to both effluent suspended solids and BOD.

Activated Sludge Systems

A further reduction in treatment plant size can be accomplished by resorting to an activated sludge system. The basic process flow sheet is shown in Figure 12.11 and consists of two units: an aeration tank and a settling basin.

Many of the organisms produced during the metabolism of organics stick together in clumps, forming floc particles that settle at rates high enough to

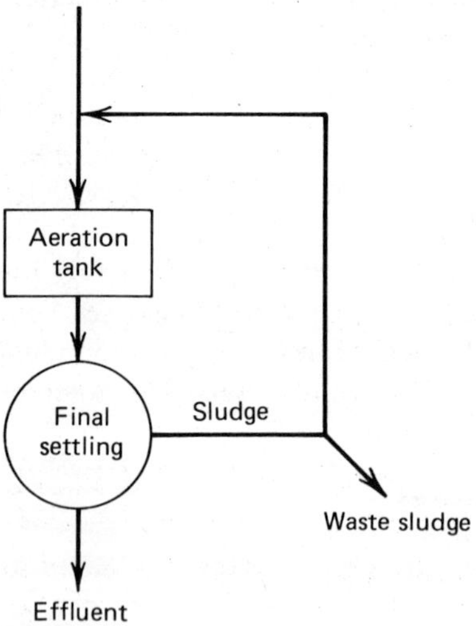

FIGURE 12.11
Flow sheet for the activated sludge process.

allow removal from the wastewater in an hour or two. The final settling tank separates the organisms from plant effluent before its discharge, and the active (living) sludge is removed from the bottom of the tank and returned to the aeration basin influent. Over a period of days or weeks, it is possible to build up the concentration of live organisms in an aeration system to many hundreds of times their numbers in the influent sewage. This accelerates the rate at which they can metabolize organics in the wastewater and makes it possible to accomplish in a few hours the same removal that otherwise might require several days in an aerated lagoon or weeks in a storage lagoon or in the receiving stream.

Aeration is accomplished by using bubble diffusers, surface aerators, or various other types of devices. It accomplishes two goals: (1) maintenance of DO in the aeration tank to satisfy the needs of the organisms and avoid the generation of odors, and (2) mixing to keep the sludge in suspension so that the organisms contact the organics to be removed.

Aeration detention time varies widely among the many modifications of activated sludge encountered in practice, ranging from less than an hour in some to a few days in others, most plants falling in the range of 4 to 10 hours. The required time depends on the characteristics of the wastewater being treated, degree of BOD and suspended solids removal to be attained, and several other factors that must be taken into consideration during plant design. The settling basin ("final settling") usually has a detention time of one to two hours. Sludge is returned continuously from that unit to the aeration tank.

Because there is growth of new cell material during metabolism of the organics, there is a gradual accumulation of excess sludge in the system. That must be removed ("wasted") from the system periodically and treated separately before disposal.

Although the combination of units in a municipal activated sludge plant varies with the needs of the system, the flow sheet in Figure 12.12 is typical of many plants. It includes screens and grit removal for preliminary treatment and the settling step used in primary treatment. These are followed by the activated sludge process itself (aeration and final settling) and, often, chlorine addition for disinfection of plant effluent. This type of plant usually can produce BOD and suspended solids removals from municipal wastewater approximating 85 to 95 percent.

Trickling Filters

The trickling filter process also is biological, but it uses a somewhat different approach. Instead of growing the organisms as floc particles suspended in the wastewater, they are attached to stone or other fixed medium and the wastewater flows over them in a thin sheet. The organics adsorb on the attached

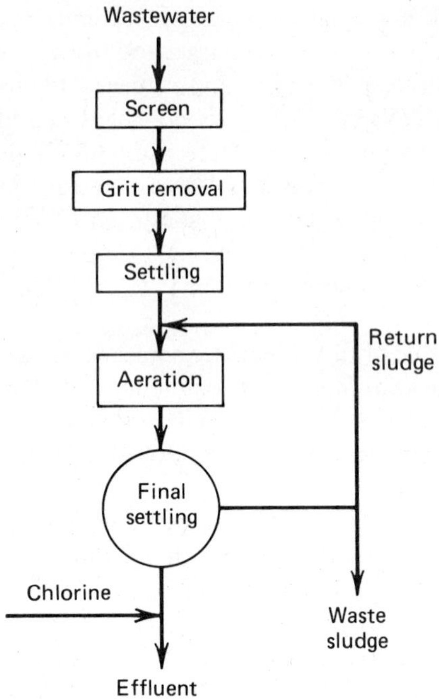

FIGURE 12.12
Flow sheet for a typical activated sludge treatment plant.

growths through physical chemical surface phenomena and are metabolized by organisms in the film; reaction by-products are released back into the flow passing through the bed. Excess accumulations of cells produced in the process wash out of the unit and are removed in final settling tanks, similar to those in the activated sludge process.

Figure 12.13 is a cutaway view of a trickling filter. Flow from primary settling passes through the feed pipe, enters the hollow distributor arms, and is discharged as a spray over the stone bed through orifices located along one face of each arm. The reaction force created by water leaving the orifices makes the distributor rotate slowly, and the combination of orifice spacings and rotation of the unit provides uniform application of wastewater over the stone surface.

The bed (filter medium) is 4 to 8 feet (1.2 to 2.4 m) deep and contains stone 2 to 4 inches (5 to 10 cm) in diameter. Clearly, the term "filter" is a gross misnomer for this unit—it is actually a biological-sorption process, as outlined above. The stone rests on a perforated underdrain system that allows wastewater to pass out of the unit through the network of channels in its bottom. The upper parts of the underdrain channels also serve as passages through which air can flow to ventilate the bed. That is the mechanism by which growths and flow are maintained in an aerobic state, minimizing the potential for odor problems.

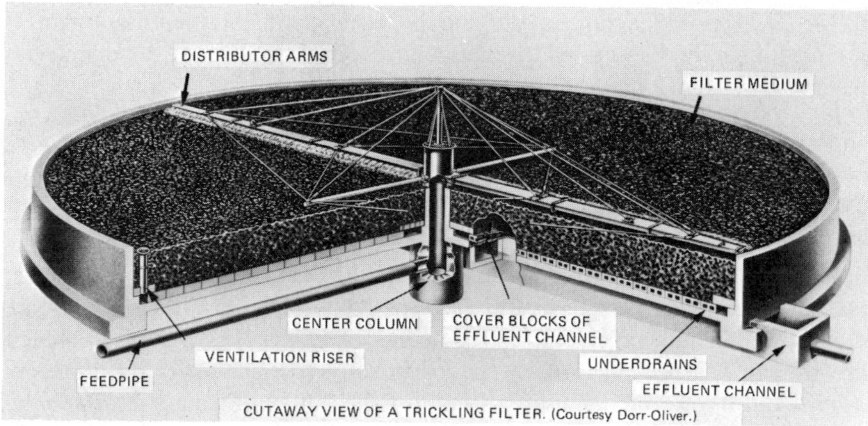

FIGURE 12.13
Cutaway view of a trickling filter. Photo courtesy of Dorr-Oliver, Inc.

Figure 12.14 shows the flow sheet for a trickling filter process, which is located within a secondary treatment plant at the position often occupied by aeration and settling in an activated sludge plant; that is, it normally follows primary settling. The "recycle" is effluent that is pumped back to a location upstream of the trickling filter to provide more than one pass through it to improve efficiency in removing organics. The patterns in which flow may be recycled in a trickling filter plant are many, two of which are shown in Figure

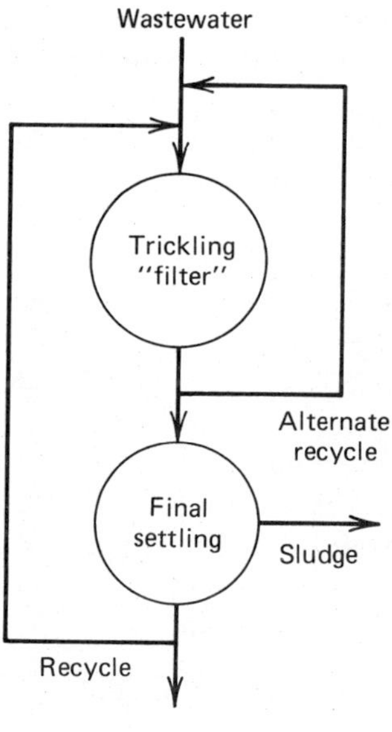

FIGURE 12.14
Flow sheet for a trickling filter process.

12.14. The recycle arrangement can be very complex when the plant is designed with two stages of trickling filters arranged in series to improve efficiency.

Performances of trickling filter plants depend on the character of the waste, volume of the stone, recycle patterns and flows, and whether the plant is arranged with one or two stages of biological units. Usually, they are somewhat less efficient than activated sludge plants, averaging perhaps 70 to 90 percent BOD and suspended solids removals.

Other treatment devices also use growth attached to a supporting structure, but they may differ substantially from the stone beds just described. Some alternatives include special plastic media, fixed vertical plates on which growths can adhere, and various types of industrial cooling tower structures. A somewhat different version uses partially submerged plastic disks on which the growths adhere, rotating about a horizontal axis perpendicular to the disks. They are called rotating biological contactors (RBCs). As the disks rotate, the growths pass through the liquid and adsorb organics in the lower part of the revolution; they then are aerated during movement through air in the upper part.

Oxidation Ponds

Although they seem simple at first sight, oxidation ponds actually are among the most complex biological processes because they combine several functions in a single basin. An oxidation pond usually consists of one or more lagoons, about 3 to 5 ft deep, with long (perhaps about 30 days) detention time.

Suspended solids in wastewater entering the pond settle and become part of the bottom deposits. The organics not removed by settling undergo an aerobic attack in biological reactions that are basically the same as those that occur in activated sludge plants, but with kinetics that are relatively slow because the population of suspended organisms is very low.

Oxygen required to keep the process aerobic is derived from photosynthetic activity of algae in the pond. Those microorganisms use carbon dioxide, trace elements in the water, and energy from sunlight to produce more cells, with oxygen as a by-product. The amount of oxygen produced is a function of algal population and the amount of sunlight received by them. That, in turn, is related to the area of pond surface available for intercepting sunlight, making pond area the prime design factor for this type of facility.

Some of the bacteria, algae, and other organisms settle to the bottom and become part of the sludge deposits, which also undergo biological attack. However, the limited exchange between aerobic water in the upper regions of the pond and deposits at the bottom prevents oxygen from penetrating in amounts sufficient to keep the sludge aerobic. Therefore, the biochemical processes in the sludge deposits are anaerobic in character.

The complexity of processes occurring in the basin become apparent when it is realized that they include (1) settling of suspended matter and organisms, (2) aerobic biological metabolism of organics in the wastewater, (3) photosynthetic production of oxygen by algae, and (4) anaerobic biological metabolism (digestion) of the sludge accumulations at the bottom. Furthermore, there are many process interactions because of exchanges of chemicals, by-products, gases, and solids between the aerobic and anaerobic zones, which greatly influence the breakdown of materials in the pond.

Oxidation ponds are widely used for municipal wastewater treatment, especially in small communities where relatively low-cost land is available. Design factors vary considerably, depending on climate, waste characteristics, and type of pond system installed. Commonly, about an acre of pond surface is provided for each 40 to 50 pounds per day of BOD in the applied wastewater, which is equivalent to an acre for each 200 to 300 persons in the contributing population.

Land Treatment

This approach to wastewater treatment sometimes is classified as an "advanced" process, which may seem strange in view of its ancient origins, but it probably is better classified as secondary treatment. Recent literature suggests that the technique, which often combines treatment and disposal into a single operation, is being rediscovered and currently is very popular.

In land application, the liquid fraction of wastewater is used for irrigation of plants, sometimes producing useful crops. Much of the flow usually is lost to the atmosphere through evaporation and transpiration. Much of the remaining liquid percolates into the soil and becomes part of the groundwater flow, some of which ultimately reaches streams and lakes. Usually surface runoff is minimal or nil; however, sometimes it may be substantial following precipitation.

Constituents of the wastewater are modified substantially through biological decomposition by soil organisms. Nutrients and many inorganic chemicals are removed through biological action, adsorption, and ion exchange, radically changing the character of the effluent.

Practical application of this approach may be limited by several factors, the most obvious of which include characteristics of the wastewater, soil character, climate, geology of the area, and cost of the land. Generally, land application is favored for small communities that discharge principally domestic wastewaters and by the presence nearby of extensive areas of low-cost land, with permeable soil for which irrigation would be desirable. Because this approach may result in little or no surface discharge, it also is attractive where especially stringent effluent standards must be met.

An obvious disadvantage of land disposal is that this highly "consumptive" use reduces the amount of water available to others downstream. As more and more communities adopt the approach under encouragement, and sometimes even pressure, from regulatory agencies, problems of excessive water consumption may be expected to arise occasionally. Eventually, it may become necessary to limit its use in some areas.

ACCOMPLISHMENTS AND LIMITATIONS OF SECONDARY TREATMENT

Table 12.4 shows typical concentrations of selected constituents in municipal wastewater after secondary treatment. The right column indicates the net increase of each constituent over its concentration in the municipal water supply before use by the community. Comparing the data with those in Table 12.1 shows typical removal efficiencies of the various constituents by secondary treatment. The wastewater BOD and suspended solids content are reduced by 80 to 90 percent, leaving about 20 to 40 mg/liter of each in the treated effluent. Much the nitrogen and phosphorus added during community use of the water, as well as virtually all of the dissolved inorganic chemicals, pass through the treatment system.

TABLE 12.4
Typical Constituents in Municipal Wastes After Secondary Treatment

Constituent	Typical Concentrations (mg/liter)	Concentration Increase During Use (mg/liter)
Suspended solids	20–40	20–40
Dissolved solids	200–300	100–200
BOD_5	20–40	20–40
TOC	30–60	30–60
Nitrogen	30–35	30–35
Phosphorus	5	5
Chlorides	90	75
Sulfates	50	25
Calcium	40	15
Magnesium	20	10
Sodium	75	60

The data make it clear that the principal accomplishments of secondary treatment lie in the removal of suspended matter, primarily by settling or floating, and in the reduction in oxygen demand (BOD) imposed on the receiving watercourse. Of course, the BOD removal is accomplished simply by making the biochemical reactions occur *before* the wastewater is discharged into the stream. With respect to specific constituents, those chemicals that are attacked by organisms in biological treatment are reduced in concentration. Nonbiodegradable chemicals are not destroyed in the process, although some may be removed by being adsorbed on sludge solids and subsequently separated by settling.

There are four major removal mechanisms at work in secondary treatment systems: settling, flotation, biochemical degradation, and adsorption on sludge produced in the process. In addition, relatively minor quantities of volatile substances may be released to the atmosphere by gas transfer during aeration. Also, minor amounts of some chemicals may react directly with oxygen or other constituents during the treatment. Anything not removed from wastewater by one of these mechanisms passes through the system into the effluent.

Although some chemicals remaining in the discharge are potentially troublesome, dilution often prevents downstream problems from developing. It should be noted, however, that sometimes water may be used several times successively by communities and industries along a stream, producing accumulations of chemicals that are not removed by secondary treatment and eventually causing increases in their concentrations to objectionable levels. In such instances, further removals of those constituents often must be accomplished through "advanced treatment."

ADVANCED WASTEWATER TREATMENT

In most practical situations today, advanced treatment of municipal wastewater is not required. It is undertaken only when necessary to provide more efficient removal of constituents remaining after secondary treatment. Various types of advanced waste treatment systems are available to satisfy any of several specific goals, including the removal of (1) suspended solids, (2) BOD, (3) NOD, (4) phosphorus, (5) nitrogen, (6) nonbiodegradable organics, and (7) dissolved inorganic chemicals. These materials may cause several different types of pollution problems, some of which were outlined in Chapter 8.

Removal of Suspended Solids

The effluent from a typical secondary treatment plant contains 20 to 40 mg/liter suspended solids, which may be objectionable in some streams. Furthermore, the suspended materials are responsible for causing some of the oxygen demand in the plant effluent; each milligram per liter of suspended solids may contribute about 0.5 mg/liter BOD (Lamb, 1970; Hanson et al, 1970).

One of the more straightforward approaches for removing suspended solids from secondary plant effluents is chemical coagulation and settling. A simple technique is to add alum or iron salts ahead of the final settling tank to help in agglomerating particles and improving its efficiency. In other instances, effluent from the final settling tank may be treated by coagulation and another stage of settling before discharge. The principles and techniques involved in this approach are virtually identical to those used in coagulation and settling in municipal water treatment plants, discussed in Chapter 11, and the process can remove most suspended solids from the effluent (Metcalf and Eddy, 1979).

Another approach that can be used instead of, or in addition to, coagulation and settling is sand filtration. This process also is essentially the same as that used in municipal water treatment plants. When effluent from secondary treatment is consistently low in suspended solids, it may be applied directly to the filters, producing an effluent containing little or no suspended matter. If the effluent sometimes contains high concentrations of solids, the filters may require backwashing so often that it would be preferable, or necessary, to treat it first by coagulation and settling (U.S. EPA, 1975b; Tchobanoglous, 1970).

Additional BOD Removal

Further reduction in BOD beyond that accomplished in conventional secondary treatment may be attained in several ways, one of which is to remove suspended matter by coagulation, settling, and filtration, as described previously. Another is to install additional treatment plant capacity and reduce loadings on plant units, producing more efficient performance, or add more stages of biological treatment for increased destruction of organic matter.

A process that sometimes may be used for removing BOD, along with nonbiodegradable organics, is adsorption on activated carbon. That may be done either by adding powdered activated carbon to the wastewater in a coagulation-settling system or, more efficiently, by passing the wastewater through columns of granular activated carbon (Metcalf and Eddy, 1979).

Reduction of NOD

Oxygen required for the biochemical conversion of ammonia to nitrite and nitrate may be substantial, as indicated in Chapter 8. There are two basic approaches for reducing this NOD: removal of the ammonia or biological conversion of the ammonia to nitrate before discharge (U.S. EPA, 1975a; Mulbarger, 1971).

Ammonia may be removed by adding lime to the wastewater to raise its pH, followed by aeration to strip the gas from the liquid and release it into the atmosphere. A different process involves passing the wastewater through a system designed to exchange other ions for the ammonia (ion exchange). In practice, ammonia is seldom removed from wastewaters by either technique because it is more economical to reduce the NOD by "nitrification," that is, conversion of the ammonia to nitrate through special biological treatment. Of course, this is precisely the same basic rationale as employed in conventional biological treatment—completion of the biochemical reactions *before* discharge of the wastewater instead of later, in the stream.

Biological nitrification may be accomplished by modifying the design of conventional systems to provide much lower loadings or by adding a second stage of biological treatment after the conventional one, as shown in the flow sheet of Figure 12.15.

Phosphorus Removal

Phosphorus is usually removed from wastewater by chemical precipitation, either in conjunction with biological treatment or in separate processes (U.S. EPA, 1976b). The precipitation may be accomplished with alum or iron com-

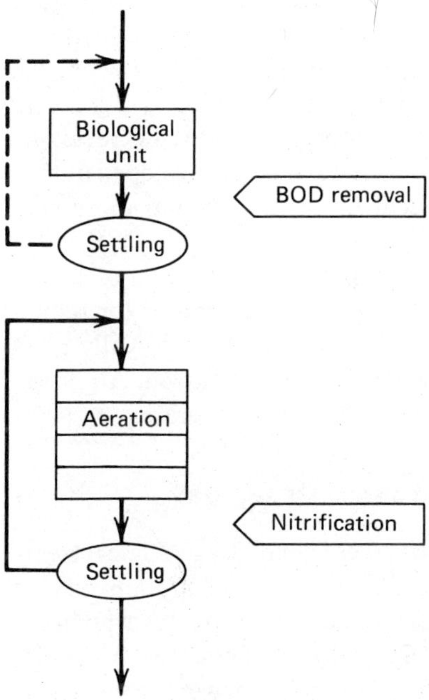

FIGURE 12.15
Flow sheet for a two-stage nitrification process.

pounds, which form insoluble products when reacted with phosphorus, or by adding lime to precipitate the phosphorus with calcium.

The choice between different approaches often depends on what processes are employed in the plant to remove other potential pollutants. Alum, for example, may be added to an existing activated sludge system for precipitation and removal of phosphorus in the final settling tank, or it may be used in a separate coagulation-precipitation process in other plants.

Biological systems for removing phosphorous have received increased attention during the past decade. These are based on sequential aerobic and anaerobic biological stages in activated sludge types of systems. With proper design and operation, it is possible to enhance phosphorous removal by the biological system by increasing the amount of that element incorporated into the cells, which subsequently are removed as waste sludge.

Nitrogen Removal

Nitrification does not remove nitrogen from the wastewater; it merely converts most of it from ammonia to nitrate and thereby reduces oxygen demand in the effluent. However, if another biological stage is added to the nitrification plant, as shown in the flow sheet of Figure 12.16, nitrogen can be removed by "denitrification" (U.S. EPA, 1975a; Mulbarger, 1971). Anaerobic conditions are generated in the third stage by adding a source of BOD (methanol in Figure 12.16) to the wastewater. With mixing but without aeration, organisms in the system metabolize the methanol and satisfy their oxygen needs by using oxygen from the nitrates. That releases the nitrogen to the atmosphere as a gas and eliminates its availability as a nutrient for microorganisms in the receiving watercourse. Other promising biological processes have been developed during the past decade that accomplish the same goal using simpler and more economical approaches.

Another method for removing nitrogen is to add alkaline material to the wastewater and aerate it to strip the ammonia, as described earlier for reduction in NOD, or to use ion-exchange systems. The fact is that no really good and firmly established technology exists today for nitrogen removal. All of the systems described are expensive and complex to design and operate.

Removal of Nonbiodegradable Organics

Most nonbiodegradable organics are not removed efficiently by biological treatment, although some removal may be possible through adsorption on the sludges. The process most commonly used to remove them is adsorption on activated carbon (Metcalf and Eddy, 1979; Culp, 1971). That may be implemented in a coagulation-settling type of system in which powdered carbon is

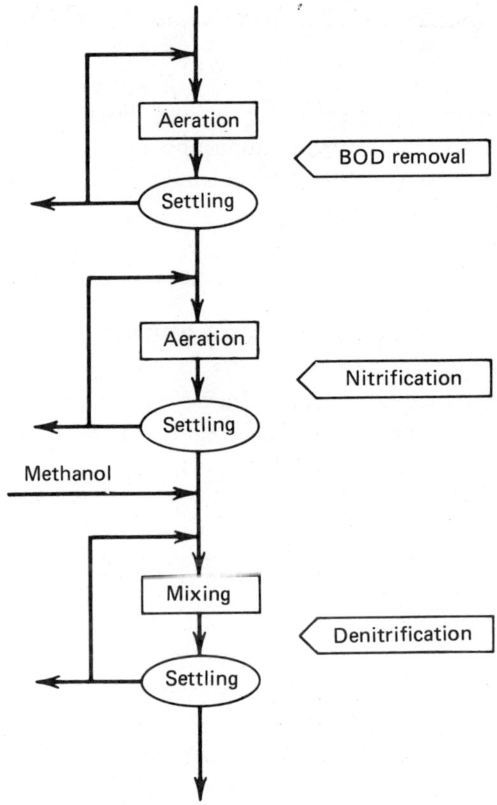

FIGURE 12.16
Flow sheet for a three-stage biological nitrogen removal process.

added to the wastewater and time allowed for sorption to occur; the carbon then is removed by settling, perhaps with the aid of chemical coagulants.

The other, more efficient, approach is to use granular activated carbon in columns through which the wastewater is passed. If the flow contains significant suspended solids, it may be necessary to subject it to sand filtration before adsorption to avoid premature clogging of the bed and the loss of carbon.

Removal of Inorganic Chemicals

Many inorganic chemicals (for example, metals) have relatively low solubility under some circumstances and may be removed efficiently through precipitation and settling, sometimes simultaneously with other potential pollutants, such as suspended solids. Examples of materials often removed by precipitation include iron, manganese, calcium, magnesium, chromium, lead, copper, zinc, and fluoride.

Other processes available for removing inorganic chemicals include ion exchange, reverse osmosis, electrodialysis, and distillation. These usually are much more expensive than precipitation and are considered principally for removing highly soluble constituents, such as sodium, chloride, and others not amenable to the simpler approaches.

ADJUNCT PROCESSES

All of the processes discussed to this point have been directed toward removing constituents from wastewaters without regard to the ultimate fate of those materials. Some of the most difficult and expensive operations in wastewater treatment plants involve treating and disposing of sludges and concentrated streams from the wastewater treatment processes. Some methods used for that purpose will be considered briefly here, in the context of sludge disposal in typical secondary treatment plants.

Sludges from secondary plants contain suspended solids removed from the wastewater by settling and growths developed in activated sludge systems, trickling filter systems, or other biological processes. They are highly putrescible and difficult to dewater, causing serious problems for plant operators. The treatment processes applied to them are directed toward (1) reducing putrescibility through the further destruction of biodegradable organics, (2) reducing the risks to health by pathogenic microorganisms in the sludge, (3) facilitating dewatering of the sludge solids, and (4) removing water to reduce the volume of material that must be handled.

Sludge Digestion

In most municipal plants, sludges are subjected to further biological treatment to accomplish the preceding goals before dewatering and disposal. The most common single process is anaerobic sludge digestion, carried out in the absence of DO by organisms different from those in the aerobic activated sludge and trickling filter systems (McCarty, 1964; Metcalf and Eddy, 1979).

Although actually very complex, the process may be viewed as involving two broad groups of organisms: acid formers and methane formers. The first group decomposes many organics in the sludge to produce relatively simple organic acids. Those subsequently are decomposed by the second group to produce methane, carbon dioxide, and smaller amounts of other gases. The methane has fuel value, which is recovered in most plants as fuel for heating the digester to enhance biological reaction rates and, in some large plants, by using it to operate engines.

Other end results of the anaerobic digestion process are the production of a humuslike residue, without major odor problems; removal of perhaps one-half of the water from the sludge solids, making the volume to be handled much smaller; production of a sludge that can be dewatered with greater ease and at lower cost; and destruction of many pathogenic organisms.

Sludge also can be digested by further aerobic biological action over many days or weeks. This involves increased energy consumption and costs and does not produce methane. Nevertheless, sometimes the approach is used because the equipment required and its operation are much simpler (U.S. EPA, 1974).

Sludge Dewatering

The disposal of sludge without nuisance or other problems often requires removing most of its water content to produce a solid end product. That is accomplished by applying one or more of several processes, including (1) further gravity settling, (2) flotation, (3) sludge centrifuges, (4) chemical treatment to facilitate separation of water and solids, (5) allowing it to drain and air dry on sand drying beds, and (6) various types of filtration systems (Metcalf and Eddy, 1979; U.S. EPA, 1974).

Some of these operations can result in the production of a sludge "cake" with solids concentrations as high as 25 percent by weight, which can be handled as a solid in spite of its 75 percent water content. Sludge with that consistency can be spread on land as a "dry" material or placed in landfills without creating moisture problems. Often, it is used as a soil conditioner, preferably on crops that are not eaten raw because of the possibility that some pathogens may remain in it.

Sludge Drying

Another step sometimes taken in removing water from sludge is the production of a product that is essentially free of water. This often is done through heating when the sludge is to be packaged for sale as a commercial fertilizer and soil conditioner. As part of the heat drying process, the sludge is brought to temperatures that also insure the destruction of all pathogenic organisms, making it safe for public use and suitable for application to edible crops.

Incineration

This exposes the sludge to very high temperatures, causing the destruction of its organic content. That eliminates its value as a fertilizer or soil conditioner, but it reduces the volume for disposal to a minimum and eliminates odor nuisances and health risks from pathogens.

Unfortunately, it is an energy-intensive process, making it less likely for use today unless it can be combined with other incineration needs of the community in which extra fuel value may be available. An example of a situation in which that sometimes is practical is the joint incineration of sludge with municipal refuse.

Ultimate Disposal

Several methods are employed in practice for the ultimate disposal of sludge. The simplest, from the point of view of preparatory treatment, is spreading it on land without dewatering. The incidence of this approach has increased in recent years, but the method requires certain caution because of the potentially harmful effects on agricultural potential of the soil and the accumulation of metals and other constituents in plants.

Other methods often used for ultimate disposal include the use of dewatered or dried sludge as a soil conditioner, disposal into landfills, subsurface injection, use for recovery of strip-mining areas, and ocean dumping, either through ships or long pipelines.

STUDY QUESTIONS

1. Compare and contrast the water distribution system and wastewater collection system for a typical municipality.
2. Sketch a flow sheet showing the main components of a *complete* municipal water system, beginning with a river source of water and ending with the outfall for discharging treated wastewater back into the stream. In a sentence or two, explain the function of each major element in the system.
3. Sketch a reasonable water supply hydrograph for a municipality over a 24-hour period, showing the flow as percent of average flow on the ordinate and time of day as the abcissa. On the same graph, sketch a reasonable hydrograph for total wastewater flow from the same municipality. Briefly explain the most important differences between the two.
4. List reasonable per capita municipal wastewater flows for the average day, the maximum day, and the peak. What would be a reasonable range of average per capita wastewater flows among various U.S. communities?
5. Identify several factors that affect municipal wastewater flows and outline the nature of those effects.

6. Give reasonable estimates for concentrations of the following constituents in municipal wastewater.
   (a) Total solids (mg/liter).
   (b) Suspended solids (mg/liter).
   (c) BOD_5 (mg/liter).
   (d) Nitrogen (mg/liter).
   (e) Phosphorus (mg/liter).
   (f) Water (%).
7. A municipality of 15,000 population has a wastewater discharge of 1.8 million gal/day (79 liters/sec) and a BOD for untreated wastewater of 175 mg/liter. What is the total discharge of BOD in pounds per day (kilograms per day)? What is the BOD discharged in pounds per person per day (kilograms per person per day)?
8. Based on data in Table 12.1 and average wastewater flow, calculate the "population equivalent" (amount contributed daily per person) for BOD, suspended solids, nitrogen, and phosphorus in municipal wastewater.
9. A 2000-gal/min (126-liters/sec) industrial wastewater discharge has a BOD of 650 mg/liter. Based on the BOD per capita ("population equivalent") from question 8, what is the municipal population that would produce the same oxygen demand as this industrial wastewater?
10. Assuming that an oxidation pond could be loaded at 40 lb BOD/day/acre (45 kg/day/ha) of surface, how many acres (hectares) of pond would be required to treat the municipal wastewater in question 7? How many acres (hectares) would be required to treat the industrial waste of question 9?
11. If the municipal oxidation pond of question 10 is constructed with an average depth of 4.0 ft (1.2 m), what would be its detention time for the wastewater? What would be the detention time in the industrial pond, with a depth of 4.0 ft (1.2 m)?
12. Using data reported in Figures 12.4 and 12.5, calculate the daily range in BOD loading on the Chapel Hill wastewater treatment plant.
13. Identify at least 10 specific potential pollutants in municipal wastewater.
14. Identify two types of preliminary treatment units and describe their functions in a municipal wastewater treatment plant.
15. Compare and contrast processes occurring in polluted streams and those in activated sludge treatment systems.
16. Draw a flow sheet for a typical complete municipal activated sludge treatment plant. Identify each unit in the plant and describe its role.
17. Draw a flow sheet for a complete trickling filter treatment plant. Identify each unit and describe its role.
18. What BOD and suspended solids removals may be expected in typical primary and secondary municipal wastewater treatment plants?

19. Identify the types of processes occurring in an oxidation pond and outline the role of each in accomplishing treatment of the wastewater.
20. Explain the role of algae in oxidation ponds. What unit or equipment satisfies the same need in an activated sludge plant? In a trickling filter plant?
21. Compare and contrast processes underway in oxidation ponds and in activated sludge plants.
22. Explain the similarities and differences of aerated lagoons and activated sludge systems.
23. What is the purpose of "return sludge" in the activated sludge process?
24. Identify several constituents of municipal wastewater that are removed little or not at all during secondary treatment.
25. What are the major mechanisms for removing potential pollutants in the activated sludge process?
26. Identify several specific goals of "advanced treatment" systems and a process that can be used to accomplish each.
27. Identify four types of adjunct processes often used in municipal plants and explain the role of each in a wastewater treatment system.
28. Identify five circumstances that would favor land disposal of a wastewater.

REFERENCES

Brown, J. C., "Alum Treatment of High Rate Trickling Filter Effluent at Chapel Hill, North Carolina," *Proceedings, National Symposium on Ultimate Disposal of Wastewaters and Their Residuals,* pp. 153–177, N.C. Water Resources Research Institute, Raleigh (April 1973).

Bunch, R. L., E. F. Barth, and M. B. Ettinger, "Organic Materials in Secondary Effluents," *Journal Water Pollution Control Federation,* vol. 33, pp. 122–126 (1961).

Bunch, R. L., and M. B. Ettinger, "Water Quality Depreciation by Municipal Use," *Journal Water Pollution Control Federation,* Vol. 36, pp. 1411–1414 (1964).

Culp, R. L., and G. L. Culp, *Advanced Wastewater Treatment,* Van Nostrand Reinhold, New York (1971).

Fair, G. M., J. C. Geyer, and D. A. Okun, *Elements of Water Quality and Wastewater Disposal,* Wiley, New York (1971).

Fair, G. M., J. C. Geyer, and D. A. Okun, *Water and Wastewater Engineering: Volume 2; Water Purification and Wastewater Treatment and Disposal,* Wiley, New York (1968).

Gehm, H. W., and J. I. Bregman, *Handbook of Water Resources and Pollution Control,* Van Nostrand Reinhold, New York (1976).

Gloyna, E. F., *Waste Stabilization Ponds,* Monograph No. 60, World Health Organization, Geneva (1971).

Hanson, R. L., W. C. Walker, and J. C. Brown, *Variations in Characteristics of Wastewater Influent at the Mason Farm Wastewater Treatment Plant, Chapel Hill, North Carolina,* Report No. 13, UNC Wastewater Research Center, Department of Environmental Sciences and Engineering, Chapel Hill (1970).

Hunter, J. V., and H. Heukelekian, "The Composition of Domestic Sewage Fractions," *Journal Water Pollution Control Federation,* Vol. 37, pp. 1142–1163 (1965).

Heukelekian, H., and J. L. Balmat, "Chemical Composition of the Particulate Fractions of Domestic Sewage," *Sewage and Industrial Wastes,* Vol. 31, 413–423 (1959).

Keefer, C. E., *Sewage-Treatment Works,* McGraw-Hill, New York (1940).

Lamb, J. C., "Evaluation of Advanced Treatment Processes for Improving Plant Performance," *Journal, N.C. Section of A.W.W.A. and Water Pollution Control Association,* or *Report No. 12, UNC Wastewater Research Center,* Chapel Hill (1970).

Ligman, K., N. Hutzler, and W. C. Boyle, "Household Wastewater Characterization," *Journal of the Environmental Engineering Division,* pp. 201–213, American Society of Civil Engineers (February 1974).

McCarty, P. L., "Anaerobic Waste Treatment Fundamentals," *Public Works,* Vol. 95, Nos. 9–12 (1964).

Metcalf and Eddy, Inc., *Wastewater Engineering: Treatment, Disposal, Reuse,* McGraw-Hill, New York (1979).

Mulbarger, M. C., "Nitrification and Denitrification in Activated Sludge Systems," *Journal Water Pollution Control Federation,* Vol. 40, pp. 2059–2070 (1971).

Sawyer, C. N., "Activated Sludge Modifications," *Journal Water Pollution Control Federation,* Vol. 32, pp. 232–244 (1960).

Schroeder, E. D., *Water and Wastewater Treatment,* McGraw-Hill, New York (1977).

Tchobanoglous, G., and R. Eliassen, "Filtration of Treated Sewage Effluent," *Journal of the Sanitary Engineering Division,* Vol. 96, No. SA2, pp. 243–265, American Society of Civil Engineers (1970).

U.S. Environmental Protection Agency, *Process Design Manual for Sludge*

Treatment and Disposal, U.S. EPA Technology Transfer, Cincinnati (1974).

U.S. Environmental Protection Agency, *Process Design Manual for Nitrogen Control,* U.S. EPA Technology Transfer, Cincinnati (1975a).

U.S. Environmental Protection Agency, *Process Design Manual for Suspended Solids Removal,* Report EPA 625/1-75-003a, U.S. EPA Technology Transfer, Cincinnati (1975b).

U.S. Environmental Protection Agency, *Environmental Pollution Control Alternatives: Municipal Wastewater,* Publication No. EPA-625/5-76-012, U.S. EPA Technology Transfer, Cincinnati (1976a).

U.S. Environmental Protection Agency, *Process Design Manual for Phosphorus Removal,* Report EPA 625/1-76-001a, U.S. EPA Technology Transfer, Cincinnati (1976b).

U.S. Environmental Protection Agency, *Process Design Manual for Land Treatment of Municipal Wastewater,* Report EPA 625/1-77-008, U.S. EPA Technology Transfer, Cincinnati (1977).

U.S. Environmental Protection Agency, *Primer for Wastewater Treatment,* Publication MCD 65, Government Printing Office, Washington, D.C. (July 1980).

CHAPTER 13

Industrial Water and Wastewater

Availability of an adequate supply of water having acceptable quality is critically important in many industrial operations and often is a major factor in the selection of plant sites. Also, the impacts of industrial water uses and wastewater discharges on the quantities and qualities of water resources can be major and complex. A book on water quality would be incomplete without addressing at least some elements of industrial water and wastewater characteristics and operations.

This chapter outlines some major categories of water uses in industries, the amounts that may be required, and water quality considerations that often are important. As pointed out earlier, a corollary to water use is the need for wastewater discharge. Therefore, several categories of industrial wastewaters also are identified, selected quality characteristics are outlined, and specific examples are used to illustrate key points about them. Special attention is directed to the variability of industrial wastewaters.

TYPES OF INDUSTRIAL OPERATIONS USING WATER

Uses of water vary widely among the many different types of industrial manufacturing and processing operations. Data in Chapter 5 show that 1980 industrial demands amounted to 58 percent of the total U.S. water withdrawals. The thermoelectric power industry is by far the largest industrial user, employing virtually all of its intake for cooling to enhance the efficiency of power generation. At the other extreme, many industries use little or no water in their operations and therefore discharge no process wastewaters. Examples of these "dry" industries include operations such as blending powdered products, forming metals, shaping plastic and wooden objects, manufacturing clothes, printing documents, and plants that use nonaqueous solvent systems (for example, "dry cleaning" of clothing).

Between the extremes are many industries that use water for a wide variety of purposes, including: pulp and paper mills, textile manufacturers, steel producers, vegetable canners, printing and dyeing establishments, metal plating shops, and meat packers. Virtually all of them use water to accomplish specific tasks in their production operations and discharge most of the flow subsequently as industrial wastewater. Several broad categories of water uses in industrial operations are outlined in the paragraphs that follow.

Transportaton of Materials

Several industries use water for hydraulic transporation. For example, the lumber industry in some regions use natural streams, flumes, and ponds to transport and store logs. Food processing plants often move vegetables from storage areas through production lines in water sluices, simultaneously washing them and separating objectionable from desirable components. Mining operations use flows to transport ores, and coal now is moved over long distances by pumping it as a water slurry. The pulp and paper industry uses water to transport chips and fibers during pulping and formation of paper products.

While transporting those materials, the water usually acquires dissolved and suspended matter removed during contact with the substances. For example, transporting and washing sugarcane or sugar beets results in solution of some sugar in the carrier water. This is not desirable because it results in loss of some of the product and adds organic matter to the wastewater, but it is an integral part of the washing and transporting step. It can be avoided only to the extent that contact with water can be minimized or other processing technology can be developed for removing undesirable impurities from the raw materials.

Separation Processes

The products in some industries are derived by separating desirable substances in raw materials from other constitutents that cannot be allowed to remain in the finished product. As examples, many fruits and vegetables must be processed to remove skins, seeds, vines, or shells before canning or freezing. Pulping involves separating the cellulose fibers needed to make paper products from lignin and other constituents of wood. The extent to which the fibers must be purified is a function of the characteristics desired in the finished paper—higher-quality papers require more thorough separation of impurities to meet product specifications. In seafood processing, the edible flesh must be separated from bones, shells, and entrails—similar in principle to the separations required in the meat processing industires.

The common factor in all of these operations is that they begin with a raw material, only part of which can be retained in the final product. That requires separating the fractions that cannot be used, often by dissolving or suspending them in water. Sometimes the discarded constituents may be recovered and used, but often they cannot and constitute most, or even all, of the potential pollutants in the wastewater. Thus, wastewater discharges must be recognized as necessary by-products of some production processes, not merely as the result of carelessness or negligence, which may be suggested to some by use of the word "waste."

Media for Chemical Reactions

In many industrial operations, chemical reactions must be carried out in water-based systems to accomplish desired changes in materials. For example, wood pulping uses chemicals dissolved in water to react with some components of the wood and facilitate subsequent separation of cellulose fibers from other constituents. Here, water provides an environment in which the desired chemical reactions can occur—a role that is separate and distinct from the transportation and separation functions outlined earlier. Of course, the water acquires substantial amounts of materials dissolved from the wood in addition to residuals of chemicals added to it initially.

The use of water as a medium in which reactions can occur forms the basis for many processes in chemical manufacturing, textile dyeing, and numerous other industries. Residuals of chemicals added to the water and reaction by-products constitute the potential pollutants in discharges from those types of operations.

Dewatering and Drying

Some wastewater flows originate in the removal of water from the process or final product by evaporation or drying. For example, sugar manufacturing processes yield a strong sugar solution that must be dewatered to produce the powdered product used in homes. Many chemicals must be reduced in water content to produce the product characteristics desired by customers or necessary in subsequent manufacturing operations. Distillates from such concentrating and drying operations frequently contain large amounts of organics and other potential pollutants.

Cooling

The large flows used for cooling in the thermoelectric power industry have been discussed in Chapter 5. The principal potential pollutant added to that water is heat. Most manufacturing operations that employ heat also require cooling at some point in the process, usually by the transfer of excess heat to water. Ideally, the only potential pollutant added to water in those systems should be heat, but often heat exchangers leak and allow other potential pollutants to enter the discharge.

Cleaning and Housekeeping

Almost all industiries use water for cleaning and housekeeping. The food and dairy industries use large amounts to clean food products, storage containers, and processing equipment. The cotton textile industry uses large flows to wash and rinse cloth before dyeing. An early step in wool processing is scouring to remove grease, oils, and other impurities from the fibers before making or dyeing the cloth. In almost all industries, water is used periodically for general housekeeping and cleanup.

Sanitary Services

Finally, water is used to provide sanitary services to meet employee needs, including drinking water and the water used for flushing toilets. Many industries also require water for employee showers and food services.

INDUSTRIAL WATER USES

Table 13.1 summaries data on water intake for eight major U.S. manufacturers and the thermoelectric industry (NAS, 1972). Water used for cooling or condensing represented 94 percent of the total industrial intake, and 87 percent of the cooling water was accounted for by the thermoelectric power industry. Among other industries, the proportions of water intake used for cooling, processing, and other purposes vary widely. In industries other than thermoelectric, 66 percent of the total intake was used for cooling, 28 percent for processes, and 6 percent for boiler feed waters, sanitary services, and other purposes. The proportions used for cooling vary from 6 percent in the leather industry to 86 percent in the petroleum and coal industries.

Water intake figures tell only part of the story, however, because substantial

TABLE 13.1
Industrial Water Intakes for Major Industries (Billions of Gallons per Year)

Industrial Group	Cooling	Process	Other	Total
Food and allied products	427	290	93	810
Textile mills	24	109	22	155
Lumber and wood	62	37	20	119
Paper, allied products	652	1,478	123	2,253
Chemicals and allied products	3,533	733	210	4,476
Petroleum and coal	1,230	95	111	1,436
Leather and allied products	1	14	1	16
Primary metals	3,632	1,207	165	5,004
Subtotals	9,561	3,963	745	14,269
Other industries	574	332	291	1,197
Total manufacturing	10,135	4,295	1,036	15,466
Thermoelectric	68,200	—	—	68,200
Total	78,335	4,295	1,036	83,666

Source: National Academy of Sciences (1972).
Note: m^3/year = billions of gal/yr × 3.785 × 10^6.

additional water demand is met through recycled flows within industrial plants, as shown in Table 13.2. Flow recycle averaged 34 percent of water intake for all industry and 131 percent for industries other than thermoelectric generation plants. The relationships between recycle and intake varied widely among the industries, ranging from 12 percent recycle in thermoelectric plants to over 400 percent of intake in petroleum and coal products plants. The trend now is upward for recycle flows in virtually all industry because of increasing water consciousness, regulatory pressures, rapid growth in the use of cooling towers, and potential for savings by reducing water intakes and wastewater discharges.

Water demands vary widely from industry to industry and also among plants within the same industry. Table 13.3 summarizes wastewater flows in gallons per ton of product that have been reported for different plants in the pulp and paper industry, for example (Gehm, 1965). The data show that flows in various pulping plants extend over a range of 15:1 and that those from different paper manufacturing operations vary as much as 50:1. Furthermore, effluent flows even from any single category of the pulp and paper industry may vary substantially, as shown in Table 13.3.

The variations can be attributed to many factors, including differences in processes used, even to make similar products. The age of the mill also can be very important because facilities designed recently have received more attention to controlling water demands, both to conserve water and to reduce costs for wastewater treatment and disposal.

TABLE 13.2
Water Intake and Recycle Figures for Major Industries (Billions of gallons per year)

Industrial Group	Water Intake	Water Recycle	Total Use
Food and applied products	810	535	1,345
Textile mills	155	174	329
Lumber and wood	119	87	206
Paper and allied products	2,253	4,270	6,523
Chemicals and allied products	4,476	4,940	9,416
Petroleum and coal	1,436	5,855	7,291
Leather and allied products	16	4	20
Primary metals	5,004	2,780	7,784
Subtotal	14,269	18,645	32,914
Other Industries	1,197	1,589	2,786
Total Industry	15,466	20,234	35,700
Thermoelectric plants	68,200	8,525	76,725
Total	83,666	28,759	112,425

Source: National Academy of Sciences (1972).
Note: m^3/year = billions of gal/yr × 3.785 × 10^6.

TABLE 13.3
Effluent Flows from Manufacturing Pulp and Paper Products

Process	Gal/Ton of Product[a]
Pulp manufacture	
Kraft and soda pulps	15,000–35,000
Sulfite pulp	40,000–60,000
Semichemical pulp	30,000–40,000
Groundwood pulp	4,000–10,000
Deinked pulp	20,000–35,000
Pulp bleaching	
Kraft and soda pulp	15,000–60,000
Sulfite pulp	30,000–50,000
Neutral sulfite pulp	40,000–60,000
Paper manufacture	
White papers	20,000–40,000
Tissues	8,000–35,000
Kraft papers	2,000–10,000
Paperboard	2,000–15,000
Specialty papers	20,000–100,000

Source: Industrial Wastewater Control, C. F. Gurnham, Copyright © 1965 by Academic, New York. Reprinted by permission.
[a] 1 U.S. gal/ton = 4.17 liters/metric ton

INDUSTRIAL WATER QUALITY NEEDS

Table 13.4 presents data compiled by a national committee in an attempt to identify quality limitations on industrial raw water supplies (NAS, 1972). Actually, the data show the maximum concentration of each constituent that had been reported for raw water supplies in the specified industry, not the maximum concentration that could be tolerated.

That point is illustrated by cooling water data; those in the table relate only to plants in which freshwater supplies were used for cooling. Other data compiled by the committee for industries using brackish cooling water supplies, for example, show a maximum chloride concentration of 22,000 mg/liter and dissolved solids of 35,000 mg/liter. Obviously, then, chloride and dissolved solids concentrations much higher than the values reported in Table 13.4 could be, and actually have been, used for industrial cooling.

As pointed out by the committee, modern water treatment technology makes it possible to adjust the quality of virtually any raw water to meet the needs of any industrial use. The treatment may require an expensive system and cover a large area but, in most instances, its cost remains only a small

TABLE 13.4
Maximum Concentrations of Constituents in Raw Waters for Various Industrial Operations (mg/liter)

Characteristic	Boiler Water[a]	Cooling Water	Textile Plants	Pulp and Paper	Chemical Industry	Petroleum
Silica (SiO_2)	150	50	—	50	—	85
Aluminum	3	3	—	—	—	—
Iron	80	14	0.3	2.6	10	15
Manganese	10	2.5	1.0	—	2	—
Calcium	—	500	—	—	250	220
Magnesium	—	—	—	—	100	85
Ammonia	—	—	—	—	—	40
Bicarbonate	600	600	—	—	600	480
Sulfate	1,400	680	—	—	850	900
Chloride	19,000	600	—	200	500	1,600
Nitrate	—	30	—	—	—	8
Dissolved solids	35,000	1,000	150	1,080	2,500	3,500
Suspended solids	15,000	5,000	1,000	—	10,000	5,000
Hardness	5,000	850	120	475	1,000	900
Alkalinity	500	500	—	—	500	500
Color (units)	1,200	—	—	360	500	25

Source: National Academy of Sciences (1972).
[a]To 1500 lbs/in^2 pressure

portion of the total for the industrial product being produced. The committee concluded that water quality at its point of use within the plant is critical for many industrial processes, but quality requirements for industrial intake water are not as stringent as they are for municipal systems, recreation, irrigation, or support of aquatic life. In short, it is not feasible to identify maximum permissible concentrations of constituents in industrial raw water supplies.

Turning to quality requirements at the point of use within process industries, we find that they often are of major importance, or even essential, to the success of the industrial operation. Table 13.5 summarizes recommendations for quality at point of use in a few industries, compiled from various sources by the same committee (NAS, 1972). There are major differences among the industries selected for this example, as might be expected. What is not shown is that there also are wide variations in quality requirements among plants within each industrial category because of differences in characteristics of finished products and processes used to produce them.

Furthermore, a single industrial plant often may include several operations and processes. Each may have its own water quality requirements, which sometimes may be satisfied by using different treatment processes and water distribution systems within the industrial plant. As an example, consider a plant that produces a chemical product in steam-heated reactors and subsequently must

TABLE 13.5 Examples of Maximum Concentrations of Constituents at Point of Use in Industrial Operations (mg/liter, unless otherwise noted)

Characteristic	Boiler Water[a]	Cooling Water	Textile Plants	Kraft Paper	Tanning Industry	Soft Drinks
Silica (SiO_2)	0.7	50	25	50	—	—
Aluminum	0.01	—	—	—	—	—
Iron	0.05	—	0.3	0.2	0.3	0.3
Manganese	0.01	—	0.05	0.1	0.2	0.05
Calcium	—	200	—	20	150	—
Ammonia	0.1	—	—	—	—	—
Bicarbonate	48	600	—	—	—	—
Sulfate	—	680	100	—	250	500
Chloride	—	600	100	200	250	500
Dissolved solids	200	1000	200	300	—	—
Suspended solids	0.5	5000	5	—	—	—
Hardness	0.07	850	50	100	40	—
Alkalinity	40	500	200	75	—	85
Color, units	—	—	5	25	5	5
Turbidity, units	—	—	5	40	0	—
Coliform bacteria (per 100 ml)	—	—	—	—	1	1

Source: National Academy of Sciences (1972).
[a]To 1500 lb/in^2 pressure.

cool it. This facility could require several different water qualities, including (1) boiler water, (2) cooling water, (3) water to satisfy the needs of the chemical process, (4) water for drinking, and (5) water for housekeeping and sanitary services for the employees. Sometimes all these needs might be satisfied by a single source and treatment system but, under other circumstances, two or more separate sources and treatment systems may be used. Of course, in manufacturing plants with many different processes, each could require somewhat different quality, compounding the variety of water quality needs.

GENERAL CATEGORIES OF INDUSTRIAL WASTEWATERS

It is convenient in this discussion to divide the many different types of industrial wastewater discharges into a few broad categories that differ substantially, both in origins and characteristics. In some instances, all of the flows may be combined for joint collection, treatment, and disposal; in others their differences in character favor or require segregation and separate handling. Some of the categories most commonly encountered are outlined in the following paragraphs.

Sanitary Wastewaters

The sanitary wastes have characteristics similar to domestic sewage from a small municipality, but the flows per person working in the plant are much smaller—perhaps 25 to 50 percent. They include body wastes, flows from employee showers and perhaps from food preparation and dishwashing if dining facilities are provided. In rare instances, their characteristics may be influenced somewhat by materials laundered from clothing worn by the workers, but that usually is negligible.

Generally, the sanitary flows offer no special problems and may be handled with typical municipal wastewater technology or by combining them with the industrial wastewaters for joint treatment. In many plants they are collected separately for treatment apart from the industrial wastes or for separate discharge to municipal sewerage systems, when available.

Process Wastes

This wastewater category includes the discharges from processing of materials within industrial operations. Their constituents usually include raw materials used in the manufacturing processes as well as finished products and intermediates produced in the processes. In most industries, process wastes contain the bulk of potential pollutants discharged from the plant, but often they represent only a small portion of the total hydraulic flow. Usually, they offer the greatest potential for causing stream pollution problems and require treatment to satisfy regulatory requirements. In plants where several different processes are employed, it may be necessary to install and operate more than one type of system for collecting and treating process wastes.

Cooling Waters

As pointed out earlier, the principal function of much industrial water is to remove heat from process operations, usually by flowing through equipment in which the heat source is on one side of an exchange surface and cooling water is on the other. If the heat-exchange equipment is well maintained, there should be little or no addition of potential pollutants to the cooling water, except for heat, often making it possible to discharge this flow directly to the watercourse without treatment. However, leakage sometimes does occur, allowing process reactants to escape into the cooling water. Poorly maintained equipment can result in highly contaiminated discharges and the need for extensive treatment to remove various constituents.

In many plants, especially where cooling towers are used, substantial amounts of chemicals are added to cooling waters to control corrosion and to inhibit biological growths in the system and on heat exchange equipment. The discharges may contain chlorine, other biocides, phosphates, inhibitory organics, and metals. Often they could have significant effects on aquatic organisms in the watercourse, requiring treatment before release.

Cooling waters frequently constitute a large fraction of the total wastewater flow. The cooling towers used to remove heat, when needed, differ so greatly from the units used to remove other potential pollutants from process wastes that little would be gained, usually, by combining cooling and process wastes for treatment. Therefore, they often are collected, treated, and discharged separately.

Boiler Waters

Many industrial plants operate boilers to produce steam or electricity and, as pointed out earlier, quality requirements for water added to them usually are high. Chemicals added to the raw water to prepare it for use in boilers, as well as constituents removed from the raw water, constitute another wastewater flow. Also, the "blowdown," or concentrate left in the boiler after generation of steam, must be discharged periodically, releasing more chemicals into the watercourse.

The chemical and biological characteristics of these flows differ greatly from process wastes, sanitary flows, and cooling waters. They often are considered and handled separately.

Storm Runoff

The runoff from rain on plant property may have characteristics that vary widely, depending on the type of industrial operations being carried out and the opportunity for escape of process chemicals to areas where they can be removed by the precipitation. When operations are contained entirely within buildings, it may be possible to discharge storm runoff directly to the watercourse without treatment. On the other hand, where spills could occur periodically in exposed areas, or where process wastewaters discharge on the ground occasionally, storm runoff may be highly contaminated, making it necessary to collect and treat it.

Sometimes process wastewater treatment facilities are also suitable for treating storm runoff if a holding basin is provided to store it for gradual discharge into the treatment system. In other instances, the types of treatment needed may differ greatly, making it necessary to install entirely separate facilities.

INDUSTRIAL WASTEWATER FLOWS AND CHARACTERISTICS

Most industrial intake water usually is discharged subsequently as wastewater. However, flow may be divided among the categories outlined previously in very different proportions, depending on the nature of the industrial operations. Therefore, the amounts of industrial wastewater that can be discharged without treatment and those requiring treatment vary widely among different industries. This makes it necesary to consider wastewater flows separately for each plant and allows only crude generalizations about them.

The same types of differences restrict our ability to generalize about the quality of industrial wastewater discharges. Table 13.6 summarizes some characteristics of wastewaters from a few industries and illustrates quality variability among them for the four parameters cited (Water Pollution Control Federation, 1976). The ranges of parameters within each industrial group show the scope of variations commonly observed in overall discharges from those types of plants, but it must be emphasized that far wider ranges often are seen!

That is illustrated by data in Table 13.7, showing wastewater characteristics in the metal finishing industry (U.S. EPA, 1977). Special note should be made of the range of concentrations for each constituent—sometimes exceeding 50:1. Of course, it is unlikely that any single metal finishing establishment would perform operations leading to the discharge of all of the potential pollutants listed in the table. On the other hand, wastewaters from some plants may contain potential pollutants not listed in the table, perhaps because of unusual operations within that plant.

Data found in tables such as these must be used with caution because of the wide variations found in practice. Nevertheless, they may be helpful by iden-

TABLE 13.6
Characteristics of Selected Industrial Wastewaters

Industry	BOD (mg/liter)	Suspended Solids (mg/liter)	pH	Metals
Meat	300–1000	300–1000	Neutral	None
Milk	200–1000	200–300	Acid-alkaline	None
Cheese	1000	1000	Acid-alkaline	None
Malt drinks	300–1000	200–1000	Acid-Neutral	None
Wool	300–1000	300–1000	Alkaline	Yes
Cotton	200–1000	200–300	Alkaline	Yes
Chromium tanning	1000	1000	Acid-alkaline	Yes
Vegetable tanning	1000	1000	Acid-alkaline	None
Petroleum	200–300	200	Acid-alkaline	Yes
Metal finishing	Low	200–1000	Acid	High
Fruits	200–1000	200–1000	Acid-alkaline	None
Mechanical pulp	300–1000	300–1000	Neutral	None
Chemical pulp	300–1000	200–1000	Acid-alkaline	Yes

Source: Operation of Wastewater Treatment Plants, C. F. Guarino, et al., Copyright © 1976 by Water Pollution Control Federation, Washington, D.C., Reprinted by permission.

tifying some of the constituents that are likely to be in those types of industrial wastewaters and the range of variations often encountered in them. More information may be found in the thousands of reports and articles that describe the characteristics, effects, and treatment of wastewaters from different industries. Detailed evaluation of industrial wastewater discharges and their probable impacts on water resources require individual study of the specific establishments, or a prototype plant when one exists.

PROBLEMS CAUSED BY INDUSTRIAL WASTEWATERS

The effects of an industrial wastewater on a receiving stream depend on the types and amounts of potential pollutants in the discharge and on many factors about the stream and other wastewaters entering it. Preliminary estimates often can be made based on information in the literature dealing with similar industries, but those sources should be considered as guidelines only and must be used with caution.

The potential impacts of industrial discharges may include all of those discussed in earlier chapters and, often, various combinations of them. Evaluation of those impacts in specific situations seldom is as straightforward as it may seem upon first examination of the industrial operation, its discharges, and pertinent literature.

TABLE 13.7
Characteristics of Raw Wastewater from Metal Finishing Industries

Parameter	Concentration (mg/liter, unless otherwise noted)
Flow (gal/day)	10,000–500,000
Flow type	Continuous or intermittent
Suspended solids	5–20
Dissolved solids	500–1000
pH	3–13
Zinc	10–100
Iron	10–1000
Cadmium	10–100
Nickel	10–100
Copper	10–100
Lead	10–100
Sodium	500
Aluminum	0–100
Chromium, hexavalent	8–400
Chromium, total	10–500
Cyanide	20–100
Fluoride	5–500
Phosphate	5–500

Source: U.S. Environmental Protection Agency (1977).

The complexity of industrial wastewater effects may be influenced radically by interrelationships among various constituents in the flow and in the receiving stream. For example, it often has been noted that the simultaneous presence of copper and zinc in water may be synergistic; that is they may cause toxicity to fish that is more serious than would be predicted by considering the effects of those chemicals individually. Sometimes antagonistic effects also have been observed, in which the impact of one chemical may be diminished by the presence of another. In other instances, chemical reactions between a constituent in the discharge and others in the wastewater or stream may radically increase or decrease its potential for interfering with beneficial uses of the watercourse. The effects of cyanides, ammonia, and sulfides, for example, can be altered radically by relatively small changes in the pH of the water.

The breadth and complexity of some of these matters may be illustrated by briefly reviewing some water quality problems that have been observed in connection with various segments of the pulp and paper industry. The principal potential pollutants in wastewaters from that industry generally are considered to be oxygen-demanding substances (BOD) and suspended solids, which can cause DO problems in watercourses and accumulations of benthic deposits downstream. However, as outlined by Gehm (1973), many problems other than those obvous ones also may result from the discharges.

In some instances, wood sugars and other decomposition products have stimulated growths of filamentous organisms, such as *Spherotilus natans,* in receiving streams. Those organisms attach to rocks, branches, and other submerged surfaces and create unsightly conditions. Also, suspended masses of them have caused trouble by clogging the nets of commercial fishermen and water intake screens.

Persistent organics in pulp and paper mill wastes include, among other things, chemicals that can impart an undesirable color in downstream water supplies. Its removal may be difficult and require increased expenditures for chemicals in water treatment processes in addition to control problems in plant operation.

Inorganic chemicals that can cause difficulties, unless controlled properly, include acids and alkalis. The total solids content of receiving waters may be increased significantly by sulfates, chlorides, sodium, and calcium contributions from pulping and bleaching operations, perhaps impinging unfavorably on quality of the water for municipal supplies, irrigation, support of aquatic life, and other beneficial uses. Suspended inorganic solids include silt, clays, talc, calcium carbonate, and titanium dioxide, all of which can cause increased turbidity and inhibit biological action by reducing light penetration. Also, they may settle and form deposits that blanket bottom organisms that otherwise might serve as food for some types of fish.

Sulfides, organic sulfur compounds, resin acids, and terpines in kraft pulping effluents may impart objectionable tastes and odors to receiving waters and to fish and other aquatic life in them. Sometimes, tastes and odors in fish downstream from those mills have been caused by the use of kerosene for controlling foam in their wastewater treatment plants.

Pulping effluents occasionally have been associated with kills of aquatic and marine life, including fish. Some of these problems have been attributed to low DO concentrations in the watercourse, caused by high BOD loadings. In other instances, they have been caused by constituents that are toxic to aquatic life, such as sulfides, mercaptans, resin acids, terpines and metals from chemicals used in pulp processing, additives in paper making, or corrosion of equipment.

In view of the preceding outline of potential problems, wastewaters from this industry clearly require careful control to avoid water pollution problems. In this respect, they are no different basically from discharges from many other industries that may cause serious problems if not properly handled. The extent to which pulp and paper industries have been successful in preventing the deterioration of watercourses receiving their wastes is a tribute to the effectiveness of the controls that have been applied by them.

EFFECTS IN MUNICIPAL SEWERAGE SYSTEMS

In addition to possible effects on quality of watercourses, many industries discharge wastewaters into municipal sewerage systems and have the potential for causing a wide variety of problems, including the following.

1. Hydraulic overloading of sewers and treatment plants.
2. Organic (BOD) overloading of treatment plants.
3. Hazards in sewers and treatment plants because of toxic gases and explosive solvents.
4. Corrosion of sewers and treatment plant elements.
5. Excessive wear of sewers and equipment by abrasive materials.
6. Interference with biological treatment by toxic effects on organisms.
7. Overloading of sludge-handling facilities.
8. Inadequate treatment of the industrial flow.
9. Generation of objectionable odors, unsightly appearance, and excessive maintenance costs.

Effective controls over industrial wastewater discharges into municipal systems are essential if problems such as the ones listed are to be avoided. That usually is accomplished through enacting municipal ordinances that regulate the quality of wastewaters that can be discharged into the sewers. Often, pretreatment of industrial flows may be necessary to protect the sewers, treatment plant, and personnel from harmful effects by the wastewaters.

When industrial wastes are compatible in treatment needs with municipal sewage, and proper controls are exercised, joint treatment of the two can result in substantial economies for all concerned. Also, joint treatment and disposal usually result in more effective treatment and better protection of the receiving stream.

STUDY QUESTIONS

1. Identify six major categories of water uses in industrial plants and briefly outline the nature of each as it relates to industrial wastewater quality.
2. For each of the categories in question 1, identify two industries that use water for that purpose.
3. Cooling towers are used in many industrial plants as a means for increasing water recycle within the plant. Outline and briefly discuss the advantages and disadvantages of this approach.
4. Water quality criteria generally have not been published for raw waters to be used as industrial supplies. Explain why.
5. Identify several constituents of water that sometimes must be limited in

concentration at the point of industrial use. Select three of those constituents and explain why such limits might be appropriate for them.

6. Identify five major categories of aqueous discharges from industrial plants and outline typical origins of those flows.
7. Outline probable general differences in quality characteristics of discharges in each of the categories identified in question 6.
8. What is the potential for wastes in each of those categories with respect to their probability for needing treatment and likely complexity of such treatment systems, where needed?
9. Outline several types of problems that may be caused in watercourses by industrial discharges.
10. Outline several types of problems that may be caused in municipal sewerage systems and treatment plants by industrial wastes.

REFERENCES

Conway, R. A., and R. D. Ross, *Handbook of Industrial Waste Disposal,* Van Nostrand Reinhold, New York (1980).

Gehm, H. W., "Pulp and Paper," in *Industrial Wastewater Control,* by C. F. Gurnham, Academic, New York (1965).

Gehm, H. W., *State-of-the-Art Review of Pulp and Paper Waste Treatment,* Report No. EPA-R2-73-184, U.S. Environmental Protection Agency, Washington, D.C. (1973).

Gurnham, C. F., *Industrial Wastewater Control,* Academic, New York (1965).

Lamb, J. C., *Guidelines for the Control of Industrial Wastes: 5. Pulp and Paper Manufacturing Wastes,* Publication WHO/WD/72.11, World Health Organization, Geneva (1972).

National Academy of Sciences, *Water Quality Criteria 1972,* Washington, D.C. (1972).

Nemerow, N. L., Industrial Water Pollution, Addison-Wesley, Reading, Mass. (1978).

U.S. Environmental Protection Agency, *Federal Guidelines—State and Local Pretreatment Programs,* Appendix 8, Report No. EPA-430/9-76-017c, Washington, D.C. (1977).

Water Pollution Control Federation, *Operation of Wastewater Treatment Plants,* Manual of Practice No. 11, W.P.C.F., Washington, D.C. (1976).

CHAPTER 14

Standards Used in Water Quality Control

The successful development of a water quality control program requires early and clear decisions about its goals. The legal framework that forms the foundation of a regulatory program can specify its broad goals and create mechanisms to attain them, but laws enacted by legislative bodies cannot define program details or methods of implementation. Instead, they should establish guiding policies and provide legal structures that are strong enough to facilitate enforcement and flexible enough to allow the implementing agency to function efficiently and economically from day to day.

Within that legal framework, the designated agency establishes regulations that define the objectives in more detail, describe procedures for program implementation, and identify enforcement techniques to be employed. A vital part of the process is setting standards that must be met by those required to conform with the laws and regulations. Standards function as the backbone of a water quality control program and provide the language in which agencies communicate their requirements to those being regulated. They also serve as yardsticks by which performance of a regulated individual is evaluated and supply the initial basis and first line of evidence in legal action against those judged to be out of compliance with the regulations and laws. In many respects, the success or failure of a pollution control program hinges on the care, accuracy, and precision with which standards are set.

For those reasons, it is essential that one who would understand water quality control activities first must understand very clearly the nature and bases of the standards that are used. It is necessary to know just what they are and how they are derived, but it is equally important to know what they are not. One must comprehend both the value and utility of good standards and the limitations and risks involved in using poor ones.

This chapter will illuminate some of those matters. First, to avoid semantic problems, we should differentiate clearly between water quality objectives, criteria, and standards.

WATER QUALITY OBJECTIVES

The term "objective" refers to an aim or goal toward which one is proceeding. An objective in water quality control may represent either a realistic goal that probably will be attained or an ideal condition toward which one wishes to proceed but has little or no expectation of actually reaching. An example of the latter is the objective specified in 1972 Amendments to the Federal Water Pollution Control Act: " . . . it is the national goal that the discharge of pollutants into the navigable waters be eliminated by 1985. . . ." The impracticability of reaching that objective was discussed in Chapter 5 and now is recognized by most legislators and regulatory officials. Nevertheless, it remains in the law as a philosophical national "objective" and serves as an ideal toward which pollution control programs can be aimed.

Frequently, water quality objectives established by agencies are general statements that do not lend themselves effectively to enforcement insofar as any single discharger might be concerned. For example, an objective specified for a stream may state simply that its quality should be suitable for the survival and propagation of certain types of fish. This identifies the goal toward which the regulatory program is directed and defines, in readily understandable terms, its end result if it is to be fully successful. It does not identify specifically the water quality characteristics that must be maintained in the stream to make that goal possible or the types of facilities that must be installed to do so.

WATER QUALITY CRITERIA

As discussed in Chapter 6, "criteria" represent the best efforts of knowledgeable groups to define specific quality characteristics that are necessary to support various water uses, based on state-of-the-art information. They provide bases for judging whether water quality in a resource is suitable for its proposed uses and, if not, what specific changes might be necessary to make it so.

Examples of published water quality criteria have been summarized in Table 6.1, and their limitations and uses are discussed in the accompanying text of Chapter 6. They need not be repeated here except to remind the reader that criteria are not absolute values, but judgment calls that include safety margins and are subject to change as new knowledge and conditions develop.

WATER QUALITY STANDARDS

Water quality standards are different from objectives or criteria and should not be confused with either (NAS, 1972; U.S. EPA, 1976). Standards are the means by which regulatory agencies define their requirements for watercourses and for each individual regulated. The development of standards must include careful attention to program objectives and quality criteria for the designated uses; but they also must consider several additional facets of the matter, including economic aspects, public needs and desires, political reactions, and practical attainability. As contrasted with criteria, standards are relatively rigid reflections of laws and regulations and usually are viewed as absolute—they are either met or violated.

Water quality standards may not always relate directly to the specific quality criteria for uses to be protected. In many instances, they are constructed in close agreement with published criteria, but sometimes they may include additional safety margins because of a special need for conservatism, as perceived by the responsible agency. On the other hand, local conditions occasionally permit adequate protection of designated uses with standards less restrictive than published criteria, which could have been developed to meet more stringent needs imposed by special factors encountered elsewhere.

One should recognize that, under some circumstances, standards might not be based on published criteria at all but on other sources of information, on erroneous data, on excessive zeal in setting safety margins, on political and public pressures, or even on an arbitrary whim of some official or group. Today, many standards in wastewater treatment, to be described later, are not related to quality in the receiving stream but are based solely on the types of technology available for treatment of the wastewater in question. These are referred to as "technology-based" standards.

A factor that should be recognized as important in setting standards is what Phelps (1948) referred to as the "principle of expediency;" that is, the level at which a standard is set often depends heavily on what is attainable in the real world, technically and at a cost viewed as reasonable. McGauhey (1965) described an example of this relative to the total solids limitations in current drinking water standards (500 mg/liter). He stated that an effort some years ago to trace the origin of that requirement was unsuccessful, causing him to conclude that it was set at that level only because it seemed to be both desirable and attainable in most drinking water supplies.

A more modern example in water pollution control is the technology-based effluent standards, referred to previously, that now are being imposed on most dischargers. In municipal wastewater treatment, for example, the minimum accepted standard requires at least secondary treatment, regardless of flow or condition of either waste or stream. That is not based on the needs of any stream or on any considerations of costs and benefits, but only on the ability of agencies operating in today's technological and social environments to require and enforce that level of treatment.

It should be clear at this point that water quality criteria and water quality standards are not interchangeable terms. They are established by different groups for different purposes and may or may not be closely related to each other in practice. Sometimes standards are set at concentrations that are identical to applicable criteria but that may be inappropriate under other circumstances.

BASIC TYPES OF STANDARDS

For discussion here, the standards used in water quality control are divided into three broad classifications.

1. Design standards.
2. Performance standards.
3. Procedural standards.

Design standards specify certain characteristics that are required in water supply or wastewater disposal facilities. They may indicate types of plants that must be used, loadings or minimum sizes of units in those plants, or characteristics of materials used in their construction. In short, they establish how something must be built or done as contrasted with identifying the type of performance that must be attained subsequently.

Performance standards, on the other hand, specify the end result that must be produced, whatever types or sizes of facilities are used. They may indicate, for example, the maximum concentration of some constituent in a municipal water, leaving it to the planner and designer to work out details of whether facilities are needed to change its concentration from that in the raw water and, if so, what characteristics and sizes of units should be provided.

Procedural standards define various approaches and methods that must be followed in water quality control activities. They include requirements for permits to construct or operate treatment plants, methods for monitoring their performance, and many other administrative aspects of the relationships between those who control water quality and the regulatory agencies that oversee them. Procedural standards do not specify either the facilities that must be used or the results to be attained through their operation.

All of the different types of standards can be and are used effectively in water supply and pollution control activities. They are not mutually exclusive, but often can be most valuable and effective when applied simultaneously to regulate various aspects of water quality control as part of a flexible program. Each has an important role to play and offers advantages and disadvantages in specific applications that should be understood and carefully considered. They will be discussed briefly in the following sections, and examples will be given of their uses in various aspects of water quality control.

DESIGN STANDARDS

Design standards may be divided into three categories, each of which follows a somewhat different approach in regulating facility design and construction. The three may be used individually or in various combinations, depending on the situation and needs.

Specifying the Type of System to Be Employed

The broadest application of design standards is in defining the types of facilities that must be used. For example, current standards for wastewater collection systems commonly require separate sewers for sanitary flow and storm runoff from the community. Usually, the storm flow is discharged directly to a stream without treatment, while sanitary sewers deliver their flows to treatment plants for removal of potential pollutants before discharge. Today, there is occasional interest in requiring treatment of storm runoff as well, but the installation of separate sewerage systems permits the use of different types of treatment for the two wastewaters. Industries often have more complex systems with separate sewers for sanitary wastes, storm runoff, cooling waters, and one or more types of process wastes, as discussed in Chapter 13. Each flow may be treated in different fashion before discharge.

Another example of applying design standards is in specifying the type of treatment facility to be provided. Most water supply regulatory agencies require certain minimum treatment of water from surface sources before its use in municipal systems; typical requirements might be for coagulation, settling, filtration, and chlorination. A more indirect example, discussed earlier, is the design standard specifying that a drinking water supply should be taken from the best source that can be developed feasibly.

An example of a design standard in wastewater treatment is the current national requirement that municipalities must install a minimum of secondary treatment. Another is the requirement often imposed by agencies that effluents from wastewater treatment plants must be chlorinated to provide protection against transmitting waterborne diseases.

Size or Other Characteristics of Units

Design standards are widely used to specify sizes of units in water and sewerage systems. For example, it is customary to specify minimum pipe diameter in municipal water distribution networks to insure adequate capacity for fire flows. Also, minimum diameters and grades to be used in constructing sewers usually are specified to avoid problems with accumulations of deposits and loss of capacity. Even maximum manhole spacings may be stipulated to insure ready accessibility for maintenance after construction.

Standards for designing treatment plants often define minimum sizing of some units, for example, minimum detention times in primary settling tanks, maximum surface overflow rates, BOD loadings for biological treatment units, and capacities of aeration devices. As indicated earlier, a design standard may require chlorination of the treatment plant effluent. In addition, there may be another design standard specifying minimum capacity of the chlorination facilities.

Characteristics of Materials and Equipment

Another type of design standard frequently employed specifies the important characteristics of materials and equipment used in piping and treatment systems. Technical specifications have been developed by trade and professional organizations for various types of pipe, valves, and other elements in water and wastewater systems. They may include types of materials to be used in their construction or certain mechanical features that should be included in them. For example, fire hydrants may be required to have drains to prevent freezing when not in use. Other specifications may require that all valves open when turned in the counterclockwise direction to avoid confusion and damage during their use.

Frequently, those specifications are incorporated by reference into design standards issued by regulatory agencies. Sometimes, the agency may modify the recommendations to fit special local needs or different views, or they may even issue their own design standards.

PERFORMANCE STANDARDS

A wide variety of performance standards have been used in water quality control. All have been aimed, directly or indirectly, toward controlling finished water quality, but they approach the question in somewhat different ways. In the following sections, three commonly used types of performance standards are discussed briefly.

Stream Standards

The rationale behind stream standards is based on establishing regulations in pollution control programs targeted on the specific conditions that must be maintained to meet objectives for the watercourse. For many years, it was common practice for agencies to issue stream standards to some dischargers, allocating to them direct responsibility for maintaining the desired conditions downstream from their outfalls. For example, if the goal was to maintain 4.0 mg/liter DO in the stream at all times, the standard issued might contain just that specification, making the discharger responsible for installing and operating whatever facilities might be necessary to meet it. Stream standards of that type are based on performance; that is, they do not specify the nature of facilities that must be used but only the end products of whatever actions may be taken by dischargers.

For many reasons, stream standards are seldom used today for direct application to individual dischargers, but they remain an important part of regulatory programs because they are used to define specific water quality characteristics that must be maintained in the watercourse. Accordingly, they serve to identify clearly and precisely the quality goals to which the agency is committed and present them in terms that can be recognized and evaluated by all interested parties. In addition to stream water quality standards, however, performance standards now usually are issued simultaneously to specify the quality of effluent that each discharger can release into the watercourse.

Effluent Standards

These specify limitations on the quality of water to be produced by the regulated individual. In the water supply field, most of the constituent concentrations listed in the EPA drinking water standards (See Table 11-1) are performance standards. They define the quality that must be attained in finished drinking water delivered to the consumer without specifying the types of facilities that must be used to produce that quality or how they must be operated.

In water pollution control, effluent standards are used to specify the quality of treatment plant effluent released into the watercourse, again without specifying how that might be attained. One approach that was used widely for many years was based on defining the maximum concentration of various potential pollutants in the discharge. A disadvantage in that type of standard is lack of control over the *quantity* of potential pollutants unless the rate of flow also is regulated. This control technique sometimes actually discouraged water conservation in industrial plants because reduction in flow from a plant discharging a fixed amount of chemical would raise its concentration in the effluent, making it more difficult to meet a standard based on concentration. In fact, some industries have been known actually to increase their uses of water, or to mix strong wastewaters with clean cooling waters before discharge, to provide more dilution and make it easier to meet effluent concentration standards. Although this approach might "solve" the industry's effluent concentration problem, it does nothing to alleviate stream pollution, which depends principally on quantities of potential pollutants discharged into the watercourse.

An alternative type of effluent standard, and one now commonly used in the United States, specifies the *quantities* of potential pollutants allowed in the plant effluent as opposed to their concentrations. This type may limit, for example, the pounds per day of BOD or suspended solids that can be released into a watercourse, whatever the amount of wastewater flow in which it might be contained. This provides a more meaningful method for regulating wastewater discharges because it limits the release of potential pollutant to the amount that can be added to the stream without contravening its water quality standards.

Removal Standards

Another type of performance standard is based on specifying the degree of removal to be attained in the treatment plant. For example, it might specify a percentage of raw waste BOD that must be removed by the facility.

This type of standard was used often several years ago, but it suffers from basic weaknesses. It does not control the parameters that are of concern in determining water quality as directly as do other types of standards described before. Also, it fails to recognize advances that might be made by industries in reducing constituents in their discharges through inplant changes and improved controls on manufacturing processes.

The current U.S. standard for municipal wastewaters specifies "secondary" treatment as a minimum. That has been defined further as 85 percent removal of BOD from raw municipal wastewater. As pointed out earlier, it might be viewed as a design standard, specifying the type of plant to be built. However, it also can be viewed as a performance standard requiring 85 percent BOD removal.

PROCEDURAL STANDARDS

These standards are devoted mostly to administrative aspects of water quality control, ranging from the maintenance of proper working relationships between the discharger and the regulatory agency to procedures that must be followed in applying for permits and obtaining approval of proposals for treating and monitoring discharges. Several specific items for which procedures and standards often are established include the following.

1. Review and approval of plans for water and wastewater treatment facilities by state and federal agencies to insure that they conform with pollution control requirements and state and federal grant mechanisms.
2. Issuance of construction permits, authorizing building of the facilities in accordance with design standards and approved plans and specifications.
3. Definition of NPDES (National Pollutant Discharge Elimination System) performance standards, with which the plant must conform during its operation.
4. Issuance of operating permits, including NPDES standards, authorizing the owner to start up and operate the plant upon its completion.
5. Design and operation of effluent- and stream-monitoring programs, under which sampling and analyses must be carried out during plant operation.
6. Methods and timing for reporting to the regulatory agency on plant operation and results of the monitoring program.
7. Procedures to be followed in the event of violations of permits.
8. Requirements and methods for licensing of plant operating personnel.

COMPARISON OF DESIGN AND PERFORMANCE STANDARDS

Design standards offer the advantage of being easily described by regulatory agencies and readily understood by plant designers. Using specific design standards for unit sizes and loadings makes it easy for regulatory personnel to check on plans and specifications before approving them for plant construction and leaves less room for argument as to whether or not a design meets the standards. They usually insure that the plant design will conform with certain desirable practices and often help guarantee meeting at least some minimum level of performance desired by the regulatory agency.

Unfortunately, once design standards become established in a regulatory agency, there often is a tendency for them to be viewed and administered rigidly. Sometimes the standards must be established with poor scientific and engineering foundations—perhaps as a temporary expedient—but it is too easy for some regulatory personnel to forget the caveats subsequently and insist on applying them inflexibly, even when they might not apply to the situation under consideration.

The combination of design standards and an inflexible regulatory agency could inhibit desirable innovations because they often result in protracted discussions and disagreements between designers and agency personnel about departures from the standards. Some have expressed the view that this can retard advances in water quality control and can sometimes even result in the construction of treatment plants doomed to failure because they are designed to conform with the standards instead of the actual needs of the specific situation. In some instances, detailed design standards can become a crutch for incompetent design and regulatory personnel who may be unwilling to depart from them, even when field conditions require different approaches. As indicated by Sedgwick many years ago, standards sometimes can be the refuge of lazy minds (Phelps, 1948).

Performance standards, on the other hand, permit maximum flexibility and initiative in design and operation of wastewater treatment facilites because they specify the end result and leave it to the designer or operator to decide how best to get there. This provides the latitude to determine the best way to meet the standard, including consideration of all treatment alternatives as well as other possibilities such as inplant changes in industrial manufacturing processes, for example.

A desirable effect of performance standards is that they leave the responsibility for decisions about processes and facilities with the discharger, where it logically belongs. At the same time, it places the agency in its proper role—regulating water quality, instead of attempting to design and operate treatment plants. It is especially treacherous for agencies to risk falling into the trap of assuming part of the responsibility for a design that they must subsequently review and control—an obvious conflict of interest. Later failure of the facility could place the agency in a very embarassing position and could preclude effective regulatory actions.

For many years, there have been heated discussions between proponents of effluent standards and stream standards about which type has the advantage (Lamb, 1963). Frequently, the arguments obscure the fact that the final goal of either should be protection of water quality in the receiving stream. Sometimes it seems that the most direct regulatory approach would be to base standards on stream quality because they would conform exactly with the result being sought, whatever the local conditions of flow, discharge, temperature, or other parameters. Stream standards actually have been used successfully for many years where a relatively long stretch of stream receives input of potential pollutants from a single major contributor. In this situation, it is often possible to correlate conditions in the stream directly with discharges from that point-source (Okun et al., 1963).

Unfortunately, however, when many municipalities and industries are located in close proximity along a stream, it is impractical to use stream standards alone because one cannot differentiate clearly between the effects of the various dischargers. Under these circumstances, diffuse responsibility for stream conditions usually make it necessary for the agency to allocate assimilative capacity of the stream among present and potential future dischargers. That is accomplished by using effluent standards that define the quality of each discharge and thereby minimize the ambiguities about which dischargers are meeting their pollution control obligations and which are not.

Note that there is no valid reason why effluent standards must specify only one quality level. It is entirely feasible to establish variable effluent standards that fit the needs of a stream under different conditions of flow and temperature. That currently is being done in most states of the United States and will be discussed in more detail in Chapter 15 (Lamb et al., 1980, 1984).

A major disadvantage of all types of performance standards is that they are viewed as more difficult to administer than design standards because they require frequent checks on actual performance of the system (that is, more monitoring). Also, the development and approval of plans and specifications for a treatment plant may require more thought by designers and reviewers because its performance may be difficult to predict. However, much of the administrative difficulty lies in perception, as opposed to reality, because the same questions about suitability of design and projected performance really should be raised anyway, whatever types of standards might be employed.

STUDY QUESTIONS

1. Differentiate among water quality objectives, criteria, and standards.
2. Explain the different roles of objectives, criteria, and standards in planning and implementing pollution control programs.
3. Explain why it often could be inappropriate to adopt water quality criteria directly for use as standards in a pollution control program.
4. Distinguish the conceptual differences between design and performance standards and give three specific examples of each to illustrate those differences.
5. Outline and briefly explain the relative advantages and disadvantages of design and performance standards as tools for regulatory control of water quality.
6. Outline and briefly explain three categories of performance standards that have been used for regulating wastewater treatment and disposal and provide one example of each.
7. Identify five specific performance standards for the quality of water supplied in a municipal system.
8. Outline and briefly explain three categories of design standards used in the water and wastewater fields and give one example of each.

9. Discuss the advantages and disadvantages of stream and effluent standards in controlling water pollution.
10. What are "procedural standards?" Identify five specific areas in which procedural standards are used.

REFERENCES

American Water Works Association, "Various Standards Published for Components of Water Supply Systems" (See *Current Annual Directory,* AWWA, Denver.)

Cleary, E. J., "Effluent Standards Strategy: Rejuvenation of an Old Game Plan," *Journal Water Pollution Control Federation,* Vol. 46, pp. 9–17 (1974).

Cleary, E. J., *The ORSANCO Story,* Johns Hopkins Press, Baltimore (1967).

Hull, D., *A Survey of the Use of Flexible Standards by State Water Quality Management Agencies,* Unpublished report submitted to the faculty of the University of North Carolina, Department of Environmental Sciences and Engineering (1983).

Kimm, V. J., "Report on EPA Regulations," *Journal American Water Works Association,* Vol. 69, pp. 482–485 (1977).

Lamb, J. C., "Effluent and Stream Standards—A Rational Approach," *Proceedings of the 1963 Florida Industrial Waste Workshop,* pp. 102–112 (1963).

Lamb, J. C., "Water Quality Control—The Trouble With Standards," *ESE Notes,* Vol. 12, No. 5, p. 6, Department of Environmental Sciences and Engineering, University of North Carolina at Chapel Hill, (November 1976).

Lamb, J. C. and D. B. Hull, "Current Status in Use of Flexible Effluent Standards," presented at the 1984 conference of the Water Pollution Control Federation, New Orleans (October 1984).

Lamb, J. C., M. A. Shiffman, and S. A. Duletsky, "A Survey of the Use of Seasonal Effluent Standards," presented at the 1980 conference of the Water Pollution Control Federation, Las Vegas (September 1980).

McDermott, J. H., "Federal Drinking Water Standards—Past, Present and Future," *Journal of the Environmental Engineering Division,* proceedings of the American Society of Civil Engineers, Vol. 99, No. EE 4, pp. 469–478 (August 1973).

McGauhey, P. H., *Engineering Management of Water Quality,* McGraw-Hill, New York (1968).

McGauhey, P. H., "Folklore in Water Quality Standards," *Civil Engineering,* pp. 70–71 (1965).

National Academy of Sciences, *Water Quality Criteria 1972,* A report of the Committee on Water Quality Criteria, Government Printing Office, Washington, D.C. (1972).

Okun, D. A., J. C. Lamb, and C. C. Wells, "A Waste Control Program for a River With Highly Variable Flow," *Journal Water Pollution Control Federation,* Vol. 35, pp. 1025–1043 (1963).

Phelps, E. B., *Public Health Engineering, Volume 1; Air Contact, Water Contact,* Wiley, New York (1948).

Tate, C. H., adn R. R. Trussel, "Developing Drinking Water Standards," *Journal American Water Works Association,* Vol. 69, pp. 486–498, (1977).

U.S. Environmental Protection Agency, *Quality Criteria for Water,* Report EPA-440/9-76-023, Washington, D.C. (1976).

CHAPTER 15

Regulation of Water Quality

There are many different philosophies and methods for regulating water quality, ranging from some that are rational and sophisticated to others that are arbitrary and simplistic. Some approach the problem through systematic and scientific evaluation of alternatives; others are highly empirical. Some are rigid in outlook and implementation; others are flexible and tailored to cope with each specific situation.

This chapter discusses several aspects of practical regulatory programs in the United States, but primary attention is directed to developing a rational framework that is consistent with the facts and philosophies discussed in earlier chapters. In short, the emphasis is on concepts that could be applicable under many circumstances as opposed to details that comprise the case history of some individual regulatory organization. Some of the approaches are employed in only a few programs now but have been found to be successful and seem to be rational techniques that should be considered in future regulatory activities.

THE LEGAL STRUCTURE FOR WATER QUALITY CONTROL

The primary objective in applicable laws and regulations is to protect water quality and assure its maintenance at appropriate levels. However, the many diverse needs and interests of our society sometimes require that the primary goal be tempered to satisfy secondary objectives as well, including attention to costs, energy consumption, regulatory manpower requirements, and political pressures. The balance between primary and secondary objectives cannot always be defined precisely and changes from time to time in response to needs and pressures, as demonstrated in the recent dynamics of pollution control regulations (Marshall and Lamb, 1982).

Early U.S. laws principally provided for the control of water quality at the state level, with the federal role restricted to matters involving interstate aspects. For example, the U.S. Public Health Service Drinking Water Standards were applicable directly to supplies used by interstate carriers, but their use to control other water supplies depended on the adoption of the same, or similar, standards by responsible state regulatory agencies—usually state health departments. Pollution control laws and regulations also were formulated, implemented, and enforced mainly by state agencies, with the federal role limited to support activities and participation in interstate pollution matters.

This pattern has changed gradually through a succession of laws and amendments to provide more fundamental federal involvement in water quality control. The Water Quality Act of 1965 (P.L. 87-234) made federal controls applicable to interstate and navigable waters, which ultimately included virtually all streams, although the primary responsibility for establishing standards and their enforcement remained with the states. During the next few years, concern grew that water pollution control was proceeding too slowly and that some regulatory strategies were inequitable in their treatment of different dischargers.

The 1972 Amendments to the Federal Water Pollution Control Act (P.L. 92-500) tried to improve the effectiveness and speed of water pollution control by increasing federal responsibility for establishing standards and providing greater involvement in their implementation and enforcement. The basic regulatory approach shifted from standards based on water quality to those requiring dischargers to use the highest available and affordable technology. That involved the development of uniform "technology-based" standards, requiring similar effluent quality of all dischargers in a given classification, usually without regard to conditions in receiving streams.

The interim goal of the act was to provide for " . . . the protection and propagation of fish, shellfish, and wildlife and . . . recreation . . . ," and the ultimate goal was to attain zero discharge of wastes into streams by 1985. Water quality in streams was given little attention, except for situations in which prescribed technology-based effluent standards would not insure adequate protection of the resource. Then, more stringent limitations must be placed on the discharger.

In 1977, the 1972 amendments were further amended in the Clean Water Act (P.L. 95-217) to cope with failures to meet deadlines, ambiguous statements, and inadequate funding. Many changes were made to provide more reasonable goals, added flexibility, and more realistic time schedules. This act did not eliminate the zero discharge goal, but few persons in Congress or the EPA now consider it to be realistic, and it probably will be eliminated within the next few years (Marshall and Lamb, 1982).

The Safe Drinking Water Act of 1974 (P.L. 93-523) established the EPA as the agency responsible for regulating quality in public water supplies. It also requires the setting of specific limits on concentrations of drinking water constituents that might be harmful to the public health.

RATIONALE FOR WATER POLLUTION CONTROL PROGRAMS

The ultimate goal in pollution control should be to maintain quality characteristics that protect specified beneficial uses of the watercourse. A reasonable first step in developing the program is to compile the data needed to reach sound decisions about its objectives—that is, uses to be protected. Subsequently, water quality criteria for those uses may be applied to define the watercourse quality characteristics that will be program targets. Careful execution of these two steps provides a solid foundation for setting standards, through further detailed studies, and then implementing the program, including its monitoring requirements.

Unfortunately, this type of approach sometimes requires more data than can be collected within the time, personnel, and money constraints faced by the agency. Accordingly, it often is necessary to establish program requirements, including standards, using much less rational techniques, incomplete data, and perfunctory studies. For example, after selecting the beneficial uses to be protected, it may be necessary to set interim standards based largely on agency policies, negotiations with dischargers, information from other studies, and judgment of regulatory officials. In those instances, the standards should be reviewed at relatively short intervals and revised as additional data become available to provide better bases for decisions.

Most of the available literature is less than clear on the motives and rationale behind pollution control programs and, sometimes, the stated or implied motives actually may be misleading. For example, many projects are rationalized largely in terms of safeguarding public health by protecting municipal water supplies. That was reasonably valid many years ago, but today it is impossible to correlate the health of population groups in industrialized countries to the quality of *raw water* used for municipal supplies. That is because of sharp decreases in waterborne diseases in most of those countries, lack of definitive information on other types of health impacts related to water quality, and improved effectiveness of water treatment.

The relatively low regulatory priority on measures for controlling waterborne diseases in the United States may be illustrated by the fact that pollution control regulations require treatment of all wastewaters but often do not require disinfection, except when the stream is used nearby for drinking, swimming, or harvesting shellfish. If the destruction of waterborne disease organisms actually were a major criterion for building treatment plants, then simple, readily available, and economical disinfection certainly would be practiced much more extensively, if not universally.

Even a cursory examination of the U.S. pollution control standards indicates an overwhelming concern about DO levels in receiving streams, suspended solids concentrations, pH, color, and other aspects of water quality bearing little or no relationship to public health. Bacteriological standards often are omitted entirely, and the limited attention devoted to toxic substances is directed mostly toward impacts on aquatic life. One is forced to conclude that the rationale behind most pollution control programs is based principally on protecting aquatic life, recreation, and aesthetic qualities of streams—not the public health.

There is no single overall rationale for water pollution control in the United States. Programs now in existence vary considerably from state to state and time to time, depending on many factors. Even the outlook and requirements of the federal establishment change frequently in response to pressures, economics, political reactions, and lessons learned from experience. There are few details of any pollution control program that could be outlined today with the assurance that they would not change substantially in the near future. One must recognize the dynamic nature of U.S. pollution control as part of understanding its strengths and weaknesses at any time.

DEVELOPMENT OF A DATA BASE

The development of background data that can be used as a foundation for decisions must have a high priority in pollution control programs. The broad categories of information required for that purpose include:

1. Physical characteristics of the water resource (type, size, flow, etc.) toward which the program is directed.
2. Present uses of the resource, including types and extent to which it is used for each.
3. Present physical, chemical, and biological quality of the water.
4. Local needs and desires concerning future uses of the watercourse.
5. Quality characteristics necessary to protect it for those selected uses.
6. Social and economic aspects of the matter.

Physical Character of the Resource

There always is a need for accurate data on the quantities of water available in the region. Stream flow records over long periods of time are of great value in determining the utility of the resource for various purposes as well as the quality that must be required of discharges into it. The needs go far beyond mere average flows and include annual and seasonal variations and statistical evaluation of maximum and minimum flow occurrences.

Minimum flow ranges in a watercourse are of special importance in evaluating its assimilative capacity or suitability as a water supply source. Peak flows and their frequencies of occurrence are important in designing system elements to cope with flood conditions. The stream flow characteristics usually may be determined through hydrologic probability analyses, often performed by the U.S. Geological Survey using records for the stream in question or other streams judged to be similar to it.

It is important to consider watercourse data in the context of other resources that also may be present in the region. For example, the importance of maintaining stream quality suitable for use as a municipal water supply depends not only on conditions in that watercourse but also on alternative sources that might be available for the purpose. Sometimes, efficient planning requires that certain resources be earmarked for some uses, retaining others for different purposes. Often this can produce both substantial savings in pollution control and better assurance of safe municipal water supply sources. However, such decisions are feasible only when based on adequate information about the quantities and qualities of water resources available in the region.

Present Uses of the Resources

An important consideration in planning for the future use of a water resource is the character and extent of present uses. That does not necessarily dictate how a resource should be used in the future, but it does provide at least a solid point of departure for the decision process. Of course, it also is necessary to consider present uses carefully to assure equity in dealing with those who already are commited to the watercourse.

Quantities of water withdrawn from the resource for industrial, municipal, and irrigation supplies as well as flows and characteristics of wastewaters discharged into it should be compiled. Other uses should be documented and, whenever feasible, analyzed to determine their economic and social values. The goal is to develop a comprehensive evaluation of current utility of the resource, revealing its present value to the population affected and projecting the data to identify future trends and needs.

Quality of the Water

Similarly, information is needed on physical, chemical, and biological water quality to provide an accurate baseline that defines present conditions. These data are especially important in establishing reasonable uses of watercourses and identifying the potential pollutants that probably will require control.

The water quality studies should include data collection at different locations throughout the region and evaluation of annual, seasonal, and other variations that could be important in program planning. Sources of significant constituents and the amounts contributed by each should be identified. Sampling and analytical techniques are especially important because all subsequent decisions about water uses and water quality control activities depend heavily on the validity of data compiled in this operation. Inadequate attention to those seemingly minor details can lead ultimately to invalid decisions based on weak, or even erroneous, technical foundations.

ESTABLISHING OPTIMUM BENEFICIAL USES

After compiling the data base on hydrology, current uses, and present quality of the watercourse, it is necessary to identify the beneficial uses to be protected. Many factors enter into these decisions, forming a complex matrix of natural and other variables that require careful and thorough evaluation.

Some beneficial uses may be dropped quickly from consideration because natural flow and quality of the watercourse are unacceptable. For example, freshwater streams are not suitable for raising shellfish because the salinity is too low. On the other hand, an estuary that might be well suited for raising shellfish could be unacceptable for drinking water supply because of high salinity. Many streams have little or no value for spawning because they are inaccessible for fish migration or because the natural water quality or stream bottom conditions are unsatisfactory in some respect. Natural temperature variations may preclude the growth of certain fish, or organic matter from swamps or other natural sources may cause DO to be too low at times. The selection of beneficial uses to be maintained and protected must include careful consideration of these and other factors, which also must be evaluated before implementing programs directed toward controlling wastewater discharges.

A decision to protect some beneficial use of a watercourse through instituting control measures ultimately leads to the commitment of various resources, including labor, materials, and money. There should be a satisfactory relationship between the societal costs of water quality control, on the one hand, and the benefits derived from those activities, on the other. Furthermore, the benefits derived from pollution control programs should compare favorably with those that could be realized by making equivalent expenditures on competitive projects that also could improve society's welfare.

For example, a community that commits money to construct a wastewater treatment plant foregoes certain other facilities or services that could have been provided with those same funds, such as improved medical care, better education, more parks, or better streets. In a world with finite limits, those facilities and services compete with each other for the limited economic resources available to the community. All of this suggests two types of tests for evaluating benefit-cost relationships for pollution control operations: (1) one test to compare societal benefits of a proposed project with its costs, and (2) another test to determine whether benefits expected from the project exceed those from other activities that could be executed with the same funding.

The beneficial uses selected for a section of watercourse ultimately establish the types of facilities that must be installed upstream for water pollution control. On the other hand, the costs of those facilities can be important in determining which beneficial uses should be protected. Optimization of this system can require careful consideration of many alternatives, each involving specific pollution control activities, producing certain benefits, and incurring corresponding construction and operating costs. The final decision should be based on comparing all of the alternatives, with full consideration of environmental, aesthetic, social, political, and financial impacts and benefits.

The selection of uses to be protected sometimes may restrict, or even eliminate, some of the present ones or may include some that are not important currently. For example, excessive costs for pollution control, combined with small benefits, might make it economically infeasible to protect minor sport fishing in a stream reach. In a different situation, on the other hand, benefits to society from a major commercial or recreational fishery could justify or greatly exceed pollution control costs, making it desirable to impose severe restrictions on wastewater disposal.

As another example, municipal and industrial requirements for water along one stream may require close control, or even elimination, of irrigation. In a different area, irrigation may be important enough to justify restricting other uses of the water, especially when supplies could be obtained from alternate sources for municipal or industrial purposes.

Whatever the final decisions, the importance of careful evaluations and following rational approaches in selecting beneficial uses should be evident. This matter is fundamental to benefits, costs, and all other aspects of the water quality control program. It must not be treated lightly.

STREAM CLASSIFICATION SYSTEMS

The optimum balance of uses for watercourses can vary widely from region to region, stream to stream, and even along the length of a single stream. Sometimes, the importance of attaining the optimum may be neglected in publications that emphasize reducing stream pollution and producing the highest quality feasible. Also, it must be recognized that current use of technology-based effluent standards have reduced the motivation to search for the optimum balance in stream uses. After all, why bother when the standards to be applied are not responsive to stream quality anyway?

Regulatory agencies actually do recognize a range of acceptable water qualities, however, because various stream sections often are reserved for different uses, even though the specified ones really may not represent the optimum balance. Sometimes the selection process used by them is referred to as "stream classification," which identifies the beneficial uses to be preserved for each stream section and, ultimately, the water quality that must be maintained in that segment.

An Example

To illustrate how a stream classification system might work, consider this hypothetical example. The optimum use of a given stream near its headwaters could include sport fishing and use for municipal water supply after only minimal treatment. That would require maintenance of very high quality, which would be reflected in stringent water quality standards for the stream. Attainment of the high quality could be feasible because of a sparse contributing population, little or no industrial development, and negligible need by society to use the watercourse for other purposes. Any discharger wishing to locate along this section of stream either would be discouraged or required to meet very stringent effluent quality characteristics to insure protection of the resource.

Further downstream, the river may flow through farmlands and receive runoff from agricultural sources and occasional wastewaters from municipalities and industries. In this segment, the river is flatter in slope, slower in velocity, and likely to be lower in water quality. Accordingly, this stretch might be classified to require more extensive water treatment before use by municipalities and industries, still supporting the propagation of some types of fish life and swimming or other water sports. A sport fishery would not be appropriate here because of limitations imposed by the natural water quality, agricultural and other discharges, and physical characteristics of the stream. Even if it were possible to maintain that fishery, it might not be feasible to do so because its excessive cost could far outweigh any benefits that would accrue to society.

Lower reaches of the stream may pass through highly urbanized and industrialized areas, receiving major discharges of wastewaters and urban runoff. Those segments might be classified to prohibit the use of the stream for municipal water supply and preserve only limited, if any, propagation of fish. Beneficial uses in this zone could include bathing, industrial supplies, and irrigation.

Finally, the stream reaches an estuary that might be valuable for fish and shellfish production—perhaps extensive enough to warrant special steps for their protection. If so, it could be classified for that purpose and the water quality needed to support those activities specified in stream standards. Effluent requirements for dischargers here often could be very stringent.

The classification system outlined in this simple example could differ radically for other streams or even for tributaries of the same one. For example, a very small tributary might receive extensive discharges from manufacturing operations or municipalities already established along the watercourse many years before. The major discharges of wastewaters, combined with a lack of dilution water, could make it difficult, expensive, and impractical to maintain a broad spectrum of uses in the tributary, although subsequent dilution in the main stream could prevent adverse impacts on that watercourse. At the same time, there might be little need for using the tributary for, say, municipal water supply because of other resources available nearby that would be more suitable for the purpose. A reasonable approach might be to classify the tributary to allow only limited uses that could be satisfied with water quality attainable at reasonable cost while classifying the main stream for uses requiring high quality.

Factors in Stream Classification

The specific classification assigned to each stream segment usually represents a compromise between the desires of many individuals and groups, each perhaps wanting the watercourse to be used for different purposes. The compromise must be based on careful consideration of many factors, including the following broad categories:

1. Present and future demands on the watercourse.
2. Needs and desires of the public.
3. Costs of maintaining different stream qualities.
4. Benefits expected under different alternatives.

The classifications actually selected may vary widely along individual streams because of radical differences in societal needs and desires from place to place and time to time.

Stream Classification Procedures

Methods employed in stream classification vary widely but usually include most of the following elements:

1. Evaluation of stream flows and variations.
2. Summary of stream water quality data.

3. Compiling information on current uses.
4. Agency assignment of tentative classifications.
5. Mechanisms for public input to the process.
6. Agency review and assignment of final classifications.
7. Provision for appeals and changes in classifications.

There has been a tendency in some regulatory agencies to establish a "hierarchy" of water quality through identifying a few broad classifications of water uses, similar to those in Table 15.1. This type of approach implies that water quality can be arranged along a simple spectrum, with one end representing "high" quality and progressing downward in orderly steps to the lower-quality end of the classification system.

This concept is basically erroneous because, as discussed in Chapter 6 and elsewhere, each water use has its own unique quality requirements, making it impossible to arrange them in such a simple fashion. Quality characteristics suitable for a municipal water supply may not be acceptable for some other "lower" use, although the hierarchy in Table 15.1 suggests that it should be. For example, coliform bacterial counts suitable for swimming waters actually are lower than those that would be acceptable for municipal raw water supply, which is treated before delivery to the consumer.

A more rational and desirable approach is merely to identify each specific type of use to be protected in the watercourse. That facilitates establishing water quality standards with detailed attention to criteria for each use as well

TABLE 15.1
Example of a Stream Classification System

Class	Water Uses
A	Source of water supply for drinking, culinary, or food processing purposes, requiring principally disinfection, or any other usage requiring water of lower quality
B	Source of water supply for drinking, culinary, or food processing purposes, requiring more extensive treatment, or any other usage requiring water of lower quality
C	Bathing and any other usage requiring water of lower quality, but not as a source of water supply for drinking, culinary, or food processing
D	Fishing, boating, wading, and any other usage requiring water of lower quality, but not for bathing or as a source of water supply for drinking, culinary, or food processing
E	Agriculture, industrial cooling and process water supply, fish survival, navigation, and any other usage, but not for fishing, bathing, or as a source of water supply for drinking, culinary, or food processing

as consideration of other factors that may be important in determining the standards.

SETTING WATER QUALITY STANDARDS FOR STREAMS

Each beneficial use of water requires certain quality characteristics, which usually are specified in applicable quality criteria. When a watercourse is classified for several beneficial uses, successful utilization for those purposes requires meeting the quality needs for all of them simultaneously. After determining the beneficial uses for a stream segment, the overall quality criteria may be compiled by selecting the most stringent level of each criterion among those specified for the several uses. This compilation may be viewed as a specification that describes the overall water quality recommendations for that section of stream, including the safety margins inherent in each criterion used.

The overall criteria usually form the backbone of a regulatory program for that section of stream. That does not mean that the quality specified in standards necessarily must coincide with the criteria, however. Sometimes, even higher quality may be maintained in the stream with only a small additional effort or expense, and the standards may be set to provide more safety margin without unreasonable hardship to the discharger. Also, special local conditions could require standards more stringent than the criteria, perhaps because of the unusual importance of fisheries to the local population, for example.

Under different circumstances, on the other hand, the costs required to meet stringent criteria might greatly exceed the benefits to be produced. That could lead to a decision to operate the resource system at a lower level of quality that still would protect the use in question but, perhaps, with a smaller margin of safety than that normally included in published water quality criteria. In some instances, it even might be necessary to reconsider the question of whether the beneficial use for which that quality would be needed actually should be retained as part of the plan. Therefore, the projected costs themselves may be important inputs to the processes of stream classification and standards setting.

Sometimes, economic limitations or lack of knowledge about the system to be controlled may dictate setting interim standards at levels substantially lower than the water quality criteria as part of a staged approach to stream improvement. As new technology is developed, or with additional information about the response of the stream to the initial steps in cleanup, the standards may be upgraded to require further improvement. Properly implemented, this type of staged construction can make it possible to attain more accurate and economical solutions for stream quality problems by avoiding the risk of incurring waste through overregulation and overconstruction.

TECHNOLOGY-BASED EFFLUENT STANDARDS

The technology-based effluent standard has been referred to earlier and is determined by the state of the art in controlling wastewaters, as contrasted with the specific needs of the watercourse into which discharge occurs. These types of standards may specify, for example, the maximum allowable rates of discharge of various constituents from the facility.

The national technology-based standard for municipal wastewater plants is secondary treatment, as indicated earlier, and has been interpreted further as requiring 85 percent BOD removal. Standards issued to a municipal plant as part of its NPDES (National Pollutant Discharge Elimination System) permit may limit the flow discharged from the facility and the BOD in that discharge, usually expressed as pounds of BOD per day. In addition, the permit contains standards that limit discharges of other constituents and a variety of miscellaneous requirements, including those for monitoring the effluent and receiving stream and for reporting violations of standards to the agency when they occur.

The technology-based standards for industrial discharges are based on studies of production and wastewater handling facilities to determine discharges per unit of production in well-operated plants of the type in question. That is done by a national review of many plants in each industrial category. The acceptable discharge for a given plant is determined by multiplying the acceptable unit discharge for that type of industry by the rate of production at the plant being considered. Industrial NPDES permits include allowable flows, discharges of constituents, and the other miscellaneous and monitoring requirements previously referred to.

Clearly, this type of standard is based on available technology and the capability of regulatory agencies to require the specified level of performance. It is a performance standard that is technology based, as opposed to being derived from the assimilative capacity and quality requirements of the receiving stream. In fact, assimilative capacity does not enter into consideration at all, unless there are indications that the technology-based standard would not maintain stream quality at or above the specified water quality standards for the watercourse.

An advantage of technology-based standards is that they seem to be easily definable and highly objective. However, two very significant limitations are that (1) they do not relate to actual water quality (Kalur, 1979), and (2) it is not feasible to balance program costs and benefits (Freeman, 1980). The principal reason for pollution control activities is to protect water quality. To require a specific level of treatment without regard to the conditions in the stream may result in a commitment of capital and operating expenses without producing identifiable benefits (Lamb, 1982a; Marshall and Lamb, 1982).

Furthermore, imposing the same requirements nationally, regardless of location or local conditions, can be counterproductive to environmental protection. Impacts resulting from discharges can vary widely at different locations, and a sound regulatory structure should encourage the selection of industrial plant sites that minimize harmful effects. Blanket technology-based standards fail in this respect because the requirements that must be met do not relate to or depend on local conditions or environmental impacts.

EFFLUENT STANDARDS FOR WATER QUALITY-LIMITED STREAMS

The 1972 amendments (P.L. 92-500) prescribe a somewhat different approach when the information available to the agency indicates that the technology-based standards just described would not be adequate to maintain the stream at a quality consistent with its specified water quality standards. That frequently is determined by the agency by using mathematical models to predict stream conditions at the minimum flow and maximum temperature likely to be encountered. The rationale behind using those extreme environmental conditions is that aquatic life should be protected to prevent fish kills and other irreversible impacts, even under those infrequent occurrences.

In this situation the stream is termed "water quality limited," and the discharger must be required to meet more stringent effluent standards, calculated to insure meeting the quality standards for the watercourse. These effluent standards often are set by using the mathematical models, with inputs approximating the worst expected flow and temperature conditions for the stream.

This approach, then, also results in the use of the performance standards, but with the difference that they are based on maintaining specified stream water quality instead of being technology based. They are tailored to fit the specific situation under consideration and are responsive to estimated environmental conditions in the stream. However, sometimes they suffer from being inflexible, as applied by some agencies, and may be more stringent than necessary much, or even most, of the time. This aspect of the problem will be discussed shortly.

FLEXIBILITY IN STANDARDS

In the discussions of oxygen balance in streams in Chapter 10, it was pointed out that deoxygenation and reoxygenation rates depend on many variables, especially flows, temperatures, and oxygen-demand loadings, and that those parameters vary widely at different times for a given stream. The amount of BOD that can be discharged by a specific municipality or industry without violating DO standards for the stream (the stream's "assimilative capacity") also changes widely in response to those variables. This produces a significant range in the amount of BOD that can be assimilated without causing water in the stream to drop below the specified standard or, in other words, without adversely impacting the desired stream water quality as defined by the regulatory system.

State regulatory agencies may exercise judgment about the amount of flexibility to be used in implementing effluent standards. At one extreme, some agencies specify a single standard that must be met all year, without considering the effects of changes in flows, temperatures, or other variables on the assimilative capacity of the stream. That approach requires effluent quality higher than needed to meet stream water quality standards most of the time because it is based on conditions that occur only infrequently—perhaps once in ten years, for example.

Other agencies recognize the variability in assimilative capacity and allow the effluent quality to change in response to changes in temperatures, flows, and other stream conditions. Sometimes, the effluent quality may be allowed to change substantially from time to time, or even continuously, limited only by an overall requirement that the stream water quality standards cannot be violated. More often, limited variations are permitted in different seasons of the year to take advantage of the effects of changes in temperature and flow on the ability of the stream to receive discharges without creating quality problems.

The rationale behind seasonal standards is illustrated in Figure 15.1, which shows the assimilative capacity of a small North Carolina stream as calculated

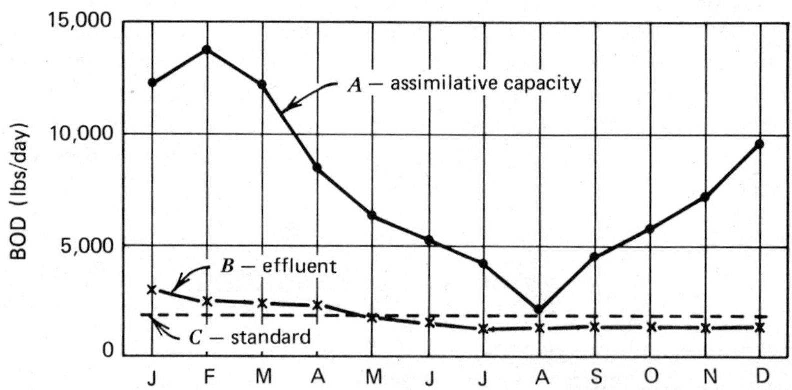

FIGURE 15.1
Illustration of monthly changes in the assimilative capacity of a stream, plant effluent quality, and a single-level standard.

with a commonly used oxygen balance model (Lamb, 1982a, 1982b, 1982c). Curve *A* shows the stream's assimilative capacity, derived by using a three-year flow and temperature data base and calculated on the basis of maintaining at least 5.0 mg/liter DO in the stream throughout each month. The assimilative capacity varies from a low of about 2000 lb/day of BOD in August to over 13,000 lb/day in February; most of the variability in this example can be attributed to changes in temperature because minimum flow in this stream is very low during all of the months. Curve *B* shows the BOD content of the plant discharge, which also varies somewhat because of temperature impacts on the biological treatment system. Curve *C* shows the single-level effluent standard under which the plant was required to operate, set at a discharge approximating the assimilative capacity in August, which is the worst month of record.

The assimilative capacity of the stream is substantially higher than the established standard during all months except August. Thus, it would be feasible for the regulatory agency to allow higher discharges of BOD into the stream during the other 11 months of the year without violating the stream water quality standard of 5.0 mg/liter DO. That would not represent a retreat from the stated goals of the program because the stipulated stream water quality (5.0 mg/l DO) still would be maintained at all times, in spite of greater discharge of oxygen-demanding materials. However, it could result in substantial savings in capital and operating costs by making it possible to defer enlargement of the treatment plant for several years (Hargett and Seagraves, 1979; Lamb et al., 1980; Lamb, 1982a, 1982b, 1982c). Various alternatives for implementing flexible standards have been outlined elsewhere (Marshall and Lamb, 1982; Lamb, 1982b).

A survey of the practices of state water pollution control agencies by Hull (1983) indicates that only two of them do not use flexible effluent standards at all. The types and degrees of flexibility employed by the others vary considerably among the states, from infrequent use for limited purposes by some to extensive issuance of flexible standards for several parameters by others. The parameters that have been allowed to vary by some in response to seasonal or other changes in environmental conditions include BOD, ammonia content, coliform bacteria, and disinfection requirements (Lamb and Hull, 1984).

VIOLATIONS OF NPDES STANDARDS

There is a growing awareness that many municipal wastewater treatment plants do not perform well enough to meet the conditions that have been imposed in their NPDES standards. A report of the General Accounting Office (GAO) in 1980 evaluated this problem using EPA data and the results of a special GAO study of 242 major plants in 10 states. The EPA data indicated that 50 to 75 percent of the plants were violating their NPDES standards at any given time, and the GAO study produced even higher figures of 87 percent violations, with 27 percent in "serious" violation (U.S. GAO, 1980b).

The GAO report attributed plant "failures" to several causes, including deficiencies in design, operation, and maintenance, but it did not consider the possibility that the standards themselves might be defective. This reveals a serious flaw in the logic of some regulatory programs and the methods employed in examining them; they are evaluated simply on whether standards are met or violated, giving little or no consideration to water quality conditions in the stream or the validity of established standards (Lamb, 1982b).

This question was studied with respect to violations of standards reported for 29 major plants in North Carolina, using data for 1978 to 1981 (King, 1982; Lamb, 1982c). An examination of stream monitoring data revealed that during those violations the downstream DO dropped below 5.0 mg/liter at only six of the plants. Furthermore, in four of those instances the upstream DO also was below 5.0 mg/liter, leaving only two of the 29 plants that logically could be charged with the responsibility for reducing downstream DO below the specified water quality standard—and the data for one of those plants was inconclusive.

Thus, during the three years only one or two of the 29 major North Carolina plants that violated their NPDES standards actually "failed," that is, probably was responsible for causing downstream DO to drop below the stipulated DO standard for the stream being protected. During those occurrences, no fish kills were reported and no other water quality problems could be attributed directly to the violations. Although that seems to be an enviable record, the fact remains that 29 plants were reported officially as "failures" in that they were not meeting their NPDES standards. This suggests that the current system for setting standards deserves a careful review because the standards and conditions in streams supposedly being protected by them do not relate very closely.

How that can happen is illustrated by Figure 15.1, introduced earlier. From January through April, the plant data (curve *B*) show violations of the standards assigned to it (curve *C*). However, curve *A* shows that in all months, including those with reported violations, the assimilative capacity exceeded the amount of BOD discharged by a substantial margin. Therefore, violations indicated by the data in Figure 15.1 are meaningless when viewed in context of the regulatory goal of maintaining at least 5.0 mg/liter DO in the stream. Another GAO report (U.S. GAO, 1980a) has addressed the question of "insignificant" violations of water quality.

This situation is recognized by many regulatory agencies, and they are moving toward building more flexibility into their systems for setting effluent standards (Hull, 1983; Lamb and Hull, 1984). Although the changes can complicate their operations to some extent and require better understanding of the streams and discharges being regulated, optimization of the regulatory systems should be far more cost effective than building or modifying treatment plants unnecesarily.

STUDY QUESTIONS

1. Describe the relationship between protection of the public health and wastewater discharge standards presently being promulgated in the United States.
2. Explain how and why that relationship might differ in some of the less developed countries of the world.
3. Outline the types of information needed for effective design of a water pollution control program and briefly justify the need for each.
4. What physical characteristics of a watercourse are important in establishing a water quality control program for it? Why is each important?
5. Outline and briefly discuss several factors that must be considered in establishing the optimum beneficial uses for a stream segment.
6. Explain the concept and utility of stream classification as part of a water quality control program.
7. What are the basic factors that should be considered in classifying streams?
8. Outline several elements that often are included in procedures for classifying streams.
9. Explain the limitations of the "hierarchy" approach to classifying streams. Describe a preferable approach and explain why it could be better in practice.
10. Distinguish between "technology-based" and "water quality-limited" effluent standards.
11. Outline the advantages and disadvantages of technology-based standards as currently employed in the United States.
12. Discuss the advantages and disadvantages of using seasonal effluent standards as opposed to single-level effluent standards that apply through the entire year.
13. Why may some violations of NPDES standards be viewed as insignificant with respect to their environmental impacts?
14. Explain the role of cost-benefit relationships in establishing water quality control regulations.
15. Discuss the relationship between water quality criteria and water quality standards.

REFERENCES

Cleary, E. J. "Effluent Standards Strategy: Rejuvenation of an Old Game Plan," *Journal Water Pollution Control Federation,* Vol. 46, pp. 291–298 (1974).

Cleary, E. J. "Frustrated Aspirations for Watershed Quality Management," *Journal Water Pollution Control Federation,* Vol. 53, p. 291 (1981).

Cleary, E. J. *The ORSANCO Story—Water Quality Management in the Ohio River Valley Under an Interstate Compact,* Johns Hopkins Press, Baltimore (1967).

Freeman, M. A.; "Technology-Based Effluent Standards: The U.S. Case," *Water Resources Research,* Vol. 16, p. 21 (1980).

Gehm, H. W., and J. I. Bregman, *Handbook of Water Resources and Pollution Control,* Van Nostrand Reinhold, New York, (1976).

Hargett, G., and J. C. Seagraves, *Benefits and Costs of Seasonal Effluent Permits in North Carolina,* Report No. 135, Water Resources Research Institute, Raleigh, N.C. (1979).

Hull, D., *A Survey of the Use of Flexible Standards by State Water Quality Management Agencies;* Unpublished report submitted in partial fulfillment of requirements for the MSPH degree, Department of Environmental Sciences and Engineering, University of North Carolina at Chapel Hill (1983).

Kalur, J. S., "Will Judicial Error Allow Industrial Point Sources to Avoid BPT and Perhaps BAT Later? A Story of Good Intentions, Bad Dictum, and Ugly Consequences," *Ecology Law Quarterly,* Vol. 7, p. 955 (1979).

King, S. D., *North Carolina Municipal Wastewater Treatment Plants: Impact of NPDES BOD_5 Permit Violations on Downstream Dissolved Oxygen,* Unpublished report submitted in partial fulfillment of requirements for the MSPH degree, Department of Environmental Sciences and Engineering, University of North Carolina at Chapel Hill (1982).

Lamb, J. C., M. A. Shiffman, and S. A. Duletsky, "A Survey of the Use of Seasonal Effluent Standards," presented at the Las Vegas meeting of the Water Pollution Control Federation (September 30, 1980a).

Lamb, J. C., "The Conflict Between Energy Conservation and Water Pollution Control Standards," *Journal Water Pollution Control Federation,* Vol. 52, pp. 656–660 (1980b).

Lamb, J. C., "Improving Regulatory Control of Water Pollution Through Industrial Collaboration," *Proceedings of the Industrial Wastes Symposia,* 54th Annual Conference, Water Pollution Control Federation; Detroit (1981).

Lamb, J. C., "Feasibility of Increased Flexibility in Water Pollution Control Standards," presented at the 1982 meeting of the Virginia Water Pollution Control Association, Charlottesville (April 28, 1982a).

Lamb, J. C., "Treatment Plant Failures: Facts, Fallacies, Solutions," *Proceedings, 14th Mid-Atlantic Industrial Waste Conference,* pp. 426–435, Ann Arbor Science Press, Ann Arbor, Mich. (1982b).

Lamb, J. C., "Assessment of the Effectiveness of Seasonal Standards," *Journal, North Carolina Section AWWA and WPCA,* 1982 meeting, Greensboro, N.C. (1982c, in press).

Lamb, J. C. and D. B. Hull, "Current Status in Use of Flexible Effluent Standards," Presented at annual meeting of the Water Pollution Control Federation, New Orleans (October, 1984).

Marshall, B. R., and J. C. Lamb, "Evolving Water Quality Regulations", presented at the A.S.C.E. Conference on Water and Energy: Technical and Policy Issues, Pittsburgh (May 23–26, 1982, in press).

U.S. General Accounting Office, "Many Water Quality Standard Violations May Not Be Significant Enough to Justify Costly Preventive Actions," *Report CED-80-86,* Washington, D.C. (1980a).

U.S. General Accounting Office, "Costly Wastewater Treatment Plants Fail to Perform as Expected," *Report CED-81-9,* Washington, D.C. (1980b).

CHAPTER 16

Progress, Problems, and Prospects

The water quality field is dynamic. Profound changes in most areas of it occur steadily and rapidly in response to public awareness of environmental matters and gains in knowledge about the technology and sociology of water resources development and uses. Any book that deals with water quality and its control must be viewed as a snapshot that, at best, can only convey a picture of this moving target at one point in time. As with many publications, by the time it is printed and reaches the eyes of a reader, it already has begun the process of obsolescence and for water quality that process is more rapid than for many other subjects. Accordingly, if the reader is to develop a sound understanding of water quality matters, it is necessary to recognize and accept their rapidly changing nature and complexity.

The rationale behind the approach followed in this book and outlined earlier in the Preface is based on trying to cast the subject in terms that illuminate basic issues and principles instead of merely reporting on current practices. Principles, at least, survive for a somewhat longer time and are applicable across a wider variety of circumstances. This makes them of greater and more lasting value in the educational armament of an individual.

The purpose of this final chapter is to convey some "feel" for the current direction and momentum of change in water quality evaluation and control. A few aspects of the field will be discussed briefly to identify important recent developments, some of the problems and issues faced today, and where we seem to be heading during the next decade. Granted that predictions in print always are risky, there seems to be no alternative if the mission of this book and my responsibility as its author are to be met.

WATER QUALITY EVALUATION

If required to identify a single area that has had the greatest influence recently on our recognition of water quality problems, understanding of key issues, and formulation of changes in policies and methods, a likely candidate would be the techniques employed to measure and evaluate water quality. The simple fact is that the validity of all decisions concerning water quality and its control hinge ultimately on the information compiled with those techniques. In the subsections that follow, we examine briefly some changes and trends in four aspects of water quality evaluation.

Identification of Chemical Parameters

Only a few years ago, those engaged in water quality control were handicapped by severe limitations in their capabilities for measuring many constituents of water—especially organics—at the very low concentrations often found in practice. In many instances, techniques were not available at all for accurate measurements in those concentration ranges. In others, the constituents could be determined only through sophisticated methods that were tedious and expensive enough to make them infeasible in practical water quality programs. As recently as the early 1970s, there was little or no reliable information on the occurrence of trace organics in watercourses or even in drinking water supplies.

Since then, there has been an explosion in analytical capability for measuring a wide array of water constituents at exceedingly low concentrations. This has been a boon in research and to those interested in identifying chemicals present in waters, tracing their movements through the environment, and evaluating their removal in treatment systems. In the short term, however, it sometimes has generated embarrassing problems for those who define water quality goals and establish pollution control standards because our analytical capability to measure constituents now often exceeds our ability to use that information meaningfully.

Today we can detect and measure chemicals that only a few years ago caused no concern simply because we did not know they were there. Now that we do know, we often do not understand their health or other significance at the part per billion or part per trillion levels at which they may be found. No longer can a regulatory agency merely stipulate "none." Now it must specify the actual permissible concentrations for many "new" constituents, often with little or no scientific basis for making such decisions.

Those problems notwithstanding, the new knowledge now being gained through improved analytical techniques will be vitally important in years to come. It can provide sound bases for decisions in many areas that before could only be ignored or evaluated through gross empirical methods. Recognizing that validity of a decision depends heavily on quality of the input data, it is clear that the impact of a quantum jump in knowledge about the constituents of water ultimately should lead to a vastly improved basis for evaluating water quality and setting standards for the control of these constituents.

Current trends are toward consolidating and capitalizing on some of those gains. Part of the effort is being directed toward improving methods for assaying the chemical content of water samples to simplify the procedures and make them more practical for routine applications. Another important avenue of attack involves developing knowledge and techniques that facilitate using the improved data base efficiently and accurately in evaluating and controlling water quality.

Use of Bioassay Techniques

The past 25 years have produced substantial increases in the use of bioassay techniques to evaluate water pollution impacts. One approach has involved using bioassays based on determining species, numbers, and diversity as tools to evaluate water quality in streams and lakes, as described in earlier discussions in Chapter 3.

Another application of bioassays has taken the form of improved and greater use of live fish and other aquatic organisms to evaluate the probable impacts of discharges on organisms in watercourses. Advances in this area have included the development of standardized procedures for short-term tests to determine acute toxic thresholds, longer-term tests to evaluate chronic effects, and special study techniques that use nonlethal responses of organisms to estimate the effects of potential pollutants.

Too little has been accomplished in evaluating the harmful effects of potential pollutants on humans, especially at very low concentrations, as discussed in Chapter 9. This is currently an area of great need and can be expected to receive substantial attention in the future. The knowledge is essential to the success of efforts to establish improved standards that define maximum permissible concentrations for the many organic constituents that now can be identified and quantified in drinking waters. This will require exceptionally sophisticated studies and will be time consuming and extremely expensive. Hopefully, the next decade or two will see major advances.

Monitoring Systems

This is an area in which modest advances have been recorded in the past two decades through the development of automatic sampling and continuous recording equipment, principally for application to wastewater discharges and watercourses. Nevertheless, most of the monitoring done today to assess the effectiveness of treatment systems and water quality in receiving streams is accomplished by relatively simple sampling techniques and manual analyses.

Furthermore, the character of analyses carried out in monitoring programs usually is directed to gross pollutional parameters, such as turbidity, color, BOD, and suspended solids. Only in relatively rare instances are they conducted for specific constituents, such as metals and organics. Usually these are done in connection with specific industrial discharges, where important constituents are known to be present from time to time, or in water supplies, where there are special grounds for concern about them.

As more specific chemicals are defined in standards for drinking waters and wastewater discharges, there will be an increased interest in and need for developing monitoring methods for those constituents. In many instances, future monitoring techniques will involve greater use of automatic sampling and analyses or continuous recording of concentrations.

Water Quality Progress and Trends

Public pronouncements have indicated that progress achieved during the decade following the passage of Public Law 92-500 has been substantial and impressive. Some accomplishments during that period are evident and easily documented, but it is not always easy to marshall the data needed to provide definitive evidence of other progress.

Between 1973 and 1982, Congress appropriated over $37 billion for grants to aid the construction of public wastewater treatment facilities, and the states and local governments spent about another $10 billion for capital improvements. During the same period, industries spent about $16 billion to improve their wastewater treatment facilities.

Currently, there are over 15,000 wastewater treatment plants in the United States, processing flows from over 160 million persons. Most of the plants use at least "secondary" treatment. Many were in existence before 1973, but the number of secondary plants increased by over 58 percent between 1976 and 1980, for example, and the trend in that direction continues even now. Considered as a whole, industries are well ahead of municipalities in meeting goals that were specified in early phases of the Public Law 92-500 program (Water Pollution Control Federation, 1983).

There is no denying that those figures constitute impressive evidence that progress has been substantial in providing treatment facilities for municipalities and industries. Also, many case histories can be identified that illustrate striking improvements in watercourse quality—the Potomac, the Houston Ship Canal, and some areas of Lake Erie, for example. It is much more difficult to document clearly improvements in watercourse quality at a national level. There are over 2 million stream channels in the United States, 75 percent of which are less than one mile long. They vary widely in physical and chemical characteristics as well as in their uses. Only limited data are available for most of the streams, and an accurate evaluation of water quality across the nation is difficult and time consuming.

Data from the National Stream Quality Accounting Network of the U.S. Geological Survey through 1980 have shown some improvements in DO, fecal coliforms, suspended solids, and phosphorus levels, but only a small proportion of the stations that were sampled produced trends that were "statistically significant" (Water Pollution Control Federation, 1983).

The 1982 National Water Quality Inventory Report to Congress, summarizing reports submitted to the EPA by the states, indicated progress in water pollution control. Twenty-one of the 33 states that reported trends indicated improvement in water quality. Twelve states reported that water quality was stable. None reported statewide degradation in quality, although 23 states did report degradation trends in localized areas (Council on Environmental Quality, 1983).

These types of qualitative or semiquantitative results are considerably less than satisfactory when trying to judge the results of a major and expensive national program over a decade. Part of the difficulty in assessing progress of the program is because water quality monitoring efforts in its early phases were given too little attention and opportunities to obtain essential baseline data were overlooked. There seems to be little help for that now, and the accurate assessment of water quality advances must wait for the program to move along further, with improved monitoring and data interpretation.

Another part of the difficulty is that there is continued growth in population and industrial development at the same time that efforts are being made to improve water quality. That situation led the Council on Environmental Quality to conclude that the little change shown by U.S. Geological Survey monitoring data, referred to before, means that:

> . . . the nation's pollution control efforts seem to be having a positive effect. Large-scale degradation of the nation's streams has stopped, even though both population and industrial activity have increased.

Additional and more sophisticated efforts are being made to evaluate trends in water quality at the national level, but time will be required to compile the data necessary for accurate determinations and to interpret the results (Council on Environmental Quality, 1983).

WATER SUPPLIES

Availability and uses of water resources have been analyzed frequently over several years, especially with respect to water-short areas. On the other hand, as outlined in Chapter 9, over the past 50 years there has been surprisingly little attention devoted to the health aspects of water supplies—partly because of complacency and partly because of the overwhelming concern about water pollution and its control. That situation is changing now because of an increased realization that many potential health problems have been overlooked too long and deserve careful attention. The following sections will outline some current trends in various aspects of the water supply field.

Water Quantities

The availability of U.S. water resources and history of their uses have been discussed in Chapters 4 and 5 and need not be repeated here. It is enough to summarize the situation by stating that water uses have been increasing much faster than the population because of the increased requirements per person and rapid growth in industrial demands. Also, the consumption of water has been increasing, causing greater inroads into our fixed resources.

When examined in a national context, the water resource-water use balance looks very good, but one must not be too glib about it. The fact is that there are serious problems in some regions, with steadily increasing uses impinging severely on limited available resources. This is especially true of areas in the West and Southwest, where authorized withdrawals from surface waters sometimes even exceed the total resource that is available reliably. Also, in some of those areas, uses of groundwaters exceed recharge rates by substantial margins and have produced an extensive decline in the elevation of the groundwater table.

Increased uses of water and the limitations of fixed resources are resulting in more reuse. This applies to the sequential use of water along streams, of course, but even more to the inplant reuse of water by industries. Increased industrial reuse can be attributed partially to the adoption of cooling towers to remove heat, which simultaneously recycle flow to production operations and reduce the water withdrawals necessary to meet industrial needs. Other important factors influencing the industrial reuse picture include higher water handling and treatment costs and tighter requirements for wastewater disposal. The latter often make it economical now to invest more in systems that facilitate water reuse and reduce costs of wastewater treatment plants by decreasing flows through them. The same factors also have made it attractive to reuse municipal wastewaters for cooling, irrigation, or other purposes, as outlined later in this chapter.

Water Quality Concerns

As pointed out in Chapter 9, the high incidence of waterborne diseases sparked actions that produced striking improvements in the quality of municipal water supplies during the first 20 to 30 years of this century. Subsequently, after the incidence of those diseases decreased almost to the vanishing point, interest in this aspect of water quality waned and remained at a low ebb until the late 1960s. Since then there has been a rapid rise in concern about the health impacts of metals and trace organics on human health.

Some of the current concern about water quality is focused on waterborne diseases, especially Giardiasis and Legionnaires disease, which only in the past few years have appeared as significant factors in the total waterborne disease picture. Much of the concern about these and some viral diseases has centered on doubts about the validity of currently used indicator organisms (coliforms) for assessing the risk of their transmission by the water route, as well as the effectiveness of current treatment systems in removing the causative organisms.

The greatest concerns now are based on the potential for adverse health effects by metals and toxic organic chemicals, much of which is an outgrowth of improved methods for detecting low concentrations of chemicals in water, discussed earlier in this chapter. Substantial activity may be expected in this area over the next decade or two, with special attention to evaluating the long-term effects of low concentrations of suspected carcinogens.

Specific chemicals that are of special interest now include the compounds formed during chlorination of water for disinfection. Activities during the next decade will include extensive investigations of the effects of those chemicals on human health and of other methods that could be used for disinfection. Of course, those alternatives also will require careful evaluation of their disinfection efficiency and potential for producing harmful by-products.

Treatment Technology

A trip through a typical water treatment plant built today usually will reveal the use of technology startlingly similar to a plant built 50 or more years ago. For the most part, the plants still are being designed and built to protect against the same waterborne diseases that caused great concern around the turn of the century. They remain appropriate for that purpose but often are not as effective as we might wish for protecting the consumer against some of the other health risks that are of rising concern today.

To minimize some of the potential risks from trace organics, there has been a trend toward increased use of activated carbon in water treatment for sorption of those chemicals. Also, many plants have begun to change their treatment approaches to avoid the formation of chlorinated organics by moving the point of addition of chlorine from the beginning of the process to its end. This allows for the removal of most organic precursors before adding chlorine, thereby reducing the formation of chlorinated organics. The approach of using disinfectants other than conventional chlorination presents additional possibilities for solving the problem.

At the time of this writing, those modifications represent the major recent changes that have occurred in water treatment technology. However, it may be anticipated that many others will be explored in the next decade or two to cope with the perceived new health risks. They will include the study and possibly the application of new treatment processes—many of which already are available—to remove materials that have been considered relatively unobjectionable in the past but now are viewed as suspicious.

To support all of those activities, there will be a major need for improving the monitoring of water to determine the types of trace chemicals present, their concentrations, and the effectiveness of various treatment schemes for their removal.

Other Water Supply Considerations

The development of groundwater and control of its quality, which have been neglected too long, also will receive concentrated attention during the next decade or two. Interest has been spurred recently by increased recognition that the improper disposal of hazardous wastes can jeopardize groundwater quality. Future activities will include extensive attention to monitoring groundwater quality, modeling underground movement of pollutants, methods for cleanup of polluted aquifers, and improved systems for treating polluted groundwaters.

As the use and consumption of water continue to increase, water reuse will become more attractive, as stated earlier. However, the direct reuse of treated wastewater to serve as a municipal water supply will remain the exception because of great resistance from a large segment of water supply professionals and the existence of better and safer approaches for augmenting water supplies. The increased reuse more often will involve the use of treated wastewaters for industrial cooling, irrigation, and other purposes involving only minimal potential for direct health impacts.

In some instances, in water-short areas there will be an increased interest in the installation of dual water systems in some communities to facilitate the safe reuse of treated wastewater. Sometimes, that could include using waters of two different qualities in households—one for potable uses and the other for flushing toilets, for example. However, the high cost of providing two distribution systems and separate plumbing systems in homes will discourage that practice in all but instances of extreme need. A more likely approach will be based on providing limited secondary distribution systems for delivering treated municipal wastewater to industries, for irrigation, and for other noncritical uses, thereby liberating better quality raw waters for potable uses.

WASTEWATER DISPOSAL

Changes in wastewater treatment and disposal have been much more pronounced than those in water supply in recent years. This can be attributed to greater pressure for water pollution control, as reflected in the passage of major federal legislation, provision of financial support for treatment facilities, and imposition of more stringent standards. The advances can be expected to continue, although perhaps at a slower rate, and seem to be easier to foresee because clear trends have been established and momentum built during the past decade. The following sections briefly review some important recent advances and directions of probable movement in the next decade or two.

Water Carriage of Wastes

In the context of approaching the goal of "zero discharge" by 1985, specified in Public Law 92-500, little progress has been made in the municipal wastewater field, and it seems that little is likely in the foreseeable future, except for improved removals of biodegradable organics (BOD) and suspended solids by treatment plants. There has been no significant attempt to reduce the use of water carriage for removing undesirable materials from the home environment, and it is difficult to envision that such changes will be initiated any time soon.

On the other hand, there have been significant efforts in that direction by some industries, where inplant reductions or changes in the character of wastewater discharges often are feasible through changes in manufacturing operations to use different raw materials or manufacturing processes. As pointed out in Chapter 13, reducing the use of water for some purposes, minimizing the contact between water and process constituents, recovering by-products, improving process efficiency, and reusing water all have proven helpful under various circumstances.

As time passes and discharge requirements become more stringent, those types of steps undoubtedly will be more and more favored because of technological difficulties in meeting standards, as well as economic considerations. It probably will remain the exception for such inplant activities to be profitable in their own right. However, when the alternatives for inplant changes and end-of-pipeline treatment are considered together, inplant approaches that seem economically unattractive if considered alone often prove to be highly desirable. Therefore, it is reasonable to anticipate that much more attention will be devoted to possibilities for solving industrial wastewater problems through inplant changes, effectively reducing water carriage of wastes in industries.

Treatment Goals

Since about 1960, there has been an extensive effort to develop new wastewater treatment processes and reduce them to practice. The result is that now a collection of treatment tools is available that makes it possible to remove virtually any constituent from wastewater discharges—at a price, of course. As the situation stands today, decisions on whether to remove constituents are made on the basis of environmental need, economics, energy consumption, or other such considerations, not on whether the technology is available to permit such action, because it is.

A few years ago, the conventional wisdom was that advanced treatment processes would be widely used to produce BOD and suspended solids removals beyond those attainable in secondary treatment systems, as well as to remove nitrogen, phosphorus, nonbiodegradable organics, and even inorganics from municipal and industrial wastewater discharges. Eventually, however, it became evident that wholesale application of those techniques would be very expensive and often would produce little or no benefits. There has been increasing caution about applying advanced technology and realization that it must be used only selectively, after careful consideration of costs and benefits.

Today, the advanced processes most often used include methods for improving BOD and suspended solids removals from discharges entering "water quality-limited" streams. The systems commonly used for those purposes include multiple stages of biological treatment, coagulation, settling, and filtration of solids from plant effluents. Other techniques used less frequently are precipitation processes to remove phosphorus from wastewaters discharged into nutrient-sensitive streams and lakes and nitrification to reduce NOD of discharges. There has been little effort in the United States outside of the industrial environment to remove nitrogen, so-called "priority pollutants" (potentially toxic constituents), or inorganics beyond the minimal removals attained in biological treatment systems.

It is anticipated that trends of the past few years, as just outlined, will continue in the next decade. We should see more attention to assessing the need for advanced treatment and, hopefully, the development of better systems for carrying out those evaluations. Efficient removals of BOD and suspended solids will be considered routine requirements in most instances, and increased attention will be devoted to the need for removing phosphorus, nitrogen and other specific constituents from industrial and municipal discharges.

Treatment Technology

As pointed out earlier, there already is a considerable inventory of treatment technology that can be used to satisfy various needs. Undoubtedly, those processes will continue to be improved through research and development, but it is unlikely that there will be major breakthroughs in this area by developing radical new processes or startling improvements in the efficiencies or economics of existing ones. The improvements are much more likely to be evolutionary than revolutionary.

An area that has received increased attention recently because of the national energy situation is the power requirements of wastewater treatment plants. This will continue as the costs and availability of energy become more important considerations. It is difficult to project what will happen in this area and its timing, of course, because of the uncertain international situation.

The quality of plant operation has received greater attention as it has become evident that many facilities constructed during the past decade are not performing up to expectations. Activity should increase in this area because plant performance often can be improved at relatively little cost through better operation, sometimes making it possible to defer or avoid major capital expenditures for enlargement or upgrading. The other side of this coin, however, is that an accusation of "poor" operation often has provided a convenient excuse to cover up for poorly designed plants or unrealistic standards (Lamb, 1982).

REGULATION OF WATER QUALITY

During the past two decades, the regulation of water quality has undergone radical changes. The governmental agencies that are responsible for planning and implementing water quality programs have shifted in identity and organization. Furthermore, the methods employed by them also have changed drastically, as discussed in Chapters 14 and 15. This section outlines probable future directions in U.S. regulatory programs.

Changing Regulatory Goals

It was pointed out earlier that most past and current regulatory activities have been directed toward protecting aquatic life, as reflected by the consuming interest in controlling oxygen-demanding substances to prevent adverse impacts on fish and other organisms. That interest has not abated but probably will be gradually supplemented by increased attention to potentially toxic chemicals (priority pollutants, for example) as part of the growing interest in human health effects.

Another change now appearing is broader consideration of various sources of potential pollutants by regulatory agencies. In the past, program activities have been directed almost exclusively to point discharges of municipal and industrial wastewaters. Currently, there is an increasing interest in evaluating the impacts of nonpoint-sources and developing methods for controlling them. The 1982 state water quality assessment report by the EPA, for example, indicated that in 20 percent of the states, nonpoint-sources are the most important cause of degradation in waters that do not meet quality requirements. Also, agricultural sources provide most of the nutrients to lakes with eutrophication problems (Ruckelshaus, 1983). The next decade will produce substantial additional attention to this aspect of the U.S. water pollution control problem.

Laws and Agencies

Principal legal and regulatory reponsibility for controlling water pollution was for many years with the states, and the federal role was restricted mostly to interstate matters. As pointed out in Chapter 15, the past decade or two has seen primary responsiblility move from the states to the federal arena, where the key laws have been passed, standards initiated, and financing provided.

Now that the basic programs are under way and momentum has been developed, there are indications that some responsibility probably will move back to state agencies over the next few years. That is reflected in recent trends for the EPA to release to the states greater authority in establishing standards and allowing increased flexibility in pollution control regulations. This reflects the realization that local input, which has been virtually ignored in setting most of the present technology-based standards, must be given more consideration if programs are to be responsive to local needs and cost effective.

That is not to suggest that all, or even most, water pollution control activities will return to the states as they were in years past. The EPA undoubtedly will continue to play a central role in establishing overall goals, guidelines, standards, and grant programs to guide and assist the states in many ways as well as to insure reasonable uniformity nationwide.

Standards

In the early 1970s, after the passage of Public Law 92-500, the systems for setting pollution control standards moved from those based on maintaining specified quality in watercourses to technology-based standards that include little or no consideration of receiving streams. Implementation of those standards has been relatively inflexible, with little attention to local circumstances, needs, or desires, in order to facilitate rapid start-up and movement of the new federal program in its early stages. It has been clear that although this approach might be good for expediting action, sometimes it can result in poor practices and wasteful expenditures by requiring construction of marginal, or even unnecessary, treatment plants. Recently there has been occasional increased interest in restructuring the federal-states program to allow or encourage more attention to impacts on affected streams in setting discharge standards.

A recent trend has involved the increased use of discharge standards that include flexibility to take advantage of variations in assimilative capacities of streams, thereby reducing capital and operating expenses for pollution control facilities without compromising water quality goals. This approach was discussed in Chapters 14 and 15. Two surveys of state regulatory agencies, in 1978 and 1983, have shown pronounced movement toward greater flexibility in setting effluent standards, and agency reactions to the results obtained through those innovations have been highly favorable (Lamb et al., 1978; Hull, 1983; Lamb and Hull, 1984). Therefore, it is anticipated that the trend in that direction should continue and grow in momentum as time passes.

The responsibility for setting and administering drinking water standards at the federal level were transferred a few years ago from the U.S. Public Health Service to the EPA. Recent actions have included some restructuring of the standards and establishing different methods for monitoring and administering water quality. Increased attention has been paid to setting limits for concentrations of various specific chemicals, especially pesticides.

This trend toward more detailed drinking water standards undoubtedly will continue during the next decade or two, for reasons that have been discussed earlier. Reviews and revisions of the regulations will be accomplished in several phases, with new standards being established periodically, for various groups of constituents, beginning in 1985.

STUDY QUESTIONS

1. Using information from the EPA or your state agency that is responsible for regulating drinking water quality, describe the processes being followed in current revisions of the drinking water standards.
2. Using information from the state water pollution control agency, outline the system used in your state in establishing effluent standards for municipal and industrial wastewater discharges.
3. What has been the trend in your state pollution control agency over the past 10 years with respect to the use of flexible standards?
4. Compare and contrast the evaluation of water quality through the use of

chemical measurements and bioassay techniques, summarizing the advantages and disadvantages of each.

5. Outline the nature of difficulties that are encountered in trying to establish permissible concentrations of a suspected toxic organic chemical in drinking water.
6. Outline several opportunities that exist for reducing water carriage of polutional materials from homes and evaluate the prospects for implementing those reductions in the practical world of the United States today. Justify your positions, whatever they may be.
7. Flow sheets for typical municipal water and wastewater treatment facilities were presented and discussed in Chapters 11 and 12. Outline the types of changes that you expect to see in those facilities during the next decade.

REFERENCES

Council on Evnironmental Quality, *Environmental Quality 1982,* 13th annual report, Government Printing Office, Washington, D.C. (1983).

Hull, D., *A Survey of the Use of Flexible Standards by State Water Quality Management Agencies;* Unpublished report submitted in partial fulfillment of requirements for MSPH Degree, Department of Environmental Sciences and Engineering, University of North Carolina at Chapel Hill (1983).

Lamb, J. C., "Treatment Plant Failures: Facts, Fallacies, Solutions," *Proceedings of the 14th Mid-Atlantic Industrial Wastes Conference,* pp. 426–435, Ann Arbor Science Press, Ann Arbor, Mich. (1982).

Lamb, J. C., M. A. Shiffman, and S. A. Duletsky, "A Survey of the Use of Seasonal Effluent Standards," presented at the Las Vegas meeting of the Water Pollution Control Federation (September 30, 1980).

Lamb, J. C. and D. B. Hull, "Current Status in Use of Flexible Effluent Standards," presented at the annual meeting of the Water Pollution Control Federation, New Orleans (October 1984).

Ruckelshaus, W. D., "Testimony Before the Subcommittee on Environmental Pollution, Committee on Environment and Public Works, United States Senate," U.S. Environmental Protection Agency, Washington, D.C. (June 14, 1983).

U.S. Environmental Protection Agency, "EPA to Revise Drinking Water Standards," *Environmental News,* press release, Thursday, October 27, 1983.

Water Pollution Control Federation, *A Decade of Progress,* Washington, D.C. (May 1983).

APPENDIX

Units, Conversions, and Other Calculations

Details about the SI system of units are discussed in other publications, which include extensive tables of conversion factors between the SI and English systems (Klingeman et al., 1976; Graham et al., 1982). In this appendix, some basic characteristics of the SI system are outlined for the benefit of those unfamiliar with it. Also, selected factors are given that can be applied to convert units commonly used in water resources practice between the English and SI systems, as may be desired. Some examples illustrate ways in which those factors may be used to compute more complex conversions and other calculations.

SOME ELEMENTS OF THE SI SYSTEM

Table A.1 lists the seven basic units of measurement and two supplementary units in the SI system, with the symbols used to abbreviate each (Graham et al., 1982). All other units used in SI are derived from these and can be readily expressed in terms of them.

TABLE A.1
SI Base Units

Quantity	SI Unit	Symbol
Base units		
Length	meter	m
Mass	kilogram	kg
Time	second	s
Electric current	ampere	A
Luminous intensity	candela	cd
Thermodynamic temp.	kelvin[a]	K
Amount of substance	mole	mol
Supplementary units		
Plane angle	radian	rad
Solid angle	steradian	sr

[a] Celsius (°C) temperature is widely used in practice and is defined as: °C = °K − 273.15.

Table A.2 shows some common derived units and a few others that are not SI but have been retained as acceptable for use. For each, its name and symbol are given, as well as the SI base, or other, units that are its equivalent (Graham et al., 1982).

Table A.3 lists approved prefixes that may be used in the SI system for numbers that are very large or very small. In written text, the prefix is merely added to the beginning of the appropriate unit name, without punctuation or space between them. For example, a distance of forty thousand meters could be written as: 40,000 meters, 4×10^4 meters, or 40 kilometers. Abbreviations for the prefixes and units are combined in similar fashion, producing for the above example: 40,000 m, 4×10^4 m, or 40 km.

SI AND ENGLISH UNITS ENCOUNTERED IN PRACTICE

Table A.4 lists some English and SI units commonly encountered in water resources literature and practice. In general, most of the data for hydrologic phenomena still are reported predominantly in English units, while the results of laboratory analyses and other scientifically based information are presented in SI—usually in metric units, actually, which may or may not conform with true SI units. The design parameters for treatment systems may be given in either or both systems, depending on authorship and purpose of the publication.

TABLE A.2
Examples of SI-Derived and Other Acceptable Units

Quantity	Unit	Symbol	Expressed in Terms of Other Units	Expressed in Terms of Base Units
SI-derived				
Acceleration	m/s/s			m/s^2
Area	m^2			m^2
Concentration	mol/m^3			mol/m^3
Density	kg/m^3			kg/m^3
Electrical potential	volt	V	W/A	$m^2kg/A\ s^3$
Electrical resistance	ohm		V/A	m^2kg/A^2s^3
Energy	joule	J	Nm	m^2kg/s^2
Force	newton	N		$m\text{-}kg/s^2$
Frequency	hertz	Hz		s^{-1}
Power	watt	W	J/s	m^2kg/s^3
Pressure	pascal	Pa	N/m^2	$m^{-1}kg/s^2$
Velocity	m/s			m/s
Viscosity, kinematic	m^2/s			m^2/s
Volume	m^3			m^3
Other acceptable units				
Angle, plane	degree	°		1.745×10^{-2}rad
Area	hectare	ha		10 000 m^2
Energy	kilowatt-hour	kWh		3.6×10^6 J
Hydraulic head	meter	m		9.806 kPa
Time	minute	min		60 s
	hour	h		3600 s
	day	d		86,400 s
	year	a		3.156×10^7 s
Volume	liter	l		$1.0 \times 10^{-3}m^3$

TABLE A.3
SI Prefixes

Prefix	Symbol	Multiplier
tera	T	10^{12}
giga[a]	G	10^9
mega	M	10^6
kilo	k	10^3
	Base Units	
milli	m	10^{-3}
micro	μ	10^{-6}
nano	n	10^{-9}
pico	p	10^{-12}

[a]Pronounced jee′ga.

TABLE A.4
Some English and SI Units Commonly Encountered in Water Resources Practice

Quantity	English Units	SI or Metric Units
Length	foot (ft)	meter (m)
	inch (in.)	centimeter (cm)
		millimeter (mm)
		micrometer (μm)
Area	square feet (ft^2)	square meter (m^2)
	square inch ($in.^2$)	square centimeter (cm^2)
	acres (ac)	hectares (ha)
Volume	cubic feet (ft^3)	cubic meters (m^3)
	cubic inches ($in.^3$)	cubic centimeters (cm^3)
	gallons (gal)	liters (1)
	million gallons	milliliters (ml)
	acre-foot (acre-ft)	
Mass or weight	pounds (lb)	grams (g)
	grains (gr)	kilograms (kg)
	ounce (oz)	milligram (mg)
		microgram (μg)
Concentration	percent	percent
	mg/liter	mg/liter
	μg/liter	μg/liter
	parts per million (ppm)	parts per million (ppm)
	parts per billion (ppb)	parts per billion (ppb)
	lbs/million gallons	moles/liter (mol/1)
	grains/gallon (gr/gal)	moles/cubic meter (mol/m^3)
		equivalents/liter

BASIC CONVERSION FACTORS

Table A.5 lists some factors that are especially helpful in making conversions between units in either system, or between the English and SI systems. More extensive lists are given in the references by Graham (1982) and Klingeman (1976).

Actually, with an understanding of methods for making the calculations, commiting only a few factors to memory makes it possible to perform almost any desired conversion without referring to tables. The entries in Table A.5 that are marked with a superscript "a" are examples of those that might be used conveniently to that end, although others could be just as suitable.

TABLE A.5
Selected Conversion Factors

Multiply Data in These Units	By	To Express Them in These Units
Acres (ac)	43,560	square feet (ft^2)
	0.405	hectares (ha)
	4047	square meters (m^2)
Acre-feet (ac ft)	1233	cubic meters (m^3)
	43,560	cubic feet (ft^3)
	325,900	gallons (gal)
Atmospheres (atm)	14.7	pounds/in.^{2} (psi)
	101	kilopascals (kPa)
British thermal units (Btu)	2.93×10^{-4}	kilowatt-hours (kW-hr)
	1.06	kilojoules (kJ)
Calories, International (cal)	3.97×10^{-3}	British thermal units (Btu)
	4.19	joules (J)
Cubic feet (ft^3)	7.48[a]	gallons (gal)
	62.4[a,b]	pounds of water (lbs)
	2.83×10^{-2}	cubic meters (m^3)
	28.3	liters
Cubic feet per second (ft^3/sec or cfs)	0.646	million gals/day (mgd)
	2.83×10^{-2}	cubic meters/sec (m^3/s)
	28.3	liters/sec (l/s)
Cubic meters (m^3)	264	gallons (gal)
Cubic meters/day (m^3/d)	2.64×10^{-4}	million gal/day (mgd)
Cubic meters/sec (m^3/s)	35.3	ft^3/sec (cfs)
	22.8	million gal/day (mgd)
Curie (Ci)	3.70×10^{10}	reciprocal seconds (s^{-1})
Degrees Celsius (°C)	To convert, use:	
	$1.8 \times °C + 32$	°F
Degrees Fahrenheit (°F)	To convert, use:	
	$(°F - 32) \times 0.556$	°C
Feet of head (ft)	0.433	lbs/in^2 (psi)
	0.305	meters (m)
	2989	pascals (Pa)
Feet (ft)	0.3048[a]	meters (m)
Feet per second (ft/sec)	0.3048	meters/sec (m/s)
Foot-pounds (ft-lb)	1.36	joules (J)
Gallons (gal)	3.785×10^{-3}	cubic meters (m^3)
	3.785[a]	liters
	8.34[a,b]	pounds of water (lbs)

TABLE A.5 (*continued*)
Selected Conversion Factors

Multiply Data in These Units	By	To Express Them in These Units
Gallons per minute (gpm)	1.44×10^{-3}	million gal/day (mgd)
	6.31×10^{-5}	cubic meters/sec (m^3/s)
Grains (gr)	64.8	milligrams (mg)
	1.43×10^{-4}	pounds (lbs)
Grains per gallon (gr/gal)	17.1	grams/cubic meter (g/m^3)
	17.1	milligrams/liter (mg/l)
Grams (g)	3.53×10^{-2}	ounces (oz)
	2.20×10^{-3}	pounds (lb)
Hectares (ha)	2.47	acres (ac)
	10,000	square meters (m^2)
Horsepower (hp)	746	watts (W)
	0.746	kilowatts (kW)
Horsepower-hours (hp-hr)	2.69	megajoules (MJ)
	0.746	kilowatt hours (kW-hr)
Inches (in.)	0.0833	feet (ft)
	25.4[a]	millimeters (mm)
Inches of mercury (in. Hg)	0.491	pounds/in^2 (psi)
	3.38	kilopascals (kPa)
Joules (J)	0.239	calories, international (cal)
	0.737	foot-pounds (ft-lb)
	2.78×10^{-4}	watt-hours (W-hr)
	9.48×10^{-4}	British thermal units (Btu)
Kelvin (K)	To convert, use:	
	(K − 273.25)	°C
Kilograms (kg)	2.20	pounds (lbs)
	6.85×10^{-2}	slugs
Kilograms, force (kgf)	9.81	newtons (N)
Kilometers (km)	0.621	miles (mi)
Kilopascals (kPa)	9.87×10^{-3}	atmospheres (atm)
	0.145	pounds/in.^{2} (psi)
Kilowatts (kW)	1.34	horsepower (hp)
Kilowatt-hours (kW-hr)	3.60	megajoules (MJ)
	1.34	horsepower-hours (hp-hr)
	3412	British thermal units (Btu)
Liters (l)	3.53×10^{-2}	cubic feet (ft^3)
	0.264	gallons (gal)
	1.06	quart (qt)

TABLE A.5
Selected Conversion Factors

Multiply Data in These Units	By	To Express Them in These Units
Liters per second (l/s)	3.53×10^{-2}	cubic feet/sec (cfs)
	15.8	gallons/min (gpm)
	2.28×10^{-2}	million gal/day (mgd)
Megajoules (MJ)	0.372	horsepower-hours (hp-hr)
	0.278	kilowatt-hours (kW-hr)
Meters (m)	3.281	feet (ft)
	39.37	inches (in.)
Micrograms per liter (μg/l)	1.00[a]	parts per billion (ppb), if specific gravity = 1.0
Micrometers (μm)	1.00	microns (μ),
Mils (mil)	1.0×10^{-3}	inches (in.)
	25.4	micrometers (μm)
Miles (mi)	5280	feet (ft)
	1.61	kilometers (km)
Milligrams per liter (mg/l)	5.85×10^{-2}	grains per gallon (gr/gal)
	1.00[b]	parts per million (ppm), if specific gravity = 1.0
Million gallons (mil gal)	3.785×10^{3}	cubic meters (m^3)
Million gal/day (mgd)	1.55	cubic feet/sec (cfs)
	694	gal/min (gpm)
	4.38×10^{-2}	cubic meters/sec (m^3/s)
	43.8	liters/sec (l/s)
Millimeters (mm)	3.94×10^{-2}	inches (in.)
Newtons (N)	0.225	pounds force (lbf)
Ounces (oz)	0.0625	pounds (lb)
	28.4	grams (g)
Parts per billion (ppb)	1.00[b]	micrograms/liter (μg/l), if specific gravity = 1.0.
	1.00	micrograms/kilogram (μg/kg)
Parts per million (ppm)	1.00[b]	milligrams/liter (mg/l), if specific gravity = 1.0.
	1.00	milligrams/kilogram (mg/kg)
Pascals (Pa)	9.87×10^{-6}	atmospheres (atm)
	3.35×10^{-4}	feet of head (ft)
	1.45×10^{-4}	pounds/in.^{2} (psi)

TABLE A.5 (*continued*)
Selected Conversion Factors

Multiply Data in These Units	By	To Express Them in These Units
Pints (pt)	0.473	liters (l)
Pounds (lb)	0.120[b]	gallons of water (gal)
Pounds-force (lbf)	4.45	newtons (N)
Pounds, mass (lb)	0.454[a]	kilograms (kg)
Pounds/in^2 (psi)	6900	pascals (Pa)
Quarts (qt)	0.25	gallons (gal)
	0.946	liters
Slugs (slug)	14.59	kilograms (kg)
Square feet (ft^2)	9.29×10^{-2}	square meters (m^2)
	144	square inches (in.^{2})
Square inches (in.^{2})	6.94×10^{-3}	square feet (ft^2)
	6.45×10^{-4}	square meters (m^2)
Square kilometers (km^2)	0.386	square miles (mi^2)
Square meters (m^2)	2.47×10^{-4}	acres (ac)
	10.76	square feet (ft^2)
Square miles (mi^2)	2.59	square kilometers (km^2)
Tons, metric (t)	1.10	tons, short
	2205	pounds (lb)
	1000	kilograms (kg)
Tons, short (ton)	907.0	kilograms (kg)
Watts (W)	3.41	British thermal units/hour (Btu/hr)
	1.34×10^{-3}	horsepower (hp)

Source: Taken in part from Klingeman (1976) and Graham et al. (1982).

[a]Convenient basic units to remember for conversions.

[b]Approximate value, assuming pure water.

EXAMPLES OF CONVERSION CALCULATIONS

The following examples illustrate methods for making calculations for conversions among units and for minimizing chances of errors. The basic procedure consists of performing the calculations for each step of the conversion with both the numbers *and* their units. Upon completion, units produced by the calculations will be correct for the numerical answer obtained through the same manipulations of the numbers.

Example No. 1

What is 225 gal expressed as milliliters?

$$225 \text{ gal} \times 3.785 \frac{\text{liters}}{\text{gal}} \times 1000 \frac{\text{ml}}{\text{liter}} = \underline{\underline{852{,}000}} \text{ ml}$$

Example No. 2

Given a flow of 8.6 million gal/day. What is that in liters per second?

$$8{,}600{,}000 \frac{\text{gal}}{\text{day}} \times 3.785 \frac{\text{liters}}{\text{gal}} \times 1 \frac{\text{day}}{86{,}400 \text{ sec}} = \underline{\underline{377}} \text{ liters/sec}$$

Example No. 3

A substance has a density of 1500 g/liter. What is that in pounds per cubic foot?

$$1500 \frac{\text{g}}{\text{liter}} \times 1 \frac{\text{lb}}{454 \text{ g}} \times 3.785 \frac{\text{liters}}{\text{gal}} \times 7.48 \frac{\text{gal}}{\text{ft}^3} = \underline{\underline{93.6}} \text{ lb/ft}^3$$

OTHER WATER QUALITY CALCULATIONS

The same numerical factors and techniques may be used to make other types of calculations involving water quantities, concentrations, and total content of constituents. They also may be used to determine the fluxes of constituents, given their concentrations and rates of water flow.

The examples that follow illustrate the methods for carrying out those calculations. It should be noted that the units and numbers again are carried through the processes together to minimize chances for errors.

Example No. 4

The concentration of a constituent in water is 1.0 mg/liter. What would be the quantity of it, in pounds, in 1 million gallons of water?

$$1.0 \frac{\text{mg}}{\text{liter}} \times 3.785 \frac{\text{liters}}{\text{gal}} \times 1 \frac{\text{g}}{1000 \text{ mg}} \times 1 \frac{\text{lb}}{454 \text{ g}} \times 10^6 \frac{\text{gal}}{\text{million gal}}$$
$$= \underline{\underline{8.34}} \text{ lb/million gal}$$

Example No. 5

The concentration of solids dissolved in water is 275 mg/liter. What is the total solids content in a tank having a volume of 300,000 gallons?

$$275 \frac{\text{mg}}{\text{liter}} \times 3.785 \frac{\text{liters}}{\text{gal}} \times 300{,}000 \text{ gal} \times 1 \frac{\text{g}}{1000 \text{ mg}} \times 1 \frac{\text{lb}}{454 \text{ g}} = \underline{\underline{688}} \text{ lb}$$

Example No. 6

A 25,000-pound spill of chemical enters a 60-million-gallon reservoir. Assuming uniform mixing, what is the resulting concentration of chemical in the water, in milligrams per liter?

$$\frac{25{,}000 \text{ lb}}{60{,}000{,}000 \text{ gal}} \times 454 \frac{\text{g}}{\text{lb}} \times 1000 \frac{\text{mg}}{\text{g}} \times 1 \frac{\text{gal}}{3.785 \text{ liters}} = \underline{\underline{50}} \text{ mg/liter}$$

Example No. 7

An industrial waste flowing at 6.2 million gal/day) contains 350 mg/liter chloride. What is the total daily flux (pounds per day) of chloride discharged in that wastewater?

$$6{,}200{,}000 \frac{\text{gal}}{\text{day}} \times 3.785 \frac{\text{liter}}{\text{gal}} \times 350 \frac{\text{mg}}{\text{liter}} \times 1 \frac{\text{g}}{1000 \text{ mg}} \times 1 \frac{\text{lb}}{454 \text{ g}}$$

$$= \underline{\underline{18{,}100}} \text{ lb/day}$$

An alternate method for working the problem, remembering that each milligram per liter is equal to 8.34 lb of constituent per million gallons (See Example 4) is as follows.

$$6.2 \frac{\text{million gal}}{\text{day}} \times 350 \frac{\text{mg}}{\text{liter}} \times 8.34 \frac{\text{lb}}{\text{million gal-mg/liter}} = \underline{\underline{18{,}100}} \text{ lb/day}$$

REFERENCES

Fair, G. M., J. C. Geyer, and D. A. Okun, *Water and Wastewater Engineering*, Vol. 2, Appendix, Wiley, New York (1968).

Graham, M. J., et al., *Units of Expression for Wastewater Management,* Manual of Practice No. 6, Water Pollution Control Federation, Washington, D.C. (1982).

Klingeman, P. C., et al., *System International d'Unites–Metric Measurement In Water Resources Engineering,* booklet prepared for the Universities Council on Water Resources Engineering, Oregon State University (June 1976).

U.S. Congress, "Metric Conversion Act of 1975," 94th Congress (1975).

Index